Genetics and Breeding of Fruit Trees

Genetics and Breeding of Fruit Trees

Editors

Dong Zhang
Libo Xing

MDPI

Basel • Beijing • Wuhan • Barcelona • Belgrade • Novi Sad • Cluj • Manchester

Editors
Dong Zhang
Northwest A&F University
Yangling, China

Libo Xing
Northwest A&F University
Yangling, China

Editorial Office
MDPI
St. Alban-Anlage 66
4052 Basel, Switzerland

This is a reprint of articles from the Special Issue published online in the open access journal *Horticulturae* (ISSN 2311-7524) (available at: https://www.mdpi.com/journal/horticulturae/special_issues/fruit_breeding_genetics).

For citation purposes, cite each article independently as indicated on the article page online and as indicated below:

Lastname, A.A.; Lastname, B.B. Article Title. *Journal Name* **Year**, *Volume Number*, Page Range.

ISBN 978-3-0365-8720-2 (Hbk)
ISBN 978-3-0365-8721-9 (PDF)
doi.org/10.3390/books978-3-0365-8721-9

Cover image courtesy of Dong Zhang

Contents

About the Editors

Dong Zhang

Dong Zhang is a professor at the College of Horticulture, Northwest A&F University, deputy chief scientist of the China Apple Research System (CARS), director of the Yangling Sub-Center of the National Center for Apple Improvement, and group leader of apple developmental biology for arid areas. His research focuses on apple developmental biology, high-density cultivation technology, and rootstock breeding. Currently, he is focusing on elucidating the physiological and molecular mechanisms underlying apple flower bud formation and adventitious root development, isolating and functionally characterizing agronomically valuable genes involved in flower bud and root development, and creating novel early-flowering, dwarf, multi-resistant germplasm and rootstocks through genetic engineering and traditional breeding. He is currently an Associate Editor of *Fruit Research*, Science Editor of *Horticulturae Advances*, and Editorial Board Member and Guest Editor of *Horticulturae*. He has published peer-reviewed papers in *The Plant Cell, Molecular Ecology, Plant Biotechnology Journal, Plant Cell and Environment, Journal of Experimental Botany, Horticulture Research, Scientia Horticulturae*, etc.

Libo Xing

Dr. Libo Xing has been an associate professor at the College of Horticulture, Northwest A&F University, since 2019. He obtained his bachelor's degree at Northwest A&F University, China, in 2010, and his master's and PhD degrees at Northwest A&F University, China, in 2013 and 2016, respectively. His research interests cover apple developmental molecular biology, and flower induction and development in particular.

horticulturae

Editorial

Genetics and Breeding of Fruit Trees

Dong Zhang * and Libo Xing

College of Horticulture, Yangling Sub-Center of the National Center for Apple Improvement, Northwest A & F University, Yangling, Xianyang 712100, China
* Correspondence: afant@nwsuaf.edu.cn

"Genetics and Breeding of Fruit Trees", published in *Horticulturae*, is a collection of 11 manuscripts focusing on commercially important fruit crops, such as apple, peach, pear, kiwi, persimmon, blueberry, and red bayberry.

Fruit trees have great economic value and, as long-living crops, pose a significant challenge for efforts aiming to produce considerable changes in many of their traits (for example, the color, shape, aroma, taste, tree structure, and environmental adaptation), particularly in comparison to their wild ancestors, over prolonged periods of rigorous cultivation and breeding practices. Fruit tree breeding research has mostly focused on the discovery of new economically valuable varieties to fulfill different consumer and industrial demands, improving the fruit tree's tolerance to environmental challenges, increasing its postharvest life, and developing cost-effective cultural approaches. The breeding of woody fruit trees is restricted by a variety of drawbacks, including long breeding cycles from the seed to fruit-bearing stage, long developmental stages that increase the costs of growing individuals to maturity in the field, and, frequently, severe heterozygosity impeding theoretical breeding methods. The expansion of biotechnology and the increasing prevalence of genomics have created new prospects to counter these limitations in the breeding of major fruit species, including the use of rapid sequencing techniques, the selection of molecular markers, whole-genome sequencing, genome-wide studies, transcriptomics, metabolomics, proteomics, etc. Because of their biological characteristics and, for some species, rigidity to plant evolution and/or rejuvenation, woody species frequently present a challenge for detailed functional analyses. These challenges severely restrict the application of conventional genetic and biotechnological techniques to functional genomic studies or plant breeding. In the past decade, the sequencing of many genomes, combined with rapid improvements in bioinformatics, has offered critical tools for in-depth molecular studies on agricultural crops, in addition to the traditional model species. The provision of sequencing data is simply a beginning, since bioinformatic techniques are insufficient for the establishment of gene roles. For this information to be useful, it is essential that we understand how thousands of genes interact with one another to establish a plant's architecture and how the metabolic pathways in which they take part affect plant growth and environmental adaptation.

The fruit sugar content is regulated by photosynthesis, but it is also influenced by the transport and accumulation of assimilates, with sugar transporter genes playing a crucial role in this process. The aim of Xiaoli Wang's [1] work was to provide insights into dwarfing rootstocks for pears (*Pyrus* spp.). 'Yunnan' quince (*Cydonia oblonga* Mill.) positively controlled the fruit sugar contents by controlling the flow of PbSWEET6-related assimilates in the scion. Breeders may use this knowledge to select pear rootstocks with the highest possible fruit sugar contents.

Sugar is a crucial component affecting the fruit's flavor quality. Fruit metabolism and sugar storage are major variables influencing fruit quality. To a considerable extent, the types and quantities of sugars impact the quality of the fruit. The aim of Junjie Tao's [2] work was to provide insights into the dynamic changes in the fruit quality, soluble sugar

Citation: Zhang, D.; Xing, L. Genetics and Breeding of Fruit Trees. *Horticulturae* **2023**, *9*, 88. https://doi.org/10.3390/horticulturae9010088

Received: 17 November 2022
Revised: 6 December 2022
Accepted: 8 December 2022
Published: 10 January 2023

contents, sucrose-metabolism-related enzyme activities, and sucrose-metabolism-related gene expressions during the growth and development of *Actinidia eriantha* 'Ganlv 1'.

During the ripening period, peaches experience textural modifications that result in tissue firmness. The importance of the role of auxin in peach post-ripening is widely accepted. The aim of Jiahui Wang's [3] work was to provide insights into the role of small auxin-up RNA (SAUR) genes in fruit post-ripening and softening. This work reported 72 SAUR family members in the peach genome. The functions of *PpSAUR43* in peaches were then established using transient overexpression. *PpSAUR43* acts as a repressor in the process of the post-ripening of peaches by inhibiting the function of the PpCMB1 protein. This study's findings will aid in the development of technologies aiming to precisely manage peach post-ripening and improve the fruit's shelf life.

The prompt and simple maturation of apples is of major interest for the determination of the harvest time and post-harvest storage conditions. The aim of Mengsheng Zhang's [4] work was to establish that quality changes result in major spectral changes during the ripening process of apples, and these changes can be monitored using low-resolution multispectral sensors. As a tool for the rapid and simple detection of the apple's maturity, cost-effective multispectral sensors could be used to identify the best harvest time and post-harvest processing approach.

The evolution of sex differentiation is a fascinating subject, because it is crucial for outbreeding strategies and the mitigation of inbreeding depression and is a major source of genetic diversity. The aim of Huimin Jia's [5] study was to identify the molecular mechanism underlying sex development and differentiation in the red bayberry (*Morella rubra*) through next-generation transcriptome sequencing and comparative analyses in order to identify differentially expressed genes in the male and female flower buds. Furthermore, ethylene is the primary hormone that interacts with other hormones and transcription factors to control sex differentiation.

Persimmon (*Diospyros kaki*) anthracnose is a fatal disease that is primarily caused by the *Colletotrichum horii* pathogen and results in fruit rotting, shoot wilting, and, sometimes, whole plant death in persimmons. The aim of Changfei Guan's [6] study was to identify a new source of resistance to anthracnose among persimmon germplasms. In total, 142 varieties were used to study anthracnose disease resistance. Eight of these highly resistant or resistant accessions exhibited the required agronomic traits. The expansion of genetic diversity and the creation of novel resistant varieties could benefit from the application of these resistance sources in the breeding of persimmons.

Endodormancy is a phase during which perennial plants prepare for the next seasonal cycle, as well as an adaptation that allows the plants to survive harsh winters. The aim of Yongqiang Li's [7] study was to comprehend the underlying molecular mechanisms of bud dormancy release in blueberries (*Vaccinium* spp.). The transcriptomes of the flower buds were examined at seven dormancy stages. RNA transport, circadian rhythm, plant hormone signal transduction, etc., are all pathways possibly related to the dormancy release process.

Kiwifruit black spot disease has increased in prevalence in many 'CuiXiang' kiwifruit planting regions. Yaming Yang's [8] study focused on the pathogenic micro-organisms of black spots using high-throughput internal transcribed spacer (ITS) sequencing to analyze the black spot disease microbial community. Seven pathogens were isolated from the infected kiwifruit. The authors predicted that the candidate pathogenic fungi, such as *Cladosporium cladosporioides*, *Diaporthe phaseolorum*, *Alternaria alternata*, and *Trichothecium roseum*, may cause black spots. The study presented a preliminary assessment of kiwifruit black spots that can be used as a foundation for future research.

Auxin response factors (ARFs) are a family of transcription factors that control the expression of auxin phase-responsive genes. Yu Zong's [9] study focused on the genome-wide identification of *ARF* genes in the tetraploid blueberry (*Vaccinium corymbosum* cv. 'Draper'). In total, 70 blueberry *ARF* genes were identified in the genome, which could be further divided into six subfamilies. *ARF* genes are crucial for the ripening of blueberry fruit, and *VcARF3*, *VcARF4*, *VcARF14*, and *VcARF52* are also important. Future studies should illuminate their function in defining the differences in firmness between the soft and firm flesh cultivars.

In the sand pear (*Pyrus pyrifolia* Nakai), ethylene production has a negative correlation with the storage life, primarily at the time of fruit harvest. The rate-limiting enzyme in the ethylene biosynthesis process, 1-aminocyclopropane-1-carboxylic acid synthase (ACS), is considered to be crucial for the fruit storage life. Jing-Guo Zhang's [10] study identified 13 *ACS* genes in the genome of the sand pear. The sand pear genome sequencing data revealed a total of 13 *ACS* genes, 9 of which were novel members. Seven of these genes seemed to be involved in the process of fruit ripening, whereas four were found to be engaged in system 1 ethylene biosynthesis, suggesting that they may play a variety of roles in ethylene biosynthesis systems.

The formation of adventitious roots (ARs) is necessary for the asexual reproduction of commercially valuable horticultural crops, such as apples. Vegetative propagation is widely used in breeding programs due to its short life cycle, high efficiency, and true-to-typeness. The inability of stem segments to generate AR limits fragment survival. Therefore, understanding the AR regulation mechanisms is critical for the long-term and appropriate use of biological resources. The aim of Muhammad Mobeen Tahir's [11] review was to present the current state of knowledge of AR formation from the physiological and molecular perspectives and highlight recent advances in research aiming to discover the underlying mechanisms involved in adventitious rooting. Although there has been progress in recent years, several questions regarding excision-induced AR creation remain unanswered. These concern the specific activities and interactions of a variety of hormonal, molecular, and metabolic components, as well as the entire structure of the stem cutting in a challenging environment.

There is no doubt that this Special Issue will provide significant knowledge to its readers and will be useful for further studies aiming to improve fruit crops.

Author Contributions: The editorial was written jointly by the D.Z. and the L.X. All authors have read and agreed to the published version of the manuscript.

Acknowledgments: We sincerely thank all the authors and reviewers for their active participation and valuable effort.

Conflicts of Interest: The authors declare no conflict of interest.

References

1. Wang, X.; Cong, L.; Pang, J.; Chen, Y.; Wang, Z.; Zhai, R.; Yang, C.; Xu, L. Dwarfing Rootstock 'Yunnan'Quince Promoted Fruit Sugar Accumulation by Influencing Assimilate Flow and PbSWEET6 in Pear Scion. *Horticulturae* **2022**, *8*, 649. [CrossRef]
2. Tao, J.; Wu, M.; Jiao, X.; Chen, S.; Jia, D.; Xu, X.; Huang, C. Dynamic changes of fruit physiological quality and sugar components during fruit growth and development of Actinidia eriantha. *Horticulturae* **2022**, *8*, 529. [CrossRef]
3. Wang, J.; Su, W.; Liu, K.; Xu, Z.; Shah, K.; Ma, J.; Zhang, D.; Hu, Y.; Zhao, C. PpSAUR43, an Auxin-Responsive Gene, Is Involved in the Post-Ripening and Softening of Peaches. *Horticulturae* **2022**, *8*, 379. [CrossRef]
4. Zhang, M.; Shen, M.; Pu, Y.; Li, H.; Zhang, B.; Zhang, Z.; Ren, X.; Zhao, J. Rapid Identification of Apple Maturity Based on Multispectral Sensor Combined with Spectral Shape Features. *Horticulturae* **2022**, *8*, 361. [CrossRef]
5. Jia, H.; Zhao, L.; Wang, Y.; Wu, H.; Zhao, H.; Zhu, Y.; Jiao, Y.; Wang, G.; Zhou, C.; Huang, C. Comparative Transcriptome Analysis Reveals Sex-Biased Expression of Hormone-Related Genes at an Early Stage of Sex Differentiation in Red Bayberry (Morella rubra). *Horticulturae* **2022**, *8*, 183. [CrossRef]
6. Guan, C.; Hu, J.; Li, Y.; Che, Q.; Yang, Y. Identification of New Sources of Resistance to Anthracnose Caused by Colletotrichum horii among Persimmon Germplasms. *Horticulturae* **2022**, *8*, 180. [CrossRef]
7. Li, Y.; Ma, R.; Li, R.; Zhao, Q.; Zhang, Z.; Zong, Y.; Yao, L.; Chen, W.; Yang, L.; Liao, F. Comparative Transcriptomic Analysis Provides Insight into the Key Regulatory Pathways and Differentially Expressed Genes in Blueberry Flower Bud Endo-and Ecodormancy Release. *Horticulturae* **2022**, *8*, 176. [CrossRef]
8. Yang, Y.; Chen, L.; Wang, C.; Peng, H.; Yin, W.; Li, R.; Liu, C.; Ren, X.; Ding, Y. Pathogenic Fungi Diversity of 'CuiXiang'Kiwifruit Black Spot Disease during Storage. *Horticulturae* **2021**, *8*, 13. [CrossRef]
9. Zong, Y.; Gu, L.; Shen, Z.; Kang, H.; Li, Y.; Liao, F.; Xu, L.; Guo, W. Genome-wide identification and bioinformatics analysis of auxin response factor genes in highbush blueberry. *Horticulturae* **2021**, *7*, 403. [CrossRef]

10. Zhang, J.-G.; Du, W.; Fan, J.; Yang, X.-P.; Chen, Q.-L.; Liu, Y.; Hu, H.-J.; Luo, Z.-R. Genome-Wide Identification of the 1-Aminocyclopropane-1-carboxylic Acid Synthase (ACS) Genes and Their Possible Role in Sand Pear (Pyrus pyrifolia) Fruit Ripening. *Horticulturae* **2021**, *7*, 401. [CrossRef]
11. Tahir, M.M.; Mao, J.; Li, S.; Li, K.; Liu, Y.; Shao, Y.; Zhang, D.; Zhang, X. Insights into Factors Controlling Adventitious Root Formation in Apples. *Horticulturae* **2022**, *8*, 276. [CrossRef]

horticulturae

MDPI

Article

Dwarfing Rootstock 'Yunnan' Quince Promoted Fruit Sugar Accumulation by Influencing Assimilate Flow and PbSWEET6 in Pear Scion

Xiaoli Wang †, Liu Cong †, Jianwen Pang, Yu Chen, Zhigang Wang, Rui Zhai, Chengquan Yang and Lingfei Xu *

College of Horticulture, Northwest A&F University, Xianyang 712100, China; Xiaoliw365@163.com (X.W.); imcongliu@163.com (L.C.); 15297303964@163.com (J.P.); cy1747023889@163.com (Y.C.); wzhg001@163.com (Z.W.); zhai.rui@nwafu.edu.cn (R.Z.); cqyang@nwsuaf.edu.cn (C.Y.)

* Correspondence: lingfeixu@nwsuaf.edu.cn; Tel.: +86-029-87081023

† These authors contributed equally to this work.

Abstract: 'Yunnan' quince (*Cydonia oblonga* Mill.) is used as the dwarfing rootstock for pear (*Pyrus* spp.). Here, we reported that the sugar contents in mature 'Zaosu' pear fruit grafted on 'Yunan' quince (Z/Q) were higher than that in 'Zaosu' pear fruit grafted on 'Duli' (*Pyrus betulifolia*) (Z/D). To investigate the underlying mechanism, the leaf photosynthetic capacity and the leaf-to-fruit assimilate transport capacity were initially analyzed. The leaf photosynthetic capacity was similar between Z/Q and Z/D, but the assimilate transport capacity was greater for Z/Q than for Z/D. Sugar transporters mediate the distribution of assimilates; therefore, changes in *PbSWEET* transcriptional patterns were examined. *PbSWEET6* was highly expressed in Z/Q fruit. Thus, the *PbSWEET6* function related to assimilate transport was further verified. Sucrose and glucose contents increased in transgenic tomato fruit and pear fruit calli overexpressing *PbSWEET6*. Taken together, these results suggest that 'Yunnan' quince positively regulated fruit sugar contents by influencing the flow of PbSWEET6-involved assimilates in the scion.

Keywords: graft; *PbSWEET6*; sugar transport; 'Zaosu' pear; *Cydonia oblonga*

Citation: Wang, X.; Cong, L.; Pang, J.; Chen, Y.; Wang, Z.; Zhai, R.; Yang, C.; Xu, L. Dwarfing Rootstock 'Yunnan' Quince Promoted Fruit Sugar Accumulation by Influencing Assimilate Flow and PbSWEET6 in Pear Scion. *Horticulturae* **2022**, *8*, 649. https://doi.org/10.3390/horticulturae8070649

Academic Editor: Stefano Giovanni La Malfa

Received: 5 June 2022
Accepted: 14 July 2022
Published: 17 July 2022

1. Introduction

Dwarfing rootstocks are used for the production of dwarf fruit trees in modern cultivation systems, which is consistent with the changes in pear cultivation technology toward dwarfing and dense planting [1–3]. Dwarfing rootstocks affect scion growth and are important for improving fruit quality and yield [4,5]. The combination of quince as the base stock and 'Hardy' (*Pyrus communis*) as the interstock leads to a better dwarfing effect and significantly improves the quality of the scion fruit [6]. The fruit sugar content is influenced by photosynthesis, but it is also affected by the transport and accumulation of assimilates, with sugar transporters playing an essential role in this process [7–9].

Sugar transporters are mainly divided into the following three superfamilies: the major facilitator superfamily, the sodium solute symporter family, and the Sugar Will Eventually be Exported Transporter (SWEET) family [10,11]. The SWEET family, which was identified relatively recently, helps to mediate the influx and efflux of sugars across the plasma membrane [12]. Additionally, SWEETs can be divided into four phylogenetic clades (Clades I–IV) [13], of which Clades I and II consist mainly of hexose transporters, whereas Clade III comprises sucrose transporters and Clade IV includes fructose transporters [14,15]. The diversity in the SWEET genes in plants enables the encoded proteins to function in various developmental and physiological processes. In *Arabidopsis thaliana*, AtSWEET1 is a low-affinity glucose transporter that contributes to glucose uptake and efflux [16]. Mutations to both *AtSWEET11* and *AtSWEET12* reportedly decrease leaf assimilate exudation and significantly inhibit sucrose loading and plant growth, resulting in the accumulation of sugar and starch in source leaves [17]. Similar findings have been reported for *Zea mays* [18]. Moreover, *AtSWEET17* and *AtSWEET16*, which encode vacuolar sugar transporters, have important functions associated with plant growth and development [19–21]. In tomato, *SlSWEET12c* may promote sucrose unloading from phloem during the tomato fruit development stage [22]. Moreover, *SlSWEET7a* and *SlSWEET14* encode hexose and sucrose transporters, and the silencing of these genes results in increases in plant height and fruit size [23]. The overexpression of *VvSWEET10* in grape calli and tomato significantly increases the glucose, fructose, and total sugar levels [24]. Previous research revealed 18 SWEETs in pear [25]. In 'Nanguo' pear (*Pyrus ussuriensis*), PuSWEET15 increases the fruit sucrose content [26].

Sugar transport and accumulation substantially affect fruit sweetness [24,27]. Different types of rootstocks can influence the transport of sugars into the scion fruit [28]. SWEETs play key roles in the accumulation of sugars in fruit [23–25]. However, the mechanism by which dwarfing rootstocks affect sugar transport in scion fruit and related sugar transporters is still unclear. In this study, 'Zaosu' (*Pyrus bretschneideri* Rehd.) pear grafted on 'Yunnan' quince (*Cydonia oblonga* Mill.) and 'Duli' (*Pyrus betulifolia*) were used as the experimental materials. A series of experiments showed that the sugar-transport-related gene *PbSWEET6* was important for mediating the effect of 'Yunnan' quince on 'Zaosu' pear fruit sugar accumulation. The results of this study may be useful for clarifying the mechanism underlying the influence of dwarfing rootstocks on fruit sugar transport in pear.

2. Materials and Methods

2.1. Plant Materials and Growth Conditions

Twelve-year-old 'Zaosu' pears were grafted on 'Yunnan' quince (Z/Q) using 'Hardy' (*Pyrus communis*) as interstock. Twelve-year-old 'Zaosu' pears were also directly grafted on 'Duli' (Z/D). Z/Q and Z/D trees were cultured in an experimental pear orchard in Meixian, Shaanxi, China (34.28° N, 108.76° E). The soil of the orchard was loam. The orchard adopted conventional management, and the management level was consistent. Mature fruits, leaves, carpopodium, and phloem of fruiting branches were harvested at 110 DAFB (days after flower bloom) in 2021. Thirty fruits from each combination were divided into three groups and brought back to the laboratory for fruit quality determination. The flesh of those fruits was immediately frozen in liquid nitrogen and stored at −80 °C. Tomato (*Solanum lycopersicum* L., 'Micro-Tom') seeds and cultured fruit calli of 'Starkrimson' pear (*Pyrus communis*) were used for genetic transformation. Tomato plants were grown in a light incubator with a 16 h/8 h (light/dark) photoperiod at 25 °C.

2.2. Photosynthetic Capacity and Assimilate Distribution

The experiment was carried out at a distance of about 1.5 m from the ground of the plant, and the biennial fruiting branch group with one fruit was selected for ^{13}C pulse labeling [29,30]. The experiment began at 9:00 a.m., when 2 mol·L^{-1} HCl solution was injected into a plastic bottle containing 1 g of $Ba^{13}CO_3$ (98%; Shanghai Research Institute of Chemical Industry, Shanghai, China) with a syringe (the HCl solution was excessive to ensure the complete reaction of $Ba^{13}CO_3$). The processing time was 7 h, during which the bag was gently shaken. At the same time, three plants not contaminated with ^{13}C were selected as blank controls (for the determination of natural ^{13}C abundance). After 24 h, the leaves, fruits, phloem, and xylem of branches were harvested for ^{13}C determination. The samples were dried for 30 min at 105 °C and then dried at 65 °C until reaching constant weight. After grinding, the samples were screened through 100 mesh and mixed well. The ^{13}C abundance of the above samples was determined with an isotope mass spectrometer (DELTA V Advantage, Thermo Fisher Scientific, Bremen, Germany).

Eight of each combination were randomly selected for photosynthesis measurement. For each tree, five mature leaves were selected from the middle of the bearing branches on the sunny side for determination. Leaf photosynthesis was measured from 9:30 to 11:00 a.m. using the LI-COR 6800 portable photosynthesis system (LI-COR, Huntington Beach, CA, USA). During the measurement, the light intensity was set at 300 and 1000 μmol m^{-2} s^{-1} (90%: 10% red: blue light), the CO_2 concentration in the leaf chamber was set at 400 umol mol^{-1} by a CO_2 mixer, the relative humidity of 60%, the leaf-to-air vapor pressure deficit was kept at 0.1 kPa, and the flow rate was 500 μmol s^{-1}. Then, 3–5 min was allowed for settling in the leaf chamber, and values were recorded after the leaf reached a steady state. The leaves were immersed in a mixed solution of 80% acetone and 95% ethanol (1:1) for 24 h in the dark to extract the chlorophyll and carotenoid, and the spectrophotometric values were obtained at 663 nm, 645 nm, and 440 nm. The corresponding chlorophyll and carotenoid contents were then calculated.

2.3. Measurements of Soluble Solids and Sugar Contents

Thirty fruits were selected from each combination, and every ten fruits were divided into one group for fruit quality determination. The content of soluble solid (Brix%, grams of soluble solid per 100 g of water) in the filtrate was determined with a refractometer (PAL-1, ATAGO, Tokyo, Japan), and titratable acid (%, percentage of mass) was measured with a fruit acidity tester (GMK-835F, G-won Hitech, Seoul, Korea). The components of soluble sugar were measured by GC–MS (ISQ & TRACE ISQ, Thermo Scientific, Waltham, MA, USA). The general steps were as follows: the sample was ground into powder in liquid nitrogen, 0.1 g of the sample was mixed with 1.4 mL of 80% (v/v) chromatographic methanol, ribitol was added (4 mg/mL) as an internal standard, and the mixture was shocked in a metal bath at 70 °C for 30 min. After centrifugation at 13,000× g for 15 min, the supernatant was taken. After that, 750 mL of chromatographic $CHCl_3$ and 1400 mL ddH_2O were added and mixed, then centrifuged at 2200× g for 15 min, and 20 μL was taken for vacuum concentration and drying. Finally, it was derivatized with methoxyamine hydrochloride and N-methyl-N-trimethylsilyl-trifluoroacetamide, then stored with brown bottles for subsequent determination.

2.4. Sequence Analysis of PbSWEET6

The sequences of SWEET gene family members were downloaded from the pear genome database (http://peargenome.njau.edu.cn, accessed on 10 December 2021) and named in reference to Li [25]. The SWEETs protein sequences of *Arabidopsis thaliana* were downloaded from Phytozome v13 (https://phytozome-next.jgi.doe.gov/, accessed on 10 December 2021), and *Solanum lycopersicum* and *Vitis vinifera*, searched on the NCBI database (https://www.ncbi.nlm.nih.gov/, accessed on 10 December 2021), were used to construct a phylogenetic tree with the maximum likelihood method and a bootstrap analysis was performed with MEGA X software. Bootstrap values were calculated from 1000 replicate analyses.

The protein accessions used were as follows: AtSWEET1 (*Arabidopsis thaliana*, AT1G21460.1), AtSWEET2 (AT3G14770.1), AtSWEET3 (AT5G53190.1), AtSWEET4 (AT3G28007.1), AtSWEET5 (AT5G62850.1), AtSWEET6 (AT1G66770.1), AtSWEET7 (AT4G10850.1), AtSWEET8 (AT5G40260.1), AtSWEET9 (AT2G39060.1), AtSWEET10 (AT5G50790.1), AtSWEET11 (AT3G48740.1), AtSWEET12 (AT5G23660.1), AtSWEET13 (AT5G50800.1), AtSWEET14 (AT4G25010.1), AtSWEET15 (AT5G13170.1), AtSWEET16 (AT3G16690.1), AtSWEET17 (AT4G15920.1); PbSWEET2 (*Pyrus bretschneideri*, XP_018501061.1), PbSWEET5 (XP_009337540.1), PbSWEET6 (XP_009349481.1), PbSWEET8 (XP_009347912.1), PbSWEET9 (XP_009340984.1), PbSWEET10 (XP_009360715.1), PbSWEET12 (XP_009376722.1), PbSWEET13 (XP_009352052.1), PbSWEET14 (XP_009377708.1), PbSWEET18 (XP_009360807.1); PuSWEET15 (*Pyrus ussuriensis*, QIJ69897.1); SlSWEET1-like (*Solanum lycopersicum*, XP_004237722.1), SlSWEET7a (XP_004245483.1), SlSWEET12c (XP_004247459.1), SlSWEET14 (XP_004235340.1); OsS-

WEET11 (*Oryza sativa*, XP_015648423.1), OsSWEET14 (XP_015615538.1), and OsSWEET15 (XP_015623673.1); VvSWEET7 (*Vitis vinifera*, XP_002263697.1), and VvSWEET10 (XP_002284244.1).

2.5. Gene Cloning and Quantitative Real-Time PCR

The total RNA was extracted from the fruits, leaves, phloem, and carpopodium of Z/D and Z/Q using an RNAprep Pure Plant Kit (TIANGEN, Beijing, China). The first-strand cDNA was synthesized from 1 μg of total RNA using the *Evo M-MLV* RT Mix Kit with gDNA Clean for qPCR (AG, Hunan, China).

The qRT-PCR was measured using the Hieff® qPCR SYBR® Green Master Mix (Yeasen, Shanghai, China) on the StepOnePlus™ Real-Time PCR Systems (Applied Biosystems, Thermo Fisher Scientific, Albany, NY, USA). The primers of selected genes and *PbActin* (an internal control) were designed on the NCBI and are listed in Table S1. At least three biological replicates were performed and analyzed using the cycle threshold ($2^{-\Delta\Delta Ct}$) method.

2.6. Subcellular Location of PbSWEET6

The full-length coding sequence (CDS) of *PbSWEET6* was amplified and cloned into the pCAMBIA2300 vector fused with the GFP reporter and driven by the CaMV35S promoter. Primers are listed in Table S1. The recombinant plasmid was transformed into tobacco leaves by injection and the empty vector expressing untargeted GFP was used as a control. The GFP green fluorescence was observed with a fluorescence microscope (BX63, OLYMPUS, Tokyo, Japan).

2.7. Agrobacterium-Mediated Tomato and Pear Fruit Calli Transformation

The full CDS of *PbSWEET6* was cloned into the gateway entry vector (pDONR222) and subsequently transferred into the pK7-203 destination vector using LR Clonase II enzymes (Invitrogen, Gibco, Grand Island, NY, USA). The transformation methods of tomato and pear fruit calli were in accordance with Zhang et al. [31].

2.8. Statistical Analysis

The data are presented as the means ± SDs (standard deviations). Significance tests were carried out using SPSS software based on Student's *t* tests at $p < 0.01$ or $p < 0.05$.

3. Results

3.1. 'Yunnan' Quince Increased the Sugar Content of Scion Fruit

The quality of Z/Q and Z/D fruits was determined after harvest at 110 DAFB. The Z/Q fruit was heavier than the Z/D fruit. Moreover, the soluble solid content (SSC) was higher in Z/Q fruit than in Z/D fruit, but there was no significant difference in the titratable acid content (Figure 1A–D). The analysis of sugar contents indicated that fructose, glucose, and sucrose contents were significantly higher in Z/Q fruit than in Z/D fruit (Figure 1E). Thus, 'Yunnan' quince increased the scion fruit sugar content.

3.2. 'Yunnan' Quince Promoted the Transport of Photoassimilates to Fruit

Fruit quality is closely related to photosynthesis. Therefore, the photosynthetic rate and chlorophyll content were analyzed. The data showed no significant difference in the photosynthetic rate between Z/Q and Z/D leaves, implying that there were no differences between 'Yunnan' quince and 'Duli' in terms of their effects on the 'Zaosu' pear photosynthetic rate (Figure 2A). There were also no differences in the chlorophyll a, chlorophyll b, and carotenoid contents between Z/Q and Z/D leaves (Figure 2B–D). The distribution of photoassimilates was determined on the basis of a ^{13}C feeding assay. After 1 day of ^{13}C feeding, the percentage of ^{13}C was higher in Z/Q fruit than in Z/D fruit (Figure 2E). Therefore, 'Yunnan' quince appeared to promote the transport of photoassimilates in the scion.

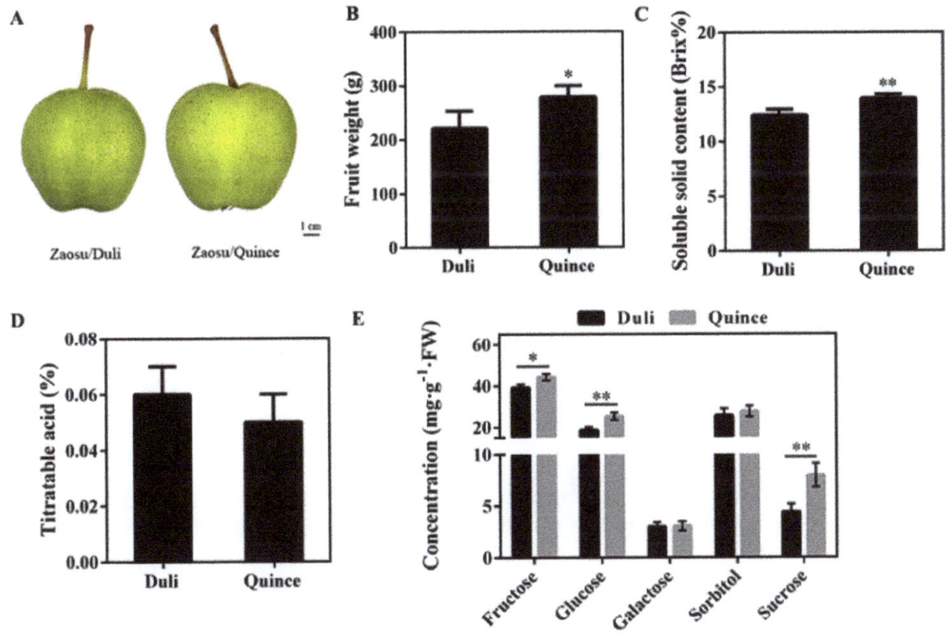

Figure 1. The fruit quality of 'Zaosu' pear grafted on 'Yunnan' quince (Z/Q) and 'Duli' (Z/D). (**A**) Z/Q and Z/D fruits at 110 (DAFB). (**B**) Fruit weight. (**C**) Soluble solid content. (**D**) Titratable acid. (**E**) Concentrations of fructose, glucose, galactose, sorbitol, and sucrose in mature Z/Q and Z/D fruits. The data represent mean values ± SDs. Asterisks indicate significant differences as determined by Student's *t* tests (* $p < 0.05$; ** $p < 0.01$).

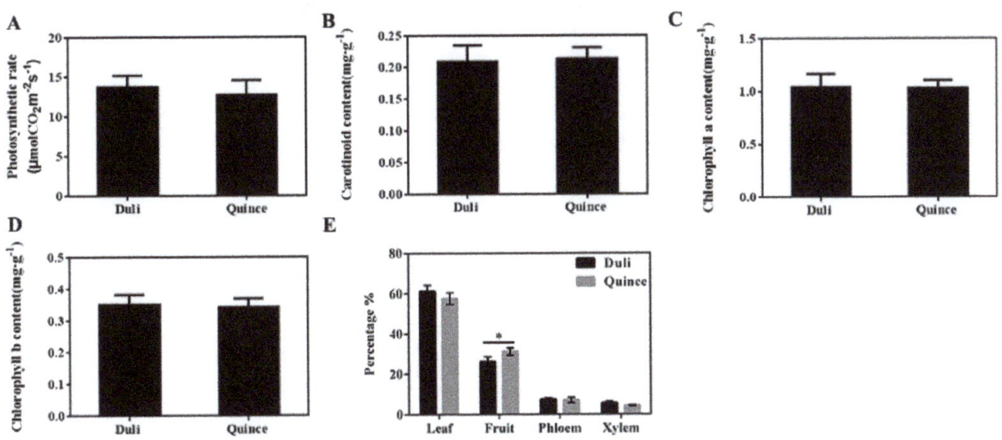

Figure 2. Photosynthetic related indexes and ^{13}C feeding experiment. (**A**) The leaf photosynthetic rate of 'Zaosu' pear grafted on 'Yunnan' quince (Z/Q) and 'Duli' (Z/D). (**B–D**) The contents of carotenoid, chlorophyll a, and chlorophyll b. (**E**) The percentage of ^{13}C in each organization. The data represent the mean values ± SDs. Asterisks indicate significant differences as determined by Student's *t* tests (* $p < 0.05$).

3.3. *PbSWEET6 May Participate in the Transport of Photoassimilates*

Photoassimilates are mainly transported in plants as sugars. A quantitative real-time PCR (qRT-PCR) analysis of 18 *PbSWEET* genes in Z/Q and Z/D fruits showed that *PbSWEET15* and *PbSWEET17* were not expressed in the fruit. The *PbSWEET6*, *PbSWEET8*, *PbSWEET10*, and *PbSWEET14* expression levels were significantly higher in Z/Q fruit than in Z/D fruit. Of these genes, the greatest difference in expression between Z/Q and Z/D fruits was detected for *PbSWEET6* (Figure 3A). Therefore, we focused on *PbSWEET6* in the subsequent analyses. An examination of tissue-specific expression indicated that *PbSWEET6* was highly expressed in the carpopodium, leaf, and fruit (Figure 3B).

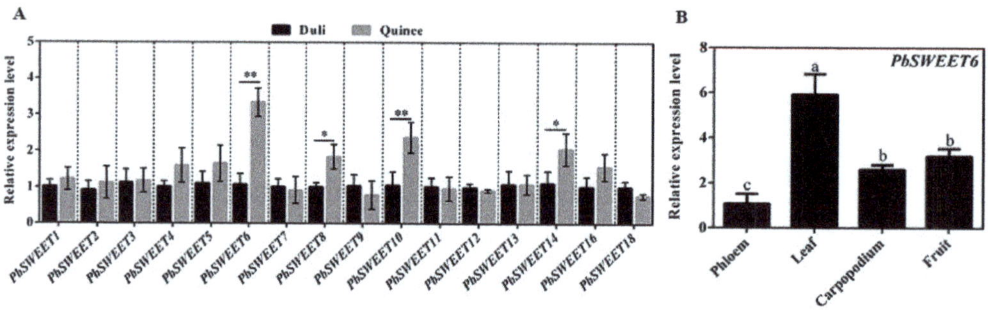

Figure 3. Selected and tissue-specific expression analysis of *PbSWEET6*. (**A**) Relative expression of *PbSWEET* genes from mature fruits of 'Zaosu' pear grafted on 'Yunnan' quince (Z/Q) and 'Duli' (Z/D) as determined by qRT-PCR. (**B**) The relative expression levels of *PbSWEET6* were detected in the phloem, leaves, carpopodium, and fruits of Z/Q during the mature fruit period. The data represent the mean values ± SDs. Different letters represent significant differences (Tukey's HSD, $p < 0.05$) and asterisks indicate significant differences as determined by Student's *t* tests (* $p < 0.05$, ** $p < 0.01$).

3.4. *Phylogenetic Analysis and Subcellular Localization of PbSWEET6*

MEGA X was used to construct a phylogenetic tree according to the maximum likelihood method and a bootstrap analysis (1000 replicates). The phylogenetic analysis revealed that PbSWEET6 is most similar to PuSWEET15 and AtSWEET15 (Figure 4A), suggesting that it is likely involved in sucrose transport. To determine the precise subcellular location of the PbSWEET6 protein, the *PbSWEET6* CDS was fused to a GFP-encoding sequence. After transforming tobacco plants with this construct, fluorescence was clearly observed near the plasmalemma of cells (Figure 4B).

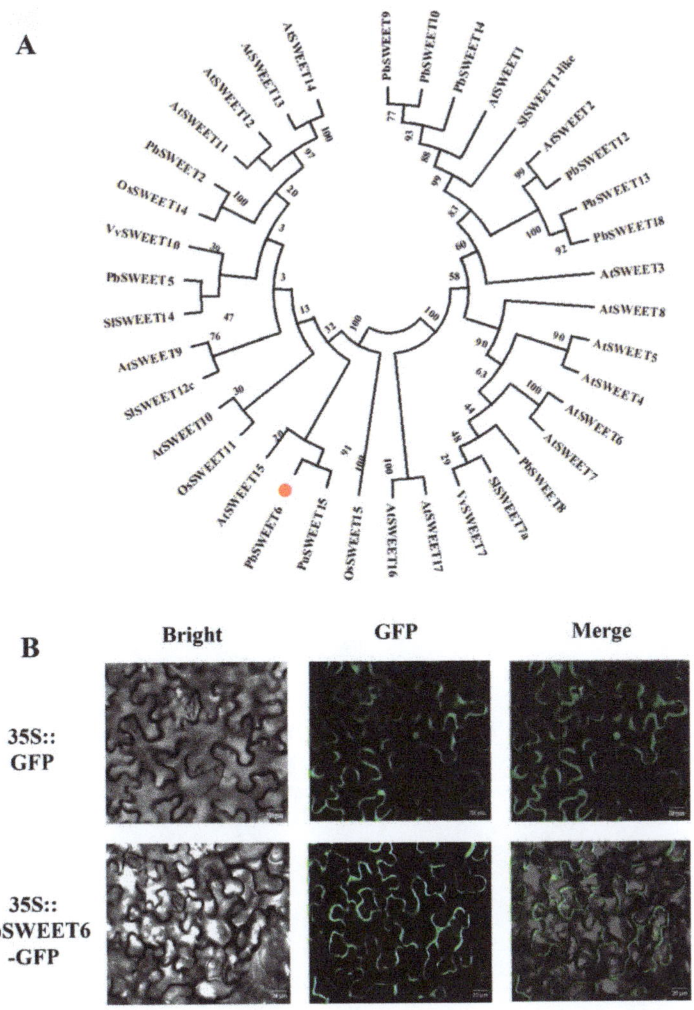

Figure 4. Phylogenetic and subcellular localization analysis of PbSWEET6. (**A**) Phylogenetic analysis of SWEET genes. The phylogenetic tree was inferred using the maximum likelihood method and JTT matrix-based model. PbSWEET6 protein is highlighted with red dots. (**B**) Localization of PbSWEET6. Bar = 20 μm.

3.5. PbSWEET6 Increased the Glucose and Sucrose Content, in Tomato and Pear Fruit Calli

PbSWEET6 was overexpressed in tomato (*PbSWEET6*-OE) to facilitate the further functional characterization of *PbSWEET6*. The PCR and qRT-PCR results confirmed that the tomato plants were successfully transformed. Two transgenic tomato lines were selected for further analyses (Supplemental Figure S1C,D). The overexpression of *PbSWEET6* resulted in significant decreases in plant height and leaf size (Supplemental Figure S1A,B). Compared with the wild-type (WT) fruit, there were no significant differences in the *PbSWEET6*-OE fruit size and weight, but the SSC was significantly higher in the *PbSWEET6*-OE fruit (Figure 5A–C). Additionally, the fructose, glucose, and sucrose contents were significantly higher in *PbSWEET6*-OE fruit than in WT fruit (Figure 5D–F).

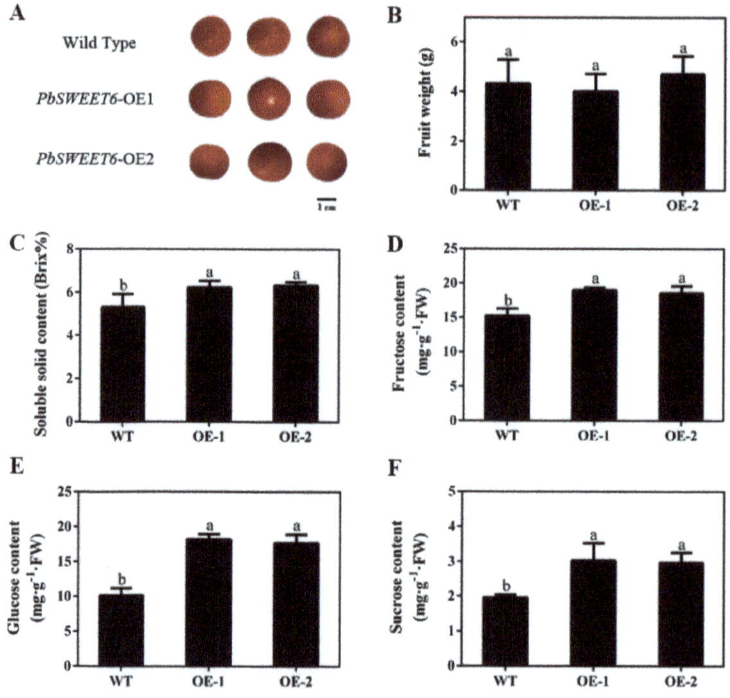

Figure 5. Fruit quality determination in wild type and *PbSWEET6*-OE tomato. (**A**) Fruits of wild type and *PbSWEET6*-OE lines at the red mature fruit stage; the scale bars correspond to 1 cm. (**B**) Single fruit weight. (**C–F**) The SSC and soluble sugar (fructose, glucose, and sucrose) contents in red mature fruits of the WT and *PbSWEET6*-OE lines. The data represent the mean values ± SDs. Different letters represent significant differences (Tukey's HSD, $p < 0.05$).

The *PbSWEET6* gene was also overexpressed in pear fruit calli, with an expression level that was significantly higher than that in the WT calli (Figure 6A,B). Furthermore, the glucose and sucrose contents were significantly higher in the *PbSWEET6*-OE fruit calli than in the WT fruit calli, whereas there were no significant differences in the fructose and sorbitol contents (Figure 6C–F).

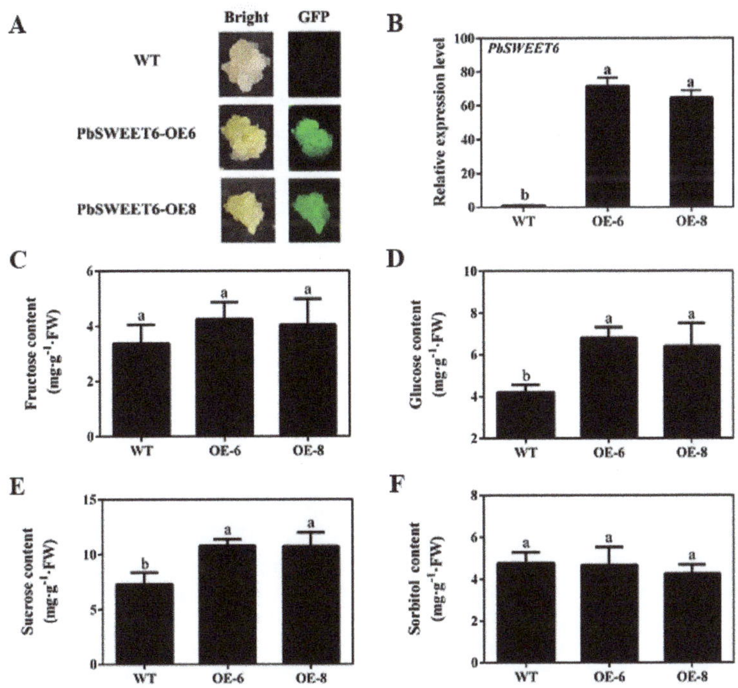

Figure 6. Effects of overexpressing *PbSWEET6* on sugar concentrations in pear fruit calli. (**A**) The transgenic pear fruit calli could be detected to express green fluorescent protein (GFP). OE-6 and OE-8 indicate the overexpressing *PbSWEET6* of pear fruit calli lines. (**B**) The qRT-PCR analysis of *PbSWEET6* expression levels in WT, OE-6 and OE-8 lines. (**C–F**) The soluble sugar (fructose, glucose, sucrose, and sorbitol) contents of WT, OE-6 and OE-8 pear fruit calli. The data represent the mean values ± SDs. Different letters represent significant differences (Tukey's HSD, $p < 0.05$).

4. Discussion

The rootstock significantly affects the vegetative and reproductive growth of the scion [32–35]. Quince is commonly used as a dwarfing rootstock for pear [36], resulting in precocious scion fruit development and increased fruit quality [4]. However, the molecular mechanism underlying the effects of quince rootstock on scion fruit quality remains largely unknown. In this study, 'Yunnan' quince significantly increased the weight and SSC of 'Zaosu' pear fruit. Moreover, the fructose, glucose, and sucrose contents were significantly greater in Z/Q fruit than in Z/D fruit. These results indicate that 'Yunnan' quince may be a better rootstock than 'Duli' for improving scion fruit quality.

Plants convert carbon dioxide and water into carbohydrates and oxygen through photosynthesis [37], which provides the energy and carbon necessary for plant growth and fruit formation. In tomato, the photosynthetic activity of green fruit influences the quality of ripe fruit [38]. Thus, photosynthesis has crucial effects on fruit quality. An examination of the leaves of 'Zaosu' pear grafted on 'Yunnan' quince and 'Duli' indicated there were no significant differences in the photosynthetic rate and chlorophyll concentration between the two combinations. Accordingly, the difference in the scion fruit sugar contents is likely unrelated to photosynthesis. In plants, photosynthates are transported from source organs to sink organs. Dwarfing rootstocks reportedly affect scion photosynthate transport. A previous study indicated that in apple, more photosynthates are transported to fruits when the scion is grafted on M9 and SH40 dwarfing rootstocks than when the scion is grafted on BC standard rootstocks [39]. Thus, we speculated that 'Yunnan' quince may facilitate the transport of photoassimilates to fruit. To assess this possibility, we performed ^{13}C

feeding experiments, which showed that 'Yunnan' quince promoted the distribution of photoassimilates to the fruit better than 'Duli', which was consistent with the results of prior research. Photoassimilates are transported from the source to the sink primarily as sugars. Hence, sugar distribution is a major determinant of fruit quality [40]. Therefore, the increase in the sugar content of 'Zaosu' pear fruit produced after grafting on 'Yunnan' quince may be associated mainly with changes in sugar transport.

Sucrose is one of the main forms of long-distance transport of assimilates. In plants, SWEET proteins can transport various sugars, including sucrose and fructose [14,26,41]. Of the four PbSWEET genes that were more highly expressed in Z/Q fruit than in Z/D fruit, *PbSWEET8* (Clade II of the *SWEET* family) as well as *PbSWEET10* and *PbSWEET14* (Clade I) likely encode hexose transporters, whereas *PbSWEET6* (Clade III) is probably involved in the transport of sucrose [9,14]. Among these four genes, the largest expression-level difference between the analyzed fruits was observed for PbSWEET6. Accordingly, it was selected for the follow-up study. Tissue-specific expression assays showed that *PbSWEET6* was highly expressed in the leaves, fruits, and carpopodium, which are involved in the transport of assimilates to fruit. These findings are in accordance with the expected sugar-transport-related function of *PbSWEET6*. Additionally, PbSWEET6 was localized in the plasma membrane, similar to PuSWEET15 and SlSWEET1a [13,26]. The phylogenetic analysis indicated that PbSWEET6 is most closely related to PuSWEET15 and AtSWEET15 (i.e., Clade III SWEETs). To further verify the *PbSWEET6* function, we overexpressed *PbSWEET6* in tomato plants and pear fruit calli. The *PbSWEET6*-OE tomato plants were shorter than the WT plants, possibly because of the increased transport of photoassimilates to fruit. An examination of sugar contents identified significant increases in the glucose and sucrose levels in tomato fruits and pear fruit calli. In contrast, the fructose content increased only in transgenic tomato fruits, which may reflect the differences in specific characteristics between tomato and pear (e.g., pear calli may have a lower glucose metabolism level). Although Clade III SWEET genes reportedly encode sucrose transporters, the clade also includes transporters of other sugars. For example, *SlSWEET14*, which belongs to Clade III, contributes to the transport of fructose, glucose, and sucrose [23]. In addition to transporting sucrose, *PbSWEET6* may also be involved in the transport of glucose. Therefore, high *PbSWEET6* expression levels might promote glucose and sucrose transport to increase the abundance of these sugars in fruit. Our findings may be useful for further clarifying how dwarfing rootstocks affect the transport of sugars into pear fruit.

5. Conclusions

'Yunnan' quince promotes the flow of photoassimilates to 'Zaosu' pear fruit, thereby increasing the fruit sugar content. The upregulated expression of *PbSWEET6* may be directly or indirectly involved in the accumulation of glucose and sucrose in 'Zaosu' pear fruit. Furthermore, overexpressed *PbSWEET6* influences the sucrose and glucose accumulation in tomato fruits and pear fruit calli. This study has expanded our understanding of how the regulation of *PbSWEET6* expression affects the fruit sugar content. This information may help breeders select pear rootstocks that can optimize fruit sugar contents.

Supplementary Materials: The following supporting information can be downloaded at: https://www.mdpi.com/article/10.3390/horticulturae8070649/s1, Table S1: Primers used for relative expression analysis, gene cloning and vector construction; Figure S1: The detection and plant phenotype of transgenic tomato (OE-1, OE-2) overexpressing PbSWEET6.

Author Contributions: Conceptualization, X.W., Z.W. and L.C.; methodology, L.C., J.P. and Y.C.; software, L.C. and J.P.; validation, X.W., L.C., Z.W., R.Z. and L.X.; formal analysis, J.P. and Y.C.; investigation, X.W., L.C., J.P. and Y.C.; resources, L.X.; data curation, X.W., L.C., Z.W. and R.Z.; writing—original draft preparation, X.W. and L.C.; writing—review and editing, X.W., L.C., Z.W. and R.Z.; visualization, Y.C. and C.Y.; supervision, L.X.; project administration, Z.W., R.Z., C.Y. and L.X.; funding acquisition, L.X. All authors have read and agreed to the published version of the manuscript.

Funding: This research was funded by the National Key Research and Development Program of China (No. 2019YFD1000100).

Institutional Review Board Statement: Not applicable.

Informed Consent Statement: Not applicable.

Data Availability Statement: The data presented in this study are available in the article and Supplementary Materials.

Acknowledgments: We thank Liwen Bianji (Edanz) (www.liwenbianji.cn, accessed on 21 April 2022) for editing the English text of a draft of this manuscript.

Conflicts of Interest: The authors declare no conflict of interest.

References

1. Einhorn, T.C. A review of recent *Pyrus*, *Cydonia* and *Amelanchier* rootstock selections for high-density pear plantings. *Acta Hortic. Sin.* **2021**, *60*, 185–196. [CrossRef]
2. Watson, A.E.; Seleznyova, A.N.; Dayatilake, G.A.; Tustin, D.S. Rootstocks affect pear (*Pyrus communis*) tree growth through extent of node neoformation and flowering wisth key differences to apple. *Funct. Plant Biol.* **2012**, *39*, 493–502. [CrossRef]
3. Donadio, L.C.; Lederman, I.E.; Roberto, S.R.; Stucchi, E.S. Dwarfing-canopy and rootstock cultivars for fruit trees. *Rev. Bras. Frutic.* **2019**, *41*, e997. [CrossRef]
4. McClymont, L.; Goodwin, I.; Whitfield, D.; O'Connell, M.; Turpin, S. Effects of rootstock, tree density and training system on early growth, yield and fruit quality of blush pear. *HortScience* **2021**, *56*, 1408–1415. [CrossRef]
5. Ou, C.; Wang, F.; Wang, J.; Li, S.; Zhang, Y.; Fang, M.; Ma, L.; Zhao, Y.; Jiang, S. A de novo genome assembly of the dwarfing pear rootstock Zhongai 1. *Sci. Data* **2019**, *6*, 281. [CrossRef] [PubMed]
6. Du Plooy, P.; van Huyssteen, P. Effect of BP1, BP3 and Quince A rootstocks, at three planting densities, on precocity and fruit quality of 'Forelle' pear (*Pyrus communis* L.). *S. Afr. J. Plant* **2000**, *17*, 57–59. [CrossRef]
7. Zhang, C.M.; Bian, Y.; Hou, S.H.; Li, X.G. Sugar transport played a more important role than sugar biosynthesis in fruit sugar accumulation during Chinese jujube domestication. *Planta* **2018**, *248*, 1187–1199. [CrossRef] [PubMed]
8. Braun, D.M.; Wang, L.; Ruan, Y.L. Understanding and manipulating sucrose phloem loading, unloading, metabolism, and signalling to enhance crop yield and food security. *J. Exp. Bot.* **2014**, *65*, 1713–1735. [CrossRef]
9. Chen, L.Q.; Cheung, L.S.; Feng, L.; Tanner, W.; Frommer, W.B. Transport of sugars. *Annu. Rev. Biochem.* **2015**, *84*, 865–894. [CrossRef]
10. Tao, Y.; Cheung, L.S.; Li, S.; Eom, J.S.; Chen, L.Q.; Xu, Y.; Perry, K.; Frommer, W.B.; Feng, L. Structure of a eukaryotic SWEET transporter in a homotrimeric complex. *Nature* **2015**, *527*, 259–263. [CrossRef]
11. Xuan, Y.H.; Hu, Y.B.; Chen, L.Q.; Sosso, D.; Ducat, D.C.; Hou, B.H.; Frommer, W.B. Functional role of oligomerization for bacterial and plant SWEET sugar transporter family. *Proc. Natl. Acad. Sci. USA* **2013**, *110*, E3685–E3694. [CrossRef] [PubMed]
12. Zhen, Q.L.; Fang, T.; Peng, Q.; Liao, L.; Zhao, L.; Owiti, A.; Han, Y.P. Developing gene-tagged molecular markers for evaluation of genetic association of apple *SWEET* genes with fruit sugar accumulation. *Hortic. Res.* **2018**, *5*, 14. [CrossRef] [PubMed]
13. Ho, L.H.; Klemens, P.A.W.; Neuhaus, H.E.; Ko, H.Y.; Hsieh, S.Y.; Guo, W.J. SlSWEET1a is involved in glucose import to young leaves in tomato plants. *J. Exp. Bot.* **2019**, *70*, 3241–3254. [CrossRef]
14. Eom, J.S.; Chen, L.Q.; Sosso, D.; Julius, B.T.; Lin, I.W.; Qu, X.Q.; Braun, D.M.; Frommer, W.B. SWEETs, transporters for intracellular and intercellular sugar translocation. *Curr. Opin. Plant Biol.* **2015**, *25*, 53–62. [CrossRef]
15. Li, J.M.; Zheng, D.M.; Li, L.T.; Qiao, X.; Wei, S.W.; Bai, B.; Zhang, S.L.; Wu, J. Genome-wide function, evolutionary characterization and expression analysis of sugar transporter family genes in pear (*Pyrus bretschneideri* Rehd). *Plant Cell Physiol.* **2015**, *56*, 1721–1737. [CrossRef] [PubMed]
16. Chen, L.Q.; Hou, B.H.; Lalonde, S.; Takanaga, H.; Hartung, M.L.; Qu, X.Q.; Guo, W.J.; Kim, J.G.; Underwood, W.; Chaudhuri, B.; et al. Sugar transporters for intercellular exchange and nutrition of pathogens. *Nature* **2010**, *468*, 527–532. [CrossRef]
17. Chen, L.Q.; Qu, X.Q.; Hou, B.H.; Sosso, D.; Osorio, S.; Fernie, A.R.; Frommer, W.B. Sucrose efflux mediated by SWEET proteins as a key step for phloem transport. *Science* **2012**, *335*, 207–211. [CrossRef]
18. Bezrutczyk, M.; Hartwig, T.; Horschman, M.; Char, S.N.; Yang, J.L.; Yang, B.; Frommer, W.B.; Sosso, D. Impaired phloem loading in *zmsweet13a,b,c* sucrose transporter triple knock-out mutants in *Zea mays*. *New Phytol.* **2018**, *218*, 594–603. [CrossRef]
19. Chardon, F.; Bedu, M.; Calenge, F.; Klemens, P.A.W.; Spinner, L.; Clement, G.; Chietera, G.; Leran, S.; Ferrand, M.; Lacombe, B.; et al. Leaf fructose content is controlled by the vacuolar transporter SWEET17 in *Arabidopsis*. *Curr. Biol.* **2013**, *23*, 697–702. [CrossRef]
20. Klemens, P.A.W.; Patzke, K.; Deitmer, J.; Spinner, L.; Le Hir, R.; Bellini, C.; Bedu, M.; Chardon, F.; Krapp, A.; Neuhaus, H.E. Overexpression of the vacuolar sugar carrier AtSWEET16 modifies germination, growth, and stress tolerance in *Arabidopsis*. *Plant Physiol.* **2013**, *163*, 1338–1352. [CrossRef]
21. Valifard, M.; Le Hir, R.; Muller, J.; Scheuring, D.; Neuhaus, H.E.; Pommerrenig, B. Vacuolar fructose transporter *SWEET17* is critical for root development and drought tolerance. *Plant Physiol.* **2021**, *187*, 2716–2730. [CrossRef] [PubMed]

22. Ru, L.; He, Y.; Zhu, Z.; Patrick, J.W.; Ruan, Y.L. Integrating sugar metabolism with transport: Elevation of endogenous cell wall invertase activity up-regulates *SlHT2* and *SlSWEET12c* expression for early fruit development in tomato. *Front. Genet.* **2020**, *11*, 592596. [CrossRef] [PubMed]
23. Zhang, X.S.; Feng, C.Y.; Wang, M.N.; Li, T.L.; Liu, X.; Jiang, J. Plasma membrane-localized SlSWEET7a and SlSWEET14 regulate sugar transport and storage in tomato fruits. *Hortic. Res.* **2021**, *8*, 186. [CrossRef]
24. Zhang, Z.; Zou, L.; Ren, C.; Ren, F.; Wang, Y.; Fan, P.; Li, S.; Liang, Z. *VvSWEET10* mediates sugar accumulation in grapes. *Genes* **2019**, *10*, 255. [CrossRef]
25. Li, J.M.; Qin, M.F.; Qiao, X.; Cheng, Y.S.; Li, X.L.; Zhang, H.P.; Wu, J. A new insight into the evolution and functional divergence of SWEET transporters in Chinese white pear (*Pyrus bretschneideri*). *Plant Cell Physiol.* **2017**, *58*, 839–850. [CrossRef] [PubMed]
26. Li, X.Y.; Guo, W.; Li, J.C.; Yue, P.T.; Bu, H.D.; Jiang, J.; Liu, W.T.; Xu, Y.X.; Yuan, H.; Li, T.; et al. Histone acetylation at the promoter for the transcription factor PuWRKY31 affects sucrose accumulation in pear fruit(1)(OPEN). *Plant Physiol.* **2020**, *182*, 2035–2046. [CrossRef]
27. Lalonde, S.; Boles, E.; Hellmann, H.; Barker, L.; Patrick, J.W.; Frommer, W.B.; Ward, J.M. The dual function of sugar carriers: Transport and sugar sensing. *Plant Cell* **1999**, *11*, 707–726. [CrossRef]
28. Morales, J.; Bermejo, A.; Navarro, P.; Forner-Giner, M.A.; Salvador, A. Rootstock effect on fruit quality, anthocyanins, sugars, hydroxycinnamic acids and flavanones content during the harvest of blood oranges 'Moro' and 'Tarocco Rosso' grown in Spain. *Food Chem.* **2021**, *342*, 128305. [CrossRef]
29. Lu, Y.H.; Watanabe, A.; Kimura, M. Input and distribution of photosynthesized carbon in a flooded rice soil. *Glob. Biogeochem. Cycles* **2002**, *16*, 1085. [CrossRef]
30. Sha, J.C.; Wang, F.; Xu, X.X.; Chen, Q.; Zhu, Z.L.; Jiang, Y.M.; Ge, S.F. Studies on the translocation characteristics of C-13-photoassimilates to fruit during the fruit development stage in 'Fuji' apple. *Plant Physiol. Biochem.* **2020**, *154*, 636–645. [CrossRef]
31. Zhang, H.Q.; Han, W.; Wang, H.B.; Cong, L.; Zhai, R.; Yang, C.Q.; Wang, Z.G.; Xu, L.F. Downstream of GA(4), *PbCYP78A6* participates in regulating cell cycle-related genes and parthenogenesis in pear (*Pyrus bretshneideri* Rehd.). *BMC Plant Biol.* **2021**, *21*, 292. [CrossRef]
32. Cantín, C.M.; Pinochet, J.; Gogorcena, Y.; Moreno, M.Á. Growth, yield and fruit quality of 'Van' and 'Stark Hardy Giant' sweet cherry cultivars as influenced by grafting on different rootstocks. *Sci. Hortic.* **2010**, *123*, 329–335. [CrossRef]
33. Kviklys, D.; Lanauskas, J.; Uselis, N.; Viskelis, J.; Viskeliene, A.; Buskiene, L.; Staugaitis, G.; Mazeika, R.; Samuoliene, G. Rootstock vigour and leaf colour affect apple tree nutrition. *Zemdirb.-Agric.* **2017**, *104*, 185–190. [CrossRef]
34. Ou, C.; Jiang, S.; Wang, F.; Tang, C.; Hao, N. An RNA-Seq analysis of the pear (*Pyrus communis* L.) transcriptome, with a focus on genes associated with dwarf. *Plant Gene* **2015**, *4*, 69–77. [CrossRef]
35. Prassinos, C.; Ko, J.H.; Lang, G.; Iezzoni, A.F.; Han, K.H. Rootstock-induced dwarfing in cherries is caused by differential cessation of terminal meristem growth and is triggered by rootstock-specific gene regulation. *Tree Physiol.* **2009**, *29*, 927–936. [CrossRef]
36. Irisarri, P.; Binczycki, P.; Errea, P.; Martens, H.J.; Pina, A. Oxidative stress associated with rootstock-scion interactions in pear/quince combinations during early stages of graft development. *J. Plant Physiol.* **2015**, *176*, 25–35. [CrossRef]
37. Okumura, M.; Inoue, S.; Kuwata, K.; Kinoshita, T. Photosynthesis activates plasma membrane H+-ATPase via sugar accumulation. *Plant Physiol.* **2016**, *171*, 580–589. [CrossRef] [PubMed]
38. Nadakuduti, S.S.; Holdsworth, W.L.; Klein, C.L.; Barry, C.S. *KNOX* genes influence a gradient of fruit chloroplast development through regulation of *GOLDEN2-LIKE* expression in tomato. *Plant J.* **2014**, *78*, 1022–1033. [CrossRef]
39. An, H.S.; Luo, F.X.; Wu, T.; Wang, Y.; Xu, X.F.; Zhang, X.Z.; Han, Z.H. Effect of rootstocks or interstems on dry matter allocation in apple. *Eur. J. Hortic. Sci.* **2017**, *82*, 225–231. [CrossRef]
40. Teo, G.; Suziki, Y.; Uratsu, S.L.; Lampinen, B.; Ormonde, N.; Hu, W.K.; DeJong, T.M.; Dandekar, A.M. Silencing leaf sorbitol synthesis alters long-distance partitioning and apple fruit quality. *Proc. Natl. Acad. Sci. USA* **2006**, *103*, 18842–18847. [CrossRef]
41. Selvana, B.; Yu, Y.C.; Chen, L.Q.; Shukla, D. Molecular basis of the glucose transport mechanism in plants. *ACS Cent. Sci.* **2019**, *5*, 1085–1096. [CrossRef] [PubMed]

horticulturae MDPI

Article

Dynamic Changes of Fruit Physiological Quality and Sugar Components during Fruit Growth and Development of *Actinidia eriantha*

Junjie Tao [1,2], Mengting Wu [1,2], Xudong Jiao [1,2], Shuangshuang Chen [1,2], Dongfeng Jia [1,2], Xiaobiao Xu [1,2] and Chunhui Huang [1,2,*]

1 College of Agronomy, Jiangxi Agricultural University, Nanchang 330045, China; taojj@jxau.edu.cn (J.T.); wmt980923@163.com (M.W.); jiaoxudong77@163.com (X.J.); css13736206317@163.com (S.C.); dongfengjia@163.com (D.J.); xbxu@jxau.edu.cn (X.X.)

2 Institute of Kiwifruit, Jiangxi Agricultural University, Nanchang 330045, China

* Correspondence: lindahch@126.com; Tel.: +86-181-7090-5629

Abstract: 'Ganlv 1' is a new cultivar of *Actinidia eriantha* selected from the wild natural population, which has the advantages of moderate taste, high yield, easy peeling and high ascorbic acid (AsA) content. In this study, 'Ganlv 1' was used to explore the changes in fruit quality, soluble sugar components, sucrose metabolism-related enzymes activities and sucrose metabolism-related enzyme genes' expression during the fruit's development. The results showed that, except for AsA, the changes in the fruit quality index and fruit growth and development during the development of 'Ganlv 1' basically exhibited the same trend. The fruit shape index was different in the different development stages of the fruit, and tended to be stable with fruit growth and development. The dynamic changes of the dry matter content indicated that the best time for fruit harvest was about 160 days after full bloom. The main sugar components in the fruit were fructose, glucose and sucrose, and sucrose and glucose were the main sugars in the soft-ripening stage. The trend of sucrose accumulation, the activities of the sucrose metabolism-related enzymes and the expression of the sucrose metabolism-related genes indicated that 130–145 days after full bloom (DAFB) might be the critical period of sucrose metabolism. The results are of great significance for clarifying the developmental characteristics and dynamic changes in the sugar components in *A. eriantha* fruits, and lay a foundation for further studying of the mechanism of sugar metabolism in *A. eriantha*.

Keywords: *Actinidia eriantha*; fruit quality; sugar component; sugar metabolism; gene expression; kiwifruit

Citation: Tao, J.; Wu, M.; Jiao, X.; Chen, S.; Jia, D.; Xu, X.; Huang, C. Dynamic Changes of Fruit Physiological Quality and Sugar Components during Fruit Growth and Development of *Actinidia eriantha*. *Horticulturae* **2022**, *8*, 529. https://doi.org/10.3390/horticulturae8060529

Academic Editor: Esmaeil Fallahi

Received: 28 April 2022
Accepted: 13 June 2022
Published: 15 June 2022

1. Introduction

Actinidia eriantha Benth. is a perennial deciduous vine of the genus, *Actinidia*. The surface of *A. eriantha* is covered with a layer of white fluff, and its emerald green flesh is rich in flavor and nutrients. Compared with *A. chinensis* and *A. deliciosa*, *A. eriantha* has stronger adaptability and stress resistance. Its high content of AsA is the most striking characteristic of *A. eriantha*, and 100 g of its fresh fruit contains as much as 596–1397 mg of AsA, which is three to four times that of *A. chinensis* [1]. Therefore, *A. eriantha* is a special germplasm resource for cultivating new varieties of kiwifruit with a high AsA content. In addition, *A. eriantha* has the characteristics of easy peeling after soft ripening, a long shelf life, and other characteristics which have high scientific research and economic value. China has abundant germplasm resources of *A. eriantha*, and many new cultivars of a good quality have been bred, such as 'White' and 'Ganmi 6' [2,3]. 'Ganlv 1' is a new cultivar selected from the natural population of wild *A. eriantha* by our group. It has the advantages of moderate taste, good yield and high AsA content, and has a high potential for development and utilization. However, the dynamic changes in the fruit quality and sugar components during the development of 'Ganlv 1' fruits are still unclear.

The fruit quality is mainly determined by the exterior and internal quality of the fruit. The exterior quality mainly includes the fruit color, fruit weight, freshness, fruit shape index and other indicators, while the internal quality is mainly determined by a variety of fruit inclusions, such as the soluble solids content, total acid, sugar components, nutrients, flavor and so on. At present, many studies have been carried out to determine and evaluate the fruit quality of *A. eriantha*. For example, by comparing the effects of different harvest maturity on the postharvest fruit quality and storage of *A. eriantha* 'White', the results showed that the fruit should be harvested, as far as possible, before 165 days after full bloom [4]. The three main components of the comprehensive evaluation of fruit quality were identified by analyzing the quality of wild *A. eriantha*, and the diversity of the fruit quality of different wild *A. eriantha* was revealed [5]. The comprehensive analysis of the fruit quality of wild A. eriantha showed that there were great differences in the fruit quality indexes, among which the coefficient of variation of single fruit weight was the largest (>33%), while the coefficient of fruit transverse and soluble solids was relatively small (11% and 11.91%, respectively), indicating that the fruit traits of wild *A. eriantha* had a rich genetic diversity [6,7]. Based on the BBCH (Biologische Bundesantalt, Bundessortenamt and Chemische Industrie) measurement method, a standard model for systematically describing the growth and development of the 'White' fruit was constructed, and the characteristics of fruit morphology, growth and development, as well as the accumulation and dynamic changes of the carbohydrate and organic acid in fruit, from anthesis to senescence, were studied [8]. In addition, the dynamic changes in the soluble sugars, titratable acid and ascorbic acid contents during the fruit development of 'Ganmi 6' and other *A. eriantha* fruits were also examined [9–11]. These studies are helpful to understand the growth and development characteristics and fruit quality of *A. eriantha*.

Sugar is an important basis for the formation of fruit flavor quality. The metabolism and accumulation of sugar in fruit are important factors affecting the fruit quality. The types and quantity of sugars determine the quality of the fruit to a large extent. The soluble sugars accumulated in the fruits of horticultural plants are mainly sucrose, fructose and glucose [12,13]. However, in the process of fruit growth and development, the types and contents of the sugars in the fruits of different horticultural plants are different to some extent. For example, the fruit of the wolfberry (*Lycium barbarum* L.) was rich in total sugars and fructose, and the predominant sugars at the maturity of the fruit were fructose and glucose [14]. In the determination of the sugar components of 86 apple varieties, it was found that the soluble sugar in the apple fruits was mainly fructose, and the content of sucrose (92.7%) in most of the varieties was higher than that of glucose [15]. The main components of sugar in grapes were glucose and fructose, and the content of sucrose was lower [16]. During the development of the 'White' fruit, the content of fructose was the highest, followed by glucose, and the content of sucrose was the lowest [8]. The sucrose content of 'Ganmi 6' was higher than that of glucose and fructose in the early stages of fruit development, while the content of fructose was the highest, followed by glucose and sucrose in the late stages of fruit development [17].

In this study, the fruit quality, sugar components and sugar metabolism-related enzyme activities of the new *A. eriantha* cultivar 'Ganlv 1' during fruit development were detected and analyzed to reveal the dynamic changes in fruit quality and sugar metabolism-related mechanism characteristics of 'Ganlv 1'. The results will lay a foundation for further understanding the fruit development characteristics, fruit quality formation mechanisms and the development and utilization of the new varieties of *A. eriantha*.

2. Materials and Methods

2.1. Experimental Materials

The experimental material of *A. eriantha* 'Ganlv 1' was collected from the germplasm resource nursery of the Kiwifruit Research Institute of Fengxin County, Jiangxi Province. The fruit samples were picked on 3 June 2019, which was 25 days after full bloom (DAFB), and the fruits were collected every 15 days for a total of 13 times. Thirty fruits of the same size and with no damage were collected randomly at each time point. The samples at the soft-ripening stage (fruit firmness reached 1.0 kg·cm^{-2}) were collected separately to the fruit ripening stage. The fruits collected each time were randomly divided into three groups, and each group contained 10 fruits, which represented the three repetitions of the experiment. After measuring the appearance quality, the collected fruit samples were placed at room temperature until naturally soft and ripe (fruit firmness reached 1.0 kg·cm^{-2}). Then, the epidermis of the soft, ripe fruit samples was removed, and the outer and inner pericarps were chopped and immediately frozen with liquid nitrogen. Finally, the samples were stored in a refrigerator at −80 °C for later use.

2.2. Measurement of Fruit Physiological Quality

The fruit shape index and single fruit weight were measured immediately after the kiwifruit samples were collected, and other fruit quality indexes were determined after the natural soft ripening at room temperature. The appearance qualities during the fruit development, including longitudinal diameter, transverse diameter and lateral diameter, were measured using vernier calipers. The fruit shape index was obtained by calculating the ratio of the longitudinal and transverse diameters. The single fruit weight was measured using an analytical balance. The soluble solids content (SSC) was determined using a digital hand-held pocket refractometer (model PAL-1, ATAGO, Tokyo, Japan). The dry matter (DM) content was determined by drying a 5 mm slice from the equatorial central transverse section of the fruit at 55 °C for about 24 h until constant weight. The content of titratable acid (TA) was determined by acid-base titration [18]. The content of vitamin C was determined by the 2,6-dichloroindophenol titration method [18]. The soluble sugar content was determined by the anthrone colorimetry method [18].

2.3. Determination of Sugar Components

The sugar components were determined by improved high-performance liquid chromatography (HPLC). About 4 g of the frozen pulp was accurately weighed from each sample, ground into homogenate under liquid nitrogen, and then transferred into a 15 mL centrifuge tube. After adding 5 mL of 80% ethanol, the centrifuge tube was placed in a water bath at 35 °C for about 25 min, then centrifuged at 1000 rpm at room temperature for 15 min, and the supernatant was transferred to a volumetric flask. The residue was extracted repeatedly three times, and finally, the supernatant was combined and the volume was fixed to 25 mL. One mL of the supernatant was dried by rotary evaporation and then 1 mL of ddH$_2$O was added to dissolve. Afterward, the fully dissolved solution was filtered by a filter with a diameter of 13 mm and an aperture of 0.44 μm. The obtained filtrate was used for the determination of the sugar content by HPLC. Each sample was repeated three times.

Analytical chromatographic separations were carried out on a Waters Spherisorb NH2 column (4.6 mm × 250 mm, 5.0 μm). The mobile phases were composed of acetonitrile and ddH$_2$O (7.8:2.2, *v*/*v*). The injection volume was 20 μL and the flow rate was set at 1.0 mL/min. The temperature of the column was kept at 35 °C. The RID (differential refraction detector) was used for detection.

2.4. Determination of Enzymes Activities Related to Sugar Metabolism

The key enzymes of the plant fruit sugar metabolism pathways were selected to determine the activities of these enzymes, including vacuolar acid invertase (VIN), neutral invertase (NI), sucrose synthase (SS), sucrose phosphate synthase (SPS) and cell wall-bound invertase (CWIN). These enzyme activities were measured using plant Enzyme-Linked Immunosorbent Assay (ELISA) kits (MEIMIAN, Yancheng, China), in accordance with the manufacturer's instructions.

2.5. Expression Analysis of Genes Related to Sugar Metabolism

The sugar metabolism-related genes *NI*, *VIN*, *SS*, *SPS* and *CWIN* of *A. eriantha* were searched from the Kiwifruit Genome Database (KGD) (http://kiwifruitgenome.org/, accessed on 8 October 2019) [19], and the corresponding gene sequences were found. The specific amplified primers were designed using Primer 3 (version 0.4.0), and synthesized by Tsingke Biological Technology (Beijing, China). The detailed information of the primers are shown in Table 1, and the kiwifruit *Actin* gene was used as a reference gene.

Table 1. Primers used for quantitative real-time PCR.

Gene	Forward Primer (5′→3′)	Reverse Primer (5′→3′)
CWIN1	GCCGAAAGGCTACATCAGTCA	TCACTGCACATAAGCACCACATAC
CWIN3	CAAGTCCAAAACCTAGCCGTG	CAAGCAGTGGTGGGGTCTCT
CWIN4	GGCTAACCTTGAAGAGTACACACC	AGCATCAGAGCACATGAGAACC
VIN3	TTCGGCAGAGAAATGGGCT	GTCATAGAATGTCTTTGATGCGTAG
VIN5	CAAGTCCAAAACCTAGCCGTG	CAAGCAGTGGTGGGGTCTCT
NIN1	ACTTTATCGGTAATGTCGGTCCT	CGGGAGTTGCCAATGACG
NIN2	TGCCGAGAGCCGTTTACTG	AGCATCTTCGCCACCAAATAG
NIN3	GGAATGTCAGCCCTGCGAG	TCAGGAGTTGCCAAAGATGCTA
SPS1	AAGCGGGGACACTGACTACG	GAAAGAGGGTAGGTTCTGTTGGC
SPS2	TTGTCTGAAGGAGAGAAGGGAG	ATTGGAAAAGTTACGCTGGAA
SPS5	CAAAGCCGAGATGAAGAAGATG	CCTCCACCACTGTCAACCCA
SS1	GCATTGCTGATACGAAGGG	CGACTATGATTTCCGCTGGT
SS2	GGGAAAACGGGTTAGAGCAG	AAACACCACGAAGAGCAGGG
SS4	TCAGAGATATTCCAGGCACCG	TCAGAGATATTCCAGGCACCG
SS5	CAAGAATCATCGCAGACGGA	GAGTGAGGGCAAGAAGTGTAAGC
Actin	GCTCCACCTGAGAGGAAATAC	CGAAATCCACATCTGTTGAAAG

The total RNA of the 'Ganlv 1' fruit at each stage was extracted by the Quick RNA isolation kit of Waryong, according to the manufacturer's instructions. The quality of RNA was detected by agarose gel electrophoresis, and the qualified RNA was used for subsequent reverse transcription. The total RNA was used as a template and the cDNA was synthesized using Takara reverse transcription kit (PrimeScriptTM RT reagent Kit with gDNA Eraser) (Takara, Dalian, China), according to the manufacturer's instructions. The gene expression analysis was performed by real-time fluorescence quantitative PCR (qRT-PCR) on the Applied Biosystems StepOne RT-PCR system. The RT-PCR reaction system was 20 μL, including TB Green Premix Ex Taq (Tli RNaseH Plus) (Takara, Dalian, China) 10 μL, forward primer (10 μmol/L) 0.5 μL, reverse primer (10 μmol/L) 0.5 μL, RNase free water 8 μL and cDNA 1 μL. The procedure was performed as follows: pre-denaturation at 95 °C for 30 s; denaturation at 95 °C for 5 s; annealing at 60 °C for 30 s; return to step 2; 40 cycles completed. Finally, the melt curve was drawn at 95 °C for 5 s and 64 °C for 30 s. Each sample was repeated in three biological replicates. The relative expression of genes was calculated by the method of $2^{-\Delta\Delta CT}$.

2.6. Data Analysis

Microsoft Excel 2010 software was used for the preliminary analysis of the experimental data and the corresponding trend chart was made. SPSS v23.0 software was used to analyze the difference significance of the data (one-way ANOVA, Duncan's test).

3. Results

3.1. Dynamic Changes of Appearance Quality-Related Indexes of 'Ganlv 1' during Fruit Development

The measurement results of the appearance quality indexes of *A. eriantha* 'Ganlv 1' at the different development stages are shown in Table 2. The vertical diameter of the fruit began to increase rapidly 25 DAFB, and the growth trend slowed down after 40 DAFB, entering the slow growth stage (Table 2). The growth patterns of the fruits' transverse diameter and lateral diameter were basically the same, and the rapid growth period of the fruits' transverse diameter and lateral diameter was from 25 DAFB to 85 DAFB, and then entered the slow growth period (Table 2). The fruit shape index increased rapidly during the period of 25 DAFB to 40 DAFB, then decreased rapidly during the period of 40 DAFB to 85 DAFB, and the fruit shape index increased slowly after 85 DAFB and tended to be stable (Table 2). During the early stages of fruit growth (25–40 DAFB), the rate of vertical growth was much higher than that of transverse growth, indicating that the vertical growth of the fruit was faster at this time and the fruit became longer. With the growth of the fruit, the trend of vertical growth slowed down and the trend of transverse and lateral growth increased, indicating that the fruit was thickening rapidly during this period. During the whole process of the fruit growth, the length of the transverse and lateral diameters was much smaller than that of the vertical diameter, which was consistent with the long cylindrical shape of the *A. eriantha*.

Table 2. Dynamic changes of fruit appearance quality-related indexes of *A. eriantha* 'Ganlv 1'.

Stages (DAFB)	Vertical Diameter (mm)	Transverse Diameter (mm)	Lateral Diameter (mm)	Single Fruit Weight (g)	Fruit Shape Index
25	28.77 ± 2.34 i	15.14 ± 2.9 h	14.7 ± 2.42 g	2.320 ± 0.841 j	1.9 ± 0.18 c
40	40.17 ± 3.58 h	17.3 ± 1.09 g	16.53 ± 0.65 f	6.597 ± 1.158 i	2.32 ± 0.28 a
55	41.56 ± 2.65 gh	19.05 ± 1.56 f	18.3 ± 1.54 e	8.538 ± 1.362 h	2.18 ± 0.13 b
70	41.27 ± 2.82 gh	22.62 ± 3.14 e	21.52 ± 2.47 d	10.349 ± 1.362 g	1.82 ± 0.22 cdef
85	42.22 ± 3.48 g	24.2 ± 3.54 cd	22.76 ± 3.40 cd	11.580 ± 2.252 g	1.74 ± 0.34 fg
100	42.59 ± 3.44 fg	24.82 ± 1.20 bcd	23.81 ± 1.39 abc	13.066 ± 2.695 f	1.72 ± 1.60 g
115	44.11 ± 2.71 ef	25.26 ± 1.69 bc	24.62 ± 2.52 ab	15.809 ± 2.223 e	1.75 ± 0.17 efg
130	44.62 ± 4.31 de	24.91 ± 2.27 bcd	23.43 ± 1.71 bc	16.555 ± 2.976 de	1.79 ± 0.24 defg
145	46.46 ± 4.46 cd	24.94 ± 2.56 bcd	23.82 ± 1.90 abc	17.560 ± 3.901 bcd	1.86 ± 0.29 cde
160	47.84 ± 2.84 bc	26.04 ± 2.52 ab	24.07 ± 1.37 abc	18.132 ± 1.196 bc	1.84 ± 0.26 cdef
175	48.5 ± 2.01 b	25.46 ± 2.32 bc	23.27 ± 2.32 bc	18.363 ± 2.832 b	1.91 ± 0.17 cd
190	51.24 ± 4.91 a	27.18 ± 1.94 a	25.16 ± 3.29 a	21.982 ± 5.741 a	1.89 ± 0.13 cd
Soft-ripening stage	49.05 ± 3.20 b	25.8 ± 2.29 bc	23.14 ± 2.11 c	19.695 ± 2.702 ab	1.93 ± 0.13 c

Letters in the table indicate the level of difference significance.

As the volume of the fruit increased, the weight of the fruit also rapidly increased. The fruit weight of 'Ganlv 1' began to increase rapidly after 25 DAFB. During the period of 145–175 DAFB, the fruit entered a slow growth period (Table 2). At the late period of fruit growth (175–190 DAFB), the fruit entered the pre-harvest expansion stage, and the vertical diameter, transverse diameter and lateral diameter of the fruit increased in varying degrees, the fruit shape index decreased, and the fruit weight also increased significantly. In the soft-ripening stage, the vertical, transverse and lateral diameter of the fruit decreased significantly, the fruit shape index increased and fruit mass decreased significantly. This may be due to more evaporation of water during the soft-ripening stage, resulting in fruit shrinkage. In addition, it could also be due to either dry weather or nutrient decomposition.

3.2. Dynamic Changes in Dry Matter and Soluble Solid Content in A. eriantha 'Ganlv 1' during Fruit Development

The content of dry matter (DM) increased during the fruit development. The DM content increased slowly in the early stages of fruit growth (85–115 DAFB), but increased sharply from 115 DAFB to 145 DAFB (Figure 1a). The DM content increased slowly during the period from 145 DAFB to 160 DAFB but decreased briefly from 160 DAFB to 175 DAFB, and then increased again, and finally reached the maximum value of 19.55% at the soft-ripening stage (Figure 1a).

The soluble solid content (SSC) also maintained an increasing trend during the whole fruit development period. Although the SSC decreased briefly in the early stage (85–100 DAFB), the SSC increased after that (Figure 1b). The SSC increased slowly from

100 DAFB to 160 DAFB, then increased rapidly after 160 DAFB, and reached the maximum value of 15.91% in the soft-ripening stage (Figure 1b).

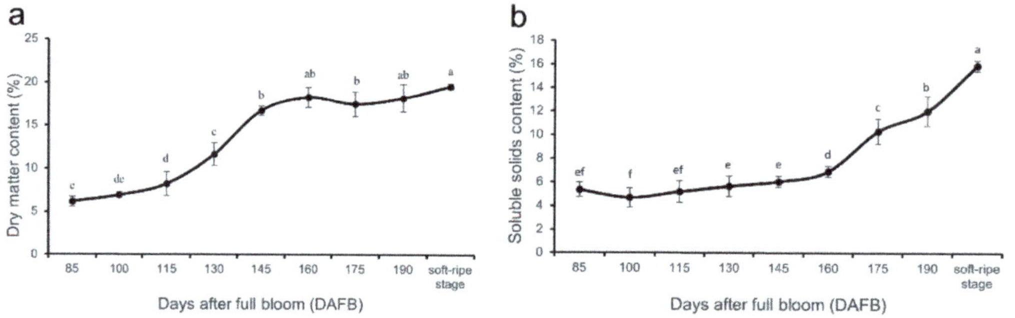

Figure 1. Dynamic changes of dry matter content and soluble solid content during fruit development of *A. eriantha* 'Ganlv 1'. (**a**) Changes of dry matter content during fruit development of 'Ganlv 1'; (**b**) Changes of soluble solid content during fruit development of 'Ganlv 1'.

3.3. Determination of Titratable Acid and AsA Contents in 'Ganlv 1' during Fruit Development

The titratable acid content of 'Ganlv 1' during fruit development is shown in Figure 2a. The titratable acid content of the fruit increased gradually from 25 DAFB to 70 DAFB, and then decreased slightly from 70 DAFB to 100 DAFB. During the period of 100 DAFB to 115 DAFB, the titratable acid content began to rise again and fluctuated in a small range from 115 DAFB to 160 DAFB. After 160 DAFB, the titratable acid content showed a rapid upward trend and reached the maximum value in the soft-ripening stage (Figure 2a). In general, although the content of titratable acid fluctuated in a small range during the fruit development, the content of titratable acid was in an overall upward trend.

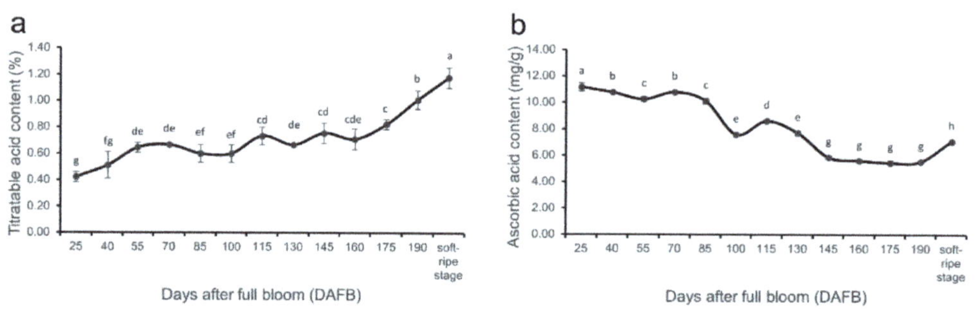

Figure 2. Dynamic changes of titratable acid content and ascorbic acid content during fruit development of *A. eriantha* 'Ganlv 1'. (**a**) Changes of titratable acid content during fruit development of 'Ganlv 1'; (**b**) Changes of ascorbic acid content during fruit development of 'Ganlv 1'.

The changes in the AsA content in 'Ganlv 1' are shown in Figure 3b. The AsA content decreased continuously from 25 DAFB to 55 DAFB, and increased briefly from 55 DAFB to 70 DAFB, followed by a sharp decrease from 70 DAFB to 100 DAFB. During the period of 100 DAFB to 115 DAFB, the AsA content increased slightly, and then began to decrease until it gradually stabilized after 145 DAFB, and there was a slight increase in the AsA content in the soft-ripening stage (Figure 2b). During the fruit development of 'Ganlv 1', the content of AsA generally showed a downward trend, but overall it can still be seen that the AsA content of *A. eriantha* is very high.

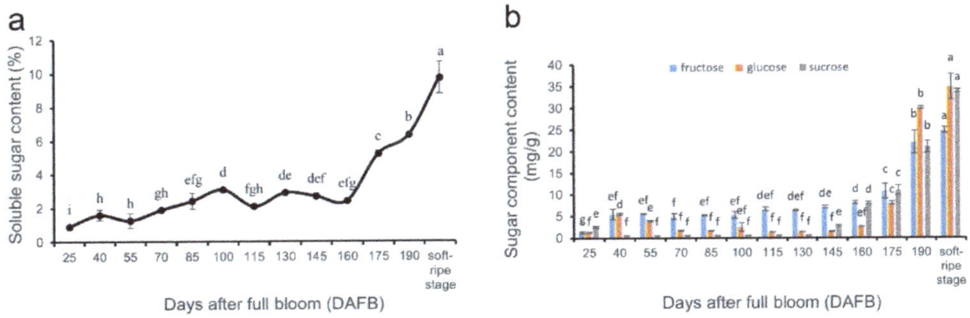

Figure 3. Dynamic changes of soluble sugar content and sugar component content during fruit development of *A. eriantha* 'Ganlv 1'. (**a**) Changes of soluble sugar content during fruit development of 'Ganlv 1'; (**b**) Changes of sugar component content during fruit development of 'Ganlv 1'.

3.4. Change of Soluble Sugar Components in 'Ganlv 1' during Fruit Development

During the fruit development, the soluble sugar content fluctuated slightly before 160 DAFB, but the overall trend was increasing. It began to grow rapidly after 160 DAFB, and reached the maximum in the soft-ripening stage (Figure 3a).

The changes in the soluble sugar components (glucose, sucrose and fructose) during fruit development were determined by HPLC. The content of glucose and fructose increased greatly from 25 DAFB to 40 DAFB, and then the fructose content increased slowly, while the glucose content decreased (Figure 3b). The fructose content began to increase rapidly after 160 DAFB and reached the maximum value in the soft-ripening stage. The glucose content increased slightly at about 100 DAFB, but then immediately decreased and tended to be stable. It began to increase rapidly after 160 DAFB, and the increase rate was much higher than the fructose, and the final glucose content was higher than fructose (Figure 3b). In the early stages of fruit development (25–40 DAFB), the sucrose content decreased slightly, then tended to be stable, and began to increase rapidly at about 130 DAFB. Finally, the sucrose content was close to the glucose content and higher than the fructose content (Figure 3b). On the whole, the content of glucose, fructose and sucrose were relatively low and changed little in the early stages of the fruit growth. The sucrose began to increase at about 130 DAFB, glucose and fructose began to increase at about 160 DAFB, at which time the fruit began to accumulate sugar rapidly.

3.5. Changes of Soluble Sugar-Related Metabolic Enzyme Activities during Fruit Development

The changes in the soluble sugar-related metabolic enzyme activities during the fruit development of 'Ganlv 1' are shown in Figure 4. The VIN activity underwent a process almost opposite to that of sucrose synthase (SS) and sucrose phosphate synthase (SPS). In the early stages of the fruit development, the activity of VIN decreased first and then increased, and reached the highest value at 130 DAFB. Then, it began to decline in a fluctuation pattern and reached the lowest value at about 190 DAFB, but there was a rapid upward trend in the soft-ripening period (Figure 4a). The activity of the soluble neutral invertase (NI) showed a gradual decline from 25 DAFB to 70 DAFB, and it then showed continuous fluctuations from 70 DAFB to 130 DAFB, and reached the lowest value at around 130 DAFB. Then, it began to fluctuate during the period of 130 DAFB to 190 DAFB, and increased significantly in the soft-ripening stage (Figure 4b). The SPS enzyme activity first increased and then decreased in the early stages of fruit development, forming a trough at 70 DAFB, and then began to rise again. It began to decrease at 100 DAFB and reached the lowest point at 130 DAFB. Then, it began to show an upward trend, and showed a slow downward trend again from 175 DAFB to the soft-ripening stage (Figure 4c). The SS enzyme activity maintained a relatively high level during the fruit growth period. From 25–130 DAFB, the SS enzyme activity showed a fluctuating downward trend and decreased to the lowest value at about 130 DAFB, and then increased rapidly from 130 DAFB to 145 DAFB. It showed a fluctuating upward trend during the period from 145 DAFB to 190 DAFB, and showed a downward trend in the soft-ripening stage (Figure 4d). The activity of CWIN decreased during the fruit development. The CWIN activity changed drastically during the 25–160 DAFB period and reached the lowest value at about 160 DAFB. During the period of 160 DAFB to the soft-ripening period, the CWIN activity increased slowly and then decreased gradually (Figure 4e).

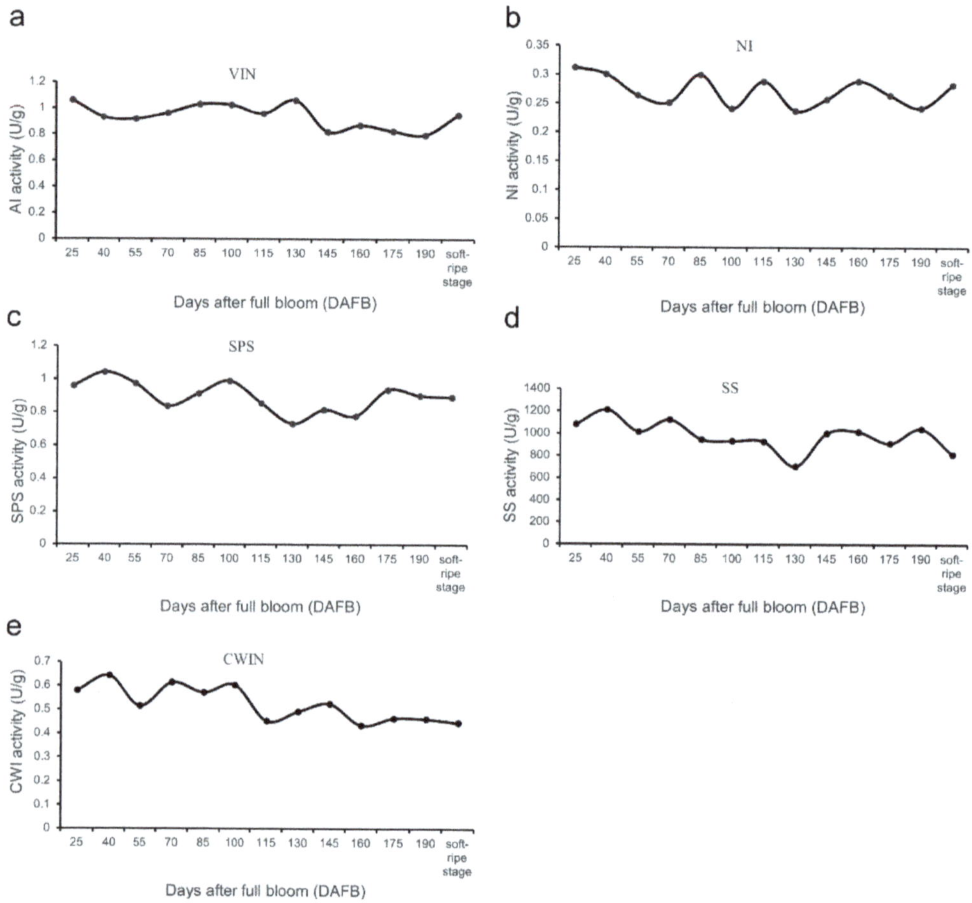

Figure 4. Changes in activities of sugar metabolism relative enzyme during fruit development of *A. eriantha* 'Ganlv 1', including VIN (**a**); NI (**b**); SPS (**c**); SS (**d**); CWIN (**e**).

3.6. Gene Expression Analysis of Soluble Sugar Metabolism-Related Enzymes during Fruit Development

The *VIN3* and *VIN5* are genes related to vacuolar acid invertase, which is also called soluble acid invertase (AI) [20]. The expression level of the *VIN3* gene was extremely low in the early stages of fruit development (25–100 DAFB), but relatively high in the middle and late stages (115 DAFB to the soft-ripening stage), and reached the maximum at 175 DAFB (Figure 5a). The expression level of the *Vin5* gene was the highest at 25 DAFB, then began to decline rapidly, followed by a rapid downward trend, and dropped to the lowest point at 55 DAFB. The expression of this gene increased slowly from 55 DAFB to 100 DAFB, then began to decrease and maintained a lower expression level in the following stages (Figure 5b).

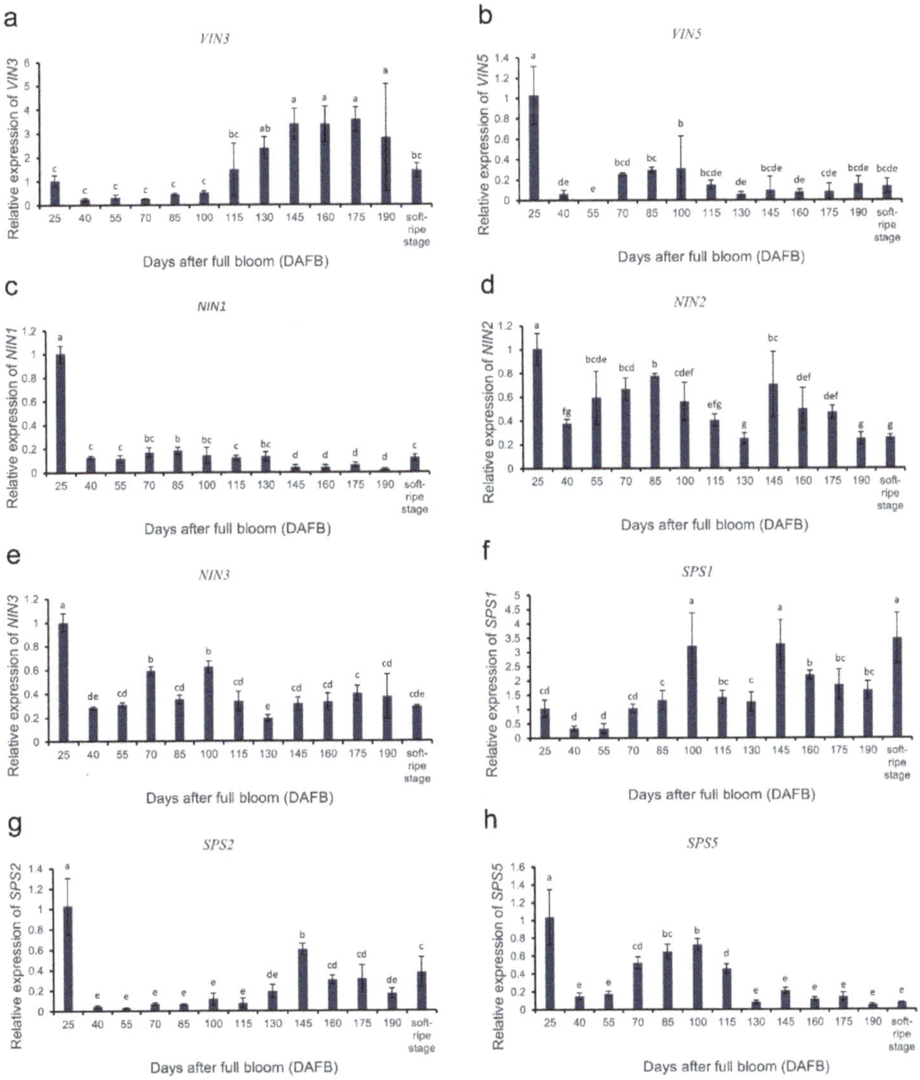

Figure 5. Expression patterns of sucrose metabolism-related genes over *A. eriantha* 'Ganlv 1' fruit development, including *VIN3* (**a**); *VIN5* (**b**); *NIN1* (**c**); *NIN2* (**d**); *NIN3* (**e**); *SPS1* (**f**); *SPS2* (**g**) and *SPS5* (**h**).

The *NIN1*, *NIN2* and *NIN3* are the NI-related genes. The expression level of *NIN1* was the highest at 25 DAFB, and then decreased rapidly afterward (Figure 5c). The *NIN2* has a high level of expression in all of the stages of fruit development, with the highest expression level at 25 DAFB, and a relatively high expression level at 85 and 145 DAFB (Figure 5d). The expression of *NIN3* was also the highest at 25 DAFB, and was relatively higher at 70 and 100 DAFB (Figure 5e).

The *SPS1* had a higher expression level in the middle and late stages of fruit development, among which the expression level was the most prominent at 100 DAFB, 145 DAFB and the soft-ripening stage (Figure 5f). The expression level of *SPS2* was the highest at 25 DAFB, then decreased rapidly, and increased in the later stages of fruit development (145 DAFB to the soft-ripening stage) (Figure 5g). The expression level of *SPS5* was also the highest at 25 DAFB, and then decreased rapidly. The expression level of *SPS5* remained relatively high at 70–115 DAFB, and then decreased rapidly (Figure 5h).

The expression level of the *SS1* gene was higher at 25 DAFB and 85 DAFB, while the expression levels in the other stages were much lower than these two periods (Figure 6a). The expression patterns of *SS2*, *SS4* and *SS5* were basically the same (Figure 6b–d). They all had the highest expression level at 25 DAFB, and then the expression level decreased rapidly and remained at a very low expression level (Figure 6b–d).

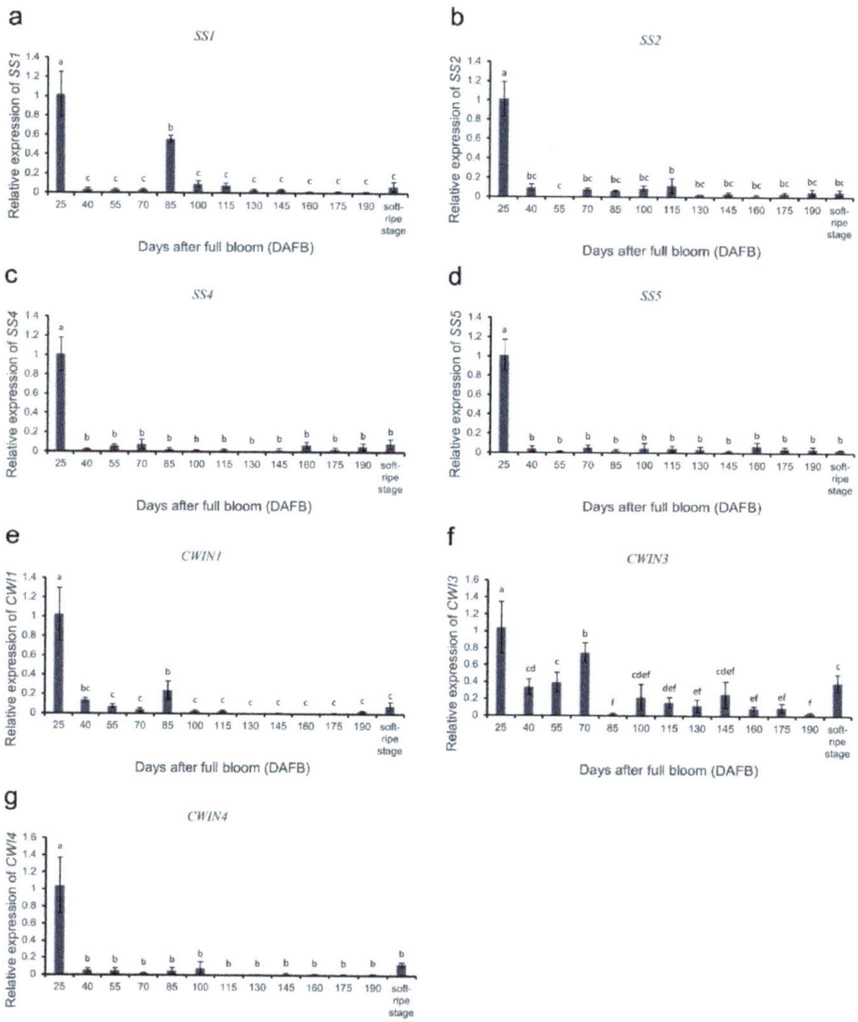

Figure 6. Expression patterns of sucrose metabolism-related genes over *A. eriantha* 'Ganlv 1' fruit development, including *SS1* (**a**); *SS2* (**b**); *SS4* (**c**); *SS5* (**d**); *CWIN1* (**e**); *CWIN3* (**f**) and *CWIN4* (**g**).

The expression levels of the three *CWIN* genes were the highest at 25 DAFB, but the expression levels were different at subsequent stages. For example, for *CWIN1*, in addition to 25 DAFB, it also had a relatively high expression level at 85 DAFB, and then the expression level decreased (Figure 6e). The *CWIN3* had a relatively high expression level at 40 DAFB, 55 DAFB, 70 DAFB and in the soft-ripening stage (Figure 6f). The expression level of *CWIN4* reached its maximum only at 25 DAFB, and then decreased rapidly and leveled off (Figure 6g).

4. Discussion

A. eriantha has strong disease resistance, high-temperature resistance, easy peeling, a long shelf life, is rich in AsA, and thus has great economic character and nutritional value. In this study, the dynamic changes in the fruit quality, sugar composition and content, sucrose synthesis-related enzyme activities and the related gene expressions of the new *A. eriantha* cultivar 'Ganlv 1' during fruit development were detected and analyzed.

The detection results of the fruit appearance quality of 'Ganlv 1' showed that the 'Ganlv 1' fruit increased rapidly in the early stages of development, and, although there was a slow increase in the middle stages, it was not obvious (Table 2). The study of the development characteristics of 'White' showed that the development of 'White' showed a 'double S' curve [8]. However, in this study, the fruit development of 'Ganlv 1' presented an approximate 'double S' curve change. The difference may be caused by the different characteristics of the varieties, or by the difference in certain factors, such as the climate and environment during fruit development. The size and weight of the fruit decreased during the soft-ripening process after picking, which might be due to fruit water loss or nutrient decomposition, suggesting the importance of postharvest preservation of *A. eriantha*. The result of the fruit shape index showed that the fruit shape index of 'Ganlv 1' was basically stable between 1.7–2.5 during the fruit development period, and the fruit shape index was relatively high in the early stages. At this time, the fruit was relatively slender, and the fruit shape index tended to be stable with the development and maturity of the fruit.

The dry matter content is an important index to evaluate the flavor quality of fruit after ripening. By measuring the dry matter content of the fruit at different developmental stages, the ripeness of the fruit can be preliminarily judged and the optimal time for harvesting can be determined [21]. In addition, as an important indicator of fruit quality, the dry matter content also affects the fruit flavor and texture of kiwifruit [22]. The dry matter content of 'Ganlv 1' increased slowly in the early stages, and then sharply from 115 DAFB to 145 DAFB (Figure 1). In the later period (145–160 DAFB), it was in a relatively stable increasing trend and reached a peak of 18.26% at 160 DAFB (Figure 1). After 160 DAFB, it decreased briefly and then began to rise again, to reach the maximum value of 19.55% in the soft-ripening period. The changes in the dry matter content at different stages of 'Ganlv 1' fruit development indicated that the best harvesting time of the 'Ganlv 1' fruit was about 160 DAFB, which is consistent with that of 'White' and 'Ganmi 6'. Previous studies have shown that 'White' harvested before 165 DAFB has a better nutritional and flavor quality [4,8], and the fruit growth period of 'Ganmi 6' fruit is 165 days [3]. The dry matter content of 'White' and 'Ganmi 6' were 17.9% and 17.3% [3,23], while the dry matter content of 'Ganlv 1' at 160 DAFB (18.26%) and the soft-ripening stage (19.55%) was higher than that of 'White' and 'Ganmi 6', indicating that 'Ganlv 1' has a very high dry matter content and its flavor and texture may also be better than these two varieties. Previous studies have shown that kiwifruit genotypes with high dry-matter content usually have higher SSC [22,24]. In this study, the SSC of 'Ganlv 1' was 15.91%, which was higher than that of 'Ganmi 6' (13.6%) [3], and was close to that of 'White' (16.0%) [23], which is consistent with the previous studies.

A. eriantha 'Ganlv 1' was rich in AsA. Although the content of the AsA decreased during fruit development, the AsA content increased in the soft-ripening stage. On the whole, the changing trend in the AsA content during the fruit development of 'Ganlv 1' was basically consistent with that of 'Ganmi 6' [9]. The content of AsA in the soft-ripening stage of 'Ganlv 1' was 7.09 mg/g, which was lower than that of 'Ganmi 6' (7.23 mg/g) [3], and higher than that of 'White' (6.28 mg/g) [23].

'Ganlv 1' had high soluble sugar content. The soluble sugar content of 'Ganlv 1' showed an overall upward trend during fruit development, but there was a large fluctuation in this process, which was somewhat inconsistent with the dynamic change trend of soluble sugar in 'White' and 'Ganmi 6' [8,9]. Before 160 DAFB, the content of the soluble sugar components in 'Ganlv 1' fruit remained at a low level with no obvious dynamic change, and the content of fructose was higher than that of sucrose and glucose during this period (Figure 3).After 160 DAFB, with the ripening of the fruit, the starch in the fruit began to degrade gradually, and the content of soluble sugar components (including sucrose, fructose and glucose) also began to accumulate gradually (Figure 3). In the soft-ripening stage, the content of glucose was the highest, followed by sucrose, and the content of fructose was the lowest. The changing trend in the soluble sugar content in 'Ganlv 1' during the fruit development and soft-ripening stage was basically similar to that of 'White' [8,25]. However, the content of total soluble sugar in 'White' and 'Ganmi 6' were 9.00% and 6.30% [3,23], respectively, which were lower than that of 'Ganlv 1' (9.74%). During the development of the 'White' fruit, the soluble sugar content was mainly accumulated by fructose and glucose, the sucrose content was always lower than glucose and fructose, and the fructose content was the highest [8]. Moreover, the study on the sugar metabolism of 'White' during storage showed that the glucose content was the highest, fructose was the second, and sucrose was the lowest [25]. However, the contents of these three sugars were relatively close, and the sucrose content was the highest in *A. chinensis* 'Hort16A' [26]. During the late growth stages of *A. deliciosa* 'Hayward', the starch began to degrade and the soluble sugar content increased rapidly, accumulating a large amount of sucrose, fructose and glucose in the late stages of the fruit development, mainly in fructose and glucose accumulation, the sucrose content was the lowest and the fructose content was the highest [27]. During the postharvest ripening process of *A. delicia*, the accumulation of glucose and fructose was always more than that of sucrose, and the accumulation of glucose was the highest [28]. The differences between the soluble sugar content and soluble sugar component content among the different varieties reflected the different sugar metabolism characteristics among the different varieties, which also reflected the diverse flavor of kiwifruit varieties.

The accumulation and changes in the sucrose content and the activities of sucrose-related metabolic enzymes and gene expression during fruit growth and development revealed the main period of sucrose metabolism in 'Ganlv 1'. During the early stages of the fruit growth and development (25–130 DAFB), the sucrose content remained at a low level, while after 130 DAFB, the sucrose content began to increase rapidly (Figure 4b), indicating that the main period of sucrose accumulation began at 130 DAFB. The sucrose synthesis-related enzymes (SS and SPS) catalyze the synthesis of sucrose, among which SPS enzyme activity is highly positively correlated with sucrose content, while sucrose invertases (VIN, NI and CWIN) degrade sucrose into glucose and fructose [29,30]. In this study, the SS and SPS activities were all lowest at 130 DAFB, and then increased rapidly during 130–145 DAFB. The sucrose invertase VIN activity decreased significantly during 130–145 DAFB (Figure 4). Although the NI and CWIN enzyme activities showed an upward trend during this period, their enzyme activities were much lower than those of other enzymes (Figure 4). The changing trend of the sucrose metabolism-related enzyme activities suggested that the period of 130–145 DAFB was favorable for the accumulation of sucrose. The expression trend of some of the sucrose metabolism-related enzyme genes was correlated with the trend of the sucrose metabolism-related enzyme activity changes. For example, the expression levels of *SPS1*, *SPS2*, *SPS5*, *NIN2* and *NIN3* were all lower at 130 DAFB, but increased at 145 DAFB to varying degrees (Figures 5 and 6).

Plant sugar metabolism-related genes mostly exist in the form of gene families. For example, 6, 5, 17 and 11 *SS* genes were identified in peach (*Prunus persica*) [31], grape (*Vitis vinifera*) [32], pear (*Pyrus bretschneideri*) [33] and apple (*Malus domestica*) [34], respectively.

Two *VIN* genes and five *CWIN* genes were identified in peach [35] At present, seven *SS* genes have been identified in *A. deliciosa* 'Hayward' [36], but the number of *SS* genes in *A. eriantha* is not clear. This study only analyzed the expression of four *SS* genes in 'Ganlv 1', which could not fully clarify the role of *SS* genes in sucrose metabolism. In future work, clarifying the members of the *SS* gene family and other sugar metabolism-related gene families will play an important role in understanding the mechanism of sugar metabolism in *A. eriantha*.

5. Conclusions

In conclusion, the dynamic changes in the fruit quality, the activities of the sucrose metabolism-related enzymes and the expression of the sucrose metabolism-related genes during the development of *A. eriantha* 'Ganlv 1' were detected and analyzed. The results clarified the dynamic characteristics of the fruit development of 'Ganlv 1', indicating that the best harvest time for 'Ganlv 1' was at about 160 DAFB. The main sugar components in the 'Ganlv 1' fruit were fructose, glucose and sucrose, among which glucose and sucrose were the most important sugars in the soft-ripening stage of the fruit. The analysis of the sucrose accumulation trend, sucrose metabolism-related enzyme activities and sucrose metabolism-related genes expression analysis indicated that 130–145 DAFB was the critical period for fruit sugar accumulation and metabolism. The results laid a foundation for clarifying the growth and development characteristics of *A. eriantha* fruit and the related mechanism of sugar metabolism.

Author Contributions: Conceptualization, C.H. and X.X.; methodology, C.H. and X.J.; validation, X.J., M.W. and S.C.; formal analysis, M.W. and X.J.; investigation, X.J. and D.J.; resources, M.W., X.J. and S.C.; data curation, M.W., X.J. and J.T.; writing—original draft preparation, J.T. and X.J.; writing—review and editing, J.T.; visualization, J.T. and X.J.; supervision, C.H.; project administration, C.H.; funding acquisition, C.H. All authors have read and agreed to the published version of the manuscript.

Funding: This research was funded by the National Natural Science Foundation of China (Grant Nos: 31760567 and 31960588).

Institutional Review Board Statement: Not applicable.

Informed Consent Statement: Not applicable.

Data Availability Statement: The data presented in this study are available on request from the corresponding author.

Conflicts of Interest: The authors declare no conflict of interest.

References

1. Huang, H.W. *Actinidia Germplasm Resources in China*; China Forestry Publishing House: Beijing, China, 2013.
2. Xie, M.; Wu, Y.J.; Jiang, G.H.; Zhang, Q.C.; Zhang, H.Q.; Peng, S.J.; Liu, K.M. A new big fruit *Actinidia eriantha* Benth. cultivar 'White'. *Acta Hortic. Sin.* **2008**, *35*, 1555.
3. Xu, X.B.; Huang, C.H.; Qu, X.Y.; Chen, M.; Zhong, M.; Lang, B.B.; Chen, C.J.; Xie, M.; Zhang, W.B. A new easy peeling *Actinidia eriantha* cultivar 'Ganmi 6'. *Acta Hortic. Sin.* **2015**, *42*, 2539–2540.
4. Xu, Y.H.; Song, Q.Q.; Hu, B.; Li, S.G.; Zheng, X.L.; Jiang, T.J. Harvest maturity affects quality and storability of *Actinidia eriantha* cv. White fruit. *J. Nucl. Agric. Sci.* **2020**, *34*, 0521–0531.
5. Qu, X.Y.; Lang, B.B.; Zhong, M.; Zhu, B.; Tao, J.J.; Huang, C.H.; Xu, X.B. Principal component analysis and comprehensive evaluation of fruit quality of *Actinidia eriantha*. *Chin. Agric. Sci. Bull.* **2016**, *32*, 92–96.
6. Lang, B.B.; Zhu, B.; Xie, M.; Zhang, W.B.; Seyrek, U.A.; Huang, C.H.; Xu, X.B. Variation and probability grading of the main quantitative characteristics of wild *Actinidia eriantha* germplasm resources. *J. Fruit Sci.* **2016**, *33*, 8–15.
7. Wang, Y.Z.; Pan, Z.M. Comprehensive evaluation of 22 Actinidia eriantha germplasm resources based on principal components analysis. *Acta Agric. Zhejiangensis* **2021**, *33*, 825–830.
8. Zhang, H.Q.; Xie, M.; Xiao, J.P.; Zhou, L.Q.; Song, G.H. Characterization of fruit development of the diploid kiwifruit, *Actinidia eriantha* 'White'. *J. Fruit Sci.* **2015**, *32*, 238–246.
9. Chen, C.J.; Tao, J.J.; Qu, X.Y.; Huang, C.H.; Xu, X.B. Dynamic variation in sugar, acid, and AsA contents of 'Gnami 6' kiwifruit (*Actinidia eriantha* Benth) fruits. *Agric. Sci. Technol.* **2015**, *16*, 2589–2591, 2595.

10. Liu, C.H.; Qiu, G.L.; Liu, Z.B.; Yang, Y.; Zhuang, Q.G.; Zhang, Q. Study on the changes of physicochemical properties of *A. eriantha* during the fruit development. *J. Sichuan Univ. (Nat. Sci. Ed.)* **2019**, *56*, 951–956.
11. Zhang, H.Q.; Xie, M.; Zhang, C.; Yang, L.Q.; Zhang, Z.; Xiao, J.P.; Zhou, L.Q. Difference in starch accumulation and characterization of sugar metabolism during fruit development of kiwifruit. *Sci. Agric. Sin.* **2014**, *47*, 3453–3464.
12. Zheng, Q.M.; Tang, Z.; Xu, Q.; Deng, X.X. Isolation, phylogenetic relationship and expression profiling of sugar transporter genes in sweet orange (*Citrus sinensis*). *Plant Cell Tissue Organ Cult.* **2014**, *119*, 609–624. [CrossRef]
13. Lee, P.R.; Tan, R.M.; Yu, B.; Curran, P.; Liu, S.Q. Sugars, organic acids, and phenolic acids of exotic seasonable tropical fruits. *Nutr. Food Sci.* **2013**, *43*, 267–276. [CrossRef]
14. Zhao, J.H.; Li, H.X.; Xi, W.P.; An, W.; Niu, L.L.; Cao, Y.L.; Wang, H.F.; Wang, Y.J.; Yin, Y. Changes in sugars and organic acids in wolfberry (*Lycium barbarum* L.) fruit during development and maturation. *Food Chem.* **2015**, *173*, 718–724. [CrossRef] [PubMed]
15. Li, Y.N.; Yan, L.Y.; Zhang, B.; Yang, S.B.; Zhao, Z.Y. A study on sugar and organic acid components in different apple cultivars. *J. Fruit Sci.* **2021**, *38*, 1877–1889.
16. Zhang, G.J. *Study on Sugars and Acids Inheritance and the Expression Pattern of β-Glucosidase Genes (VvBGs) during Grape Berry Ripening*; China Agricultural University: Beijing, China, 2013.
17. Huang, Q. *Studies on Characterization of Organic Acids Metabolism and Expression of Relative Genes during Fruit Development of Relative Genes during Fruit Development of 'Ganmi 6'*; Jiangxi Agricultural University: Nanchang, China, 2019.
18. Cao, J.K.; Jiang, W.B.; Zhao, Y.M. *Experiment Guidance of Postharvest Physiology and Biochemistry of Fruits and Vegetables*; Chinese Light Industry Press: Beijing, China, 2007.
19. Yue, J.Y.; Liu, J.C.; Tang, W.; Wu, Y.Q.; Tang, X.F.; Li, W.; Yang, Y.; Wang, L.H.; Huang, S.X.; Fang, C.; et al. Kiwifruit Genome Database (KGD): A comprehensive resource for kiwifruit genomics. *Hortic. Res.* **2020**, *7*, 117. [CrossRef]
20. Wang, X.X.; Chen, Y.; Jiang, S.; Xu, F.; Wang, H.F.; Wei, Y.Y.; Shao, X.F. PpINH1, an invertase inhibitor, interacts with vacuolar invertase PpVIN2 in regulating the chilling tolerance of peach fruit. *Hortic. Res.* **2020**, *7*, 168. [CrossRef]
21. Chen, M.Y.; Zhang, P.; Zhao, T.T.; Han, F.; Liu, X.L.; Zhong, C.H. Relationship between harvest indices and fruit quality traits in *Actinidia chinensis* 'Jintao'. *Plant Sci. J.* **2019**, *37*, 621–627.
22. Nardozza, S.; Gamble, J.; Axten, L.G.; Wohlers, M.W.; Clearwater, M.J.; Feng, J.Q.; Harker, F.R. Dry matter content and fruit size affect flavour and texture of novel *Actinidia deliciosa*. *J. Sci. Food Agric.* **2011**, *91*, 742–748. [CrossRef]
23. Wu, Y.J.; Xie, M.; Zhang, Q.C.; Jiang, G.H.; Zhang, H.Q.; Long, Q.J.; Han, W.J.; Chen, J.W.; Shong, G.H. Characteristics of 'White': A new easy-peel cultivar of *Actinidia eriantha*. *N. Z. J. Crop Hortic. Sci.* **2009**, *37*, 369–373. [CrossRef]
24. Nardozza, S.; Boldingh, H.L.; Richardson, A.C.; Costa, G.; Marsh, H.; MacRae, E.A.; Clearwater, M.J. Variation in carbon content and size in developing fruit of *Actinidia deliciosa* genotypes. *Funct. Plant Biol.* **2010**, *37*, 545–554. [CrossRef]
25. Qi, W.Y.; Zhou, C.H.; Song, L.J.; Zhong, Y.; Zheng, X.L. Study on sugar metabolism of *Actinidia eriantha* Benth 'White' during storage. *J. Fruit Sci.* **2016**, *33*, 744–751.
26. Richardson, A.C.; Boldingh, H.L.; McAtee, P.A.; Gunaseelan, K.; Luo, Z.W.; Atkinson, R.G.; David, K.M.; Burdon, J.N.; Schaffer, R.J. Fruit development of the diploid kiwifruit, *Actinidia chinensis* 'Hort16A'. *BMC Plant Biol.* **2011**, *11*, 182. [CrossRef] [PubMed]
27. Moscatello, S.; Famiani, F.; Proietti, S.; Farinelli, D.; Battistelli, A. Sucrose synthase dominates carbohydrate metabolism and relative growth rate in growing kiwifruit (*Actinidia deliciosa*, cv Hayward). *Sci. Hortic.* **2011**, *128*, 197–205. [CrossRef]
28. MacRae, E.; Quick, W.P.; Benker, C.; Stitt, M. Carbohydrate metabolism during postharvest ripening in kiwifruit. *Planta* **1992**, *188*, 314–323. [CrossRef]
29. Gomez, M.; Lajolo, F.; Cordenunsi, B. Evolution of soluble sugars during ripening of papaya fruit and its relation to sweet taste. *J. Food Sci.* **2010**, *67*, 442–447. [CrossRef]
30. Basson, C.E.; Groenewald, J.H.; Kossmann, J.; Cronjé, C.; Bauer, R. Sugar and acid-related quality attributes and enzyme activities in strawberry fruits: Invertase is the main sucrose hydrolysing enzyme. *Food Chem.* **2010**, *121*, 1156–1162. [CrossRef]
31. Zhang, C.H.; Yu, M.L.; Ma, R.J.; Shen, Z.J.; Zhang, B.B.; Korir, N.K. Structure, expression profile, and evolution of the sucrose synthase gene family in peach (*Prunus persica*). *Acta Physiol. Plant.* **2015**, *37*, 81. [CrossRef]
32. Zhu, X.D.; Wang, M.Q.; Li, X.P.; Jiu, S.T.; Wang, C.; Fang, J.G. Genome-wide analysis of the sucrose synthase gene family in grape (*Vitis vinifera*): Structure, evolution, and expression Profiles. *Genes* **2017**, *8*, 111. [CrossRef]
33. Lv, J.H.; Wang, Y.Z.; Cheng, R.; Wang, G.M.; Zhang, S.L.; Wu, J.; Zhang, H.P. Genome-wide identification and expression analysis of sucrose synthase (SUS) and sucrose phosphate synthase (SPS) gene families in pear. *Acta Hortic. Sin.* **2018**, *45*, 421–435.
34. Tong, X.L.; Wang, Z.Y.; Ma, B.Q.; Zhang, C.X.; Zhu, L.C.; Ma, F.W.; Li, M.J. Structure and expression analysis of the sucrose synthase gene family in apple. *J. Integr. Agric.* **2018**, *17*, 847–856. [CrossRef]
35. He, X.X.; Wei, Y.Y.; Kou, J.Y.; Xu, F.; Chen, Z.H.; Shao, X.F. *PpVIN2*, an acid invertase gene family member, is sensitive to chilling temperature and affects sucrose metabolism in postharvest peach fruit. *Plant Growth Regul.* **2018**, *86*, 169–180. [CrossRef]
36. Chen, C.; Wang, Y.; Yang, Y.; Yan, Y.Q. Identification of Sucrose Synthase Gene Family in Kiwifruit and Their Expression during Fruit Development. *Mol. Plant Breed.* **2021**. Available online: http://kns.cnki.net/kcms/detail/46.1068.S.20210422.1703.010.html (accessed on 17 April 2022).

horticulturae

MDPI

Article

PpSAUR43, an Auxin-Responsive Gene, Is Involved in the Post-Ripening and Softening of Peaches

Jiahui Wang, Weijing Su, Kun Liu, Ze Xu, Kamran Shah, Juanjuan Ma, Dong Zhang, Yanan Hu * and Caiping Zhao *

College of Horticulture, Northwest A&F University, Xianyang 712100, China; wangjiahui97@nwafu.edu.cn (J.W.); swj10162021@163.com (W.S.); liukun506@nwafu.edu.cn (K.L.); xuzeizi@163.com (Z.X.); kamranshah801@nwafu.edu.cn (K.S.); mjj@nwafu.edu.cn (J.M.); afant@nwsuaf.edu.cn (D.Z.)
* Correspondence: huyn_nwsuaf@163.com (Y.H.); zhcc@nwsuaf.edu.cn (C.Z.)

Abstract: Auxin's role in the post-ripening of peaches is widely recognized as important. However, little is known about the processes by which auxin regulates fruit post-ripening. As one of the early auxin-responsive genes, it is critical to understand the role of small auxin-up RNA (SAUR) genes in fruit post-ripening and softening. Herein, we identified 72 *PpSAUR* auxin-responsive factors in the peach genome and divided them into eight subfamilies based on phylogenetic analysis. Subsequently, the members related to peach post-ripening in the *PpSAUR* gene family were screened, and we targeted *PpSAUR43*. The expression of *PpSAUR43* was decreased with fruit post-ripening in melting flesh (MF) fruit and was high in non-melting flesh (NMF) fruit. The overexpression of *PpSAUR43* showed a slower rate of firmness decline, reduced ethylene production, and a delayed fruit post-ripening process. The MADS-box gene family plays an important regulatory role in fruit ripening. In this study, we showed with yeast two-hybrid (Y2H) and bimolecular fluorescence complementation (BIFC) experiments that PpSAUR43 can interact with the MADS-box transcription factor PpCMB1(PpMADS2), which indicates that PpSAUR43 may inhibit fruit ripening by suppressing the function of the PpCMB1 protein. Together, these results indicate that *PpSAUR43* acts as a negative regulator involved in the peach post-ripening process.

Keywords: peach; fruit; post-ripening; softening; *PpSAUR*

Citation: Wang, J.; Su, W.; Liu, K.; Xu, Z.; Shah, K.; Ma, J.; Zhang, D.; Hu, Y.; Zhao, C. *PpSAUR43*, an Auxin-Responsive Gene, Is Involved in the Post-Ripening and Softening of Peaches. *Horticulturae* **2022**, *8*, 379. https://doi.org/10.3390/horticulturae8050379

Academic Editor: Dilip R. Panthee

Received: 16 March 2022
Accepted: 24 April 2022
Published: 26 April 2022

Publisher's Note: MDPI stays neutral with regard to jurisdictional claims in published maps and institutional affiliations.

1. Introduction

The fruit is an organism for which metabolic activities are still in order after harvesting [1]. A series of physiological and biochemical processes in the tissues causes the degradation of certain organic macromolecules and changes in cell wall and cell membrane structures. Peaches (*Prunus persica* (L.) Batsch) are a popular commercial fruit with nutritional and pharmacological properties [2]. According to the Food and Agriculture Organization (FAO), 1,527,052 ha of peaches were cultivated worldwide in 2019, and China is home to the world's largest peach industry [3]. However, peaches generally soften rapidly, with reduced quality and nutritional value, until they become rotten after harvest. Hence, postharvest softening and the senescence regulatory mechanism have been the focus of much research.

During the period of ripening, peaches undergo textural changes that lead to the loss of tissue firmness [4]. It has already been determined in the existing literature that many different phytohormones play pivotal functions in regulating ripening and softening [5–9]. For example, ethylene has long been regarded as the main regulator of ripening in climacteric fruit [10,11]. For climacteric fruit, ethylene biosynthesis is regulated by two systems, system 1 and system 2. System 1 is only responsible for producing low concentrations of basal ethylene at the pre-climacteric stage, and when the ethylene produced by system 1 reaches a certain level, system 2 begins to produce large levels of ethylene, which is responsible for the ripening process by self-catalysis [12–14]. As a climacteric fruit, peach cultivars usually have three types: melting flesh (MF), non-melting flesh (NMF), and stony hard (SH) according to the differences in fruit firmness and texture characteristics [15,16]. Previous studies have found that a high concentration of auxin can stimulate the synthesis of system 2 ethylene through its inductive action on the expression of *PpACS1* (1-aminocyclopropane-1-carboxylic acid synthase) in MF peaches during post-ripening. In contrast, the low concentrations of auxin due to the low expression of *PpYUC11* suppress *PpACS1* expression and ethylene production in SH peaches during post-ripening [17–21]. In fact, the involvement of auxin in peach ripening has long been reported. It has been pointed out earlier that the peak of ethylene during the later stages of peach development coincides with higher levels of auxin, and the application of exogenous auxin could promote ethylene production and peach ripening, suggesting that auxin may be involved in ethylene synthesis [22,23]. Similar findings showing that auxin regulates ripening by modulating ethylene and auxin crosstalk have previously been reported in apples [23], durians [24], papayas [25], and tomatoes [26,27]. Accumulating evidence suggests that both ethylene and auxin interact in the regulation of fruit ripening. In addition to auxin indirectly regulating fruit ripening via modulating ethylene and auxin crosstalk, auxin may also play its own role in the ripening process of peaches. Our predecessors used genomic methods to determine that an increase in auxin has an independent effect on the hormones in the ripening process of climacteric peaches [28]. However, the role of auxin during fruit ripening remains obscure, although auxin accumulation appears to be critical and participates in the regulation of the ripening of fruit.

The fruit ripening regulatory network is not only dependent on the expression of genes directly involved in hormone synthesis and response but also on many transcription factor families (i.e., ERF, NAC, MADS-Box, etc.) that participate in the regulation of fruit ripening by regulating the expression of hormone synthesis and response genes to successfully complete the fruit ripening process [29–34]. Among these TFs, MADS-box genes constitute a highly conserved family of TFs and have been shown in numerous studies to be involved in fruit ripening regulation [35]. RIN, which is a member of the SEPALLATA clade (E-class) of MADS-box genes, is essential in regulating tomato fruit ripening and softening because the fruits failed to ripen in the RIN mutants [36]. In apple, a Rin-homologue of MdMADS8/9 was revealed to regulate fruit ripening by directly controlling the auxin levels. MdMADS8/9 suppressed GH3 expression, resulting in increased levels of free-state IAA during apple ripening [37]. In banana, Elitzur et al. determined that the two SEP-like MADS-box genes *MaMADS1* and *MaMADS2* are essential for fruit ripening because the silencing of either gene resulted in reduced fruit ethylene synthesis and delayed ripening [38]. In peach, Li et al. found that the suppression of *PpSEP1* could decrease the transcription of cell wall metabolism genes (i.e., *Endo-PG3*, and *PME1*) and ethylene synthesis-related genes (i.e., *ACS2* and *ACO1*) [39]. In other fruits, such as citrus, strawberries, and grapes, MADS-box transcription factors have been demonstrated to be involved in fruit ripening [40–42]. PpCMB1(PpMADS2), which belongs to the SEP clade of the MADS-box gene family, is a homolog of SlCMB1 in the tomato MADS-box family [43]. SlCMB1 has been reported to be a potential regulator of ethylene biosynthesis and carotenoid accumulation during maturation [44]. In addition, the homologous genes of PpCMB1 have been reported in apple and cherry as MdMADS6 and PaMADS2, respectively. Both MdMADS6 and PaMADS2 have been shown to play a role in fruit ripening and softening, speculating that the PpCMB1 may have a similar function [45,46]. In peach, PpCMB1 has been found to be highly expressed in fruit and is involved in fleshy fruit development and disease resistance in postharvest fruit [47–49].

As one of the three early auxin-responsive gene families, small auxin-up RNA (SAUR) can rapidly respond to auxin applications without protein synthesis and is critically involved in the auxin signaling pathway [50,51]. Since McClure and Guilfoyle [52] identified the first *SAUR* gene from soybeans, *SAUR* members have subsequently been identified in several species, including *Arabidopsis*, rice, tomatoes, and apples [53–56]. Most *SAURs* do not contain introns in their coding sequence (CDS), and SAUR proteins are small and contain a conserved SAUR-specific domain (SSD) of approximately 60 residues in the central region [57]. Previous research has shown that *SAURs* regulate a wide variety of cellular, physiological, and developmental processes that involve the hormonal and environmental control of plant growth and development [58]. For example, a few SAUR proteins have been found to be capable of binding calmodulins (CaM) and CaM-like proteins (CML) [59,60], regulating organ elongation [61] and apical hook development [62], increasing abiotic stress tolerance [63], altering stamen filament elongation [64], and promoting leaf senescence [65,66] and cell expansion [67]. Moreover, *SAUR* may be related to the polar transport of auxin. In *Arabidopsis*, both *AtSAUR19* overexpression lines and *AtSAUR19, 23*, and *24* RNA interference lines can change the transport of auxin from the hypocotyl axis to the base [68]. Kant et al. [69] reported that the *OsSAUR39* gene negatively regulates auxin synthesis and transport in rice. In addition to auxin, gibberellin [64], abscisic acid [70], brassinosteroid [71], and ethylene [72,73] are all thought to be related to the expression of the *SAUR* gene, indicating that *SAUR* may be related to other hormones that regulate plant growth and development. Studies on *SAURs* involved in ripening have seldom been reported. However, it has been shown that *SlSAUR69* is an RIN-regulated gene involved in the regulation of ripening in tomatoes. The overexpression of *SlSAUR69* represses auxin transport and enhances ethylene sensitivity, thereby resulting in the premature initiation of ripening. Conversely, the downregulation of *SlSAUR69* delays the initiation of fruit ripening [74]. However, the precise biological roles of *SAUR* in the context of peach ripening still need to be clarified.

At present, reducing the fruit softening rate has become an important breeding objective. A previous study developed by Carrasco-Valenzuela et al. [75] identified *PpSAUR43* (Prupe.8G079500) as a key candidate gene for the rate of peach softening by integrating quantitative trait loci (QTL) analyses. Additionally, the expression profile of the differentially expressed gene coding for *PpSAUR43* has been experimentally confirmed by RT-qPCR analyses and has been shown to be upregulated in low-softening-rate progenies. However, there are no direct reports on the regulation of the fruit post-ripening process by *PpSAUR* in peaches. Therefore, this study aimed to identify the regulatory pathway of *PpSAUR43* during the post-ripening of peaches, which will help to further understand the molecular mechanism of peach ripening and softening.

2. Materials and Methods

2.1. Plant Materials

Fruits of the melting flesh (MF) peach 'Yu Hua Lu' ('YHL') and the non-melting flesh (NMF) peach 'Babygold 5' ('B5') were collected as the experimental materials. The 'YHL' (27 June 2021) and 'B5' (24 July 2021) fruits were harvested in the commodity maturation period and stored at 25.0 ± 1 °C in the laboratory, from which samples were taken at two-day intervals for 'YHL' and three-day intervals for 'B5', which were then stored at −80 °C. In this study, the fruits were harvested from the orchard of the Experimental Station of the College of Horticulture of Northwest A&F University, Yangling, Shaanxi, China (34°20′ N, 108°24′ E), which contains five-year-old trees.

Transgenic tobacco (*Nicotiana benthamiana*), containing a nuclear localization signal (NLS-mCherry), used for subcellular localization analysis, was grown in a phytotron (25.0 °C; photoperiod: 16 h/8 h light/dark). *Arabidopsis*, used for a bimolecular fluorescence complementation (BiFC) assay, was grown in a growth chamber at 22.0 °C under a 16 h/8 h light/dark photoperiod.

2.2. Measurement of Fruit Firmness

Fruit firmness was measured by a GY-4 penetrometer (Top Instrument Co., Ltd., Hangzhou, China) equipped with a 7.9 mm probe. From each group, 10 fruits were randomly selected, and fruit firmness was measured around the injection site after peeling.

2.3. Determination of Ethylene Production

For ethylene production, nine fruit samples were weighed and sealed in a jar for 1 h (temperature 25.0 ± 1 °C and relative humidity of 75–85%; 0.03% of CO_2 and 21% of O_2.). Then, the sample gases (1 mL) were withdrawn from the jar and measured using a gas chromatograph (Trace GC Ultra; Thermo Fisher Scientific, Milan, Italy). The working conditions of gas chromatography were: chromatographic column (Rtx-1701) of 30 m × 0.32 mm, column temperature of 80 °C, carrier gas of nitrogen (N_2), detector (FID) temperature of 130 °C, inlet temperature of 120 °C, hydrogen flow rate of 40 mL·min^{-1}, air flow rate of 400 mL·min^{-1}, pressure of 100 kpa, tail gas flow of 30 mL·min^{-1}, column flow rate of 1.19 mL·min^{-1}, injection volume of 1 mL, and splitting ratio of 5.0. There were three replicates per treatment, and three values per replicate.

$$\text{Ethylene release}\left(\mu\text{L}\cdot\text{kg}^{-1}\cdot\text{h}^{-1}\right) = \text{c} \times \text{V} \times \text{m}^{-1} \times \text{t}^{-1} \times 0.001$$

where c is the converted ethylene content after the gas chromatographic determination in the sample gas (μL·L^{-1}), V is the volume of the closed space of the desiccator (mL), t is the measurement time (h), and m is the fruit quality (g).

2.4. Determination of IAA Concentration

The extraction and purification of IAA in the injection site after the peeling of the fruits were carried out according to the method of Tatsuki et al. [17]. The content of IAA in each sample was determined using the LC-MS/MS analysis method as described by Müller et al. [76].

2.5. Identification of the PpSAUR Genes in Peaches

To identify the *SAUR* family genes in the peach species, we first acquired the SAUR family member CDS sequences and protein sequences of *Arabidopsis thaliana* [53] and *Solanum lycopersicum* [55]. Then, the protein sequences of *Arabidopsis thaliana* and *Solanum lycopersicum* were used as queries in the Genome Database for Rosaceae (*Prunus persica* v2.0.a1 genome) to identify the potential SAUR protein sequences in peaches (GDR: https://www.rosaceae.org/species/prunus/all (accessed on 18 August 2020)). The protein candidate sequences were inspected using SMART (http://smart.embl-heidelberg.de/ (accessed on 18 August 2020)) and Pfam (the protein families database: http://pfam.xfam.org/ (accessed on 18 August 2020)) to confirm that they conserved the PF02519 domain. Subsequently, we used Expasy (https://www.expasy.org/ (accessed on 4 September 2020)) to estimate the physicochemical parameters (i.e., length, molecular weight (MW), and isoelectric point (pI)) of the PpSAURs.

2.6. Phylogenetic Analyses of the SAUR Genes in Peaches and Arabidopsis

In order to analyze the phylogenetic organization of the SAUR family, we performed multiple sequence alignments for all available SAUR full-length protein sequences of *Arabidopsis* and peaches using DNAMAN 9.0 (Lynnon Biosoft, San Ramon, CA, USA). A phylogenetic tree based on the neighbor-joining (NJ) method was constructed for SAUR proteins by using the software MEGA 7.0 (with a Jones–Taylor–Thornton (JTT) model and complete deletion). A bootstrap test was carried out using 1000 replicates to construct the reliability of the tree.

2.7. RNA Isolation and Quantitative Real-Time Polymerase Chain Reaction (RT-qPCR)

'YHL' fruits were taken after 0, 2, 4, 6, and 8 days of storage (25.0 ± 1 °C) for the RT-qPCR analysis. 'B5' fruits were taken after 0, 3, 6, 9, and 12 days of storage (25.0 ± 1 °C) for RT-qPCR analysis. Three biological replicates were designed, and each repetition was carried out by mixing flesh samples from three fruits.

The RNA extraction of the peach samples (3.5 g) was carried out using cetyltrimethylammonium bromide according to the method of Xing et al. [77]. Then, 1 μg of RNA was used to synthesize the cDNA with a PrimeScript^{TM} RT reagent Kit with a gDNA Eraser (Perfect Real Time) (TaKaRa, Beijing, China). Then, quantitative reverse transcription RT-qPCR was performed in volumes of 10 μL of reaction buffer containing 5 μL of SYBR® Premix Ex Taq^{TM} II (Tli RNaseH Plus) (2×) (TaKaRa), 1 μL of cDNA, 0.2 μL of each primer, and 3.4 μL of deionized water. The PCR program was as follows: 95 °C for 1 min, then 40 cycles of 95 °C for 15 s, 60 °C for 20 s, and 72 °C for 20 s. After this, the mixed sample was heated to 95 °C for 10 s and cooled to 65 °C for 15 s. The sample was then heated up to 95 °C at a rate of 0.1 $°C{\cdot}s^{-1}$ for melting curve analyses. RT-qPCR was run on the CFX Connect Real-Time System (Bio-Rad Laboratories, Inc., Hercules, CA, USA). The peach *PpTUA5* (Prupe.6G004100) was used as an internal control and the $2^{-\Delta\Delta CT}$ method was used to calculate the transcript accumulation [78]. Each sample was analyzed in triplicate. The primers are listed in Table S1.

2.8. Subcellular Localization Analysis

The coding regions of PpSAUR43 and PpCMB1 without the stop codon were amplified by PCR using Takara LA Taq high-fidelity DNA polymerase (TaKaRa). PCR primers (Table S1) were designed using Primer 6.0 software (PREMIER Biosoft International, CA, USA). Then, we inserted them into the pCAMBIA2300-green fluorescent protein (GFP) vector to produce the respective fusion constructs of the pCAMBIA2300-PpSAUR43-green fluorescent protein (GFP) and pCAMBIA2300-PpCMB1-green fluorescent protein (GFP) using a Seamless Cloning Kit (Sangon Biotech, Shanghai, China). The fusion plasmids and the control vector (pCAMBIA2300-GFP) were separately transformed into the *Agrobacterium tumefaciens* GV3101 chemically competent cell. Transgenic tobacco plant material containing a nuclear localization signal (NLS-mCherry) was used in this experiment [79]. The *Agrobacterium tumefaciens* GV3101 strain was agroinfiltrated into five-week-old leaves of *Nicotiana tabacum* with different vectors. Transformed leaves were analyzed at 48 h after the infection of the lower epidermal cells. Confocal imaging was performed using an inverted Leica TCS-SP8 SR laser scanning microscope. For the imaging expression of GFP constructs, excitation lines of a solid-state laser of 488 nm were used with a 498/510-nm bandpass filter in the single-track facility of the microscope. For the imaging co-expression of the GFP and mCherry constructs, excitation lines of a solid-state laser of 488 nm for GFP and 552 nm for mCherry were used alternately with line switching using the sequential scanning mode of the microscope.

2.9. Peach Injection Assays

The overexpressed *PpSAUR43* construct was generated by cloning the CDS of *PpSAUR43* into the pCAMBIA2300 vector. The empty pCAMBIA2300 vector was used as a control. The recombinant plasmids and empty vectors were respectively introduced into the *Agrobacterium tumefaciens* GV3101 chemically competent cell. The agroinfiltration of 'YHL' fruits was performed according to Jia et al. [80]. The fruits were harvested when they were commercially mature (i.e., 110 days after full bloom). At least 200 fruits per treatment were randomly collected with three biological replicates for an injection test of *Agrobacterium tumefaciens* in peaches. The infiltrated peach fruits were stored at 25.0 ± 1 °C. The ethylene production was determined with whole fruits, while the positions on fruits for the firmness measurement and for sampling with gene expression analysis were the injection site after the peeling of the fruits. The injection sites after the peeling of the fruits were then snap-frozen in liquid nitrogen and stored at −80 °C.

2.10. Yeast Two-Hybrid (Y2H) Assay

PpSAUR43 was amplified using LA Taq high-fidelity DNA polymerase (TaKaRa) and inserted into the pGBKT7 vector (Clontech, Mountain View, CA, USA) to generate pGBKT7-PpSAUR43. The CDS of PpCMB1 was inserted into the pGADT7 vector (Clontech) to generate the pGADT7-PpCMB1 recombinant plasmid using a Seamless Cloning Kit (Sangon Biotech). The recombinant plasmid pGBKT7-*PpSAUR43* was separately co-transformed with pGADT7-PpCMB1 into the Y2H-Gold yeast (Clontech) chemically competent cell. The pGADT7 empty vector was used as a negative control. A cDNA library was constructed from the RNA extracted from 'Qianjianbai' fruit during post-ripening (constructed by Takara Bio Company). The interaction assay consisted of three biological replicates. The yeast transformation experiment was conducted using the Yeastmaker Yeast Transformation System 2 User Manual (Clontech). The transformed Y2H-Gold yeast strain was cultured on SD/-Leu/-Trp and SD/-Trp/-Leu/-Ade/-His media with the indicator x-α-gal (5-Bromo-4-chloro-3-indolyl-α-D-galactoside) and cultured at 28.0 °C for 3 days.

2.11. Bimolecular Fluorescence Complementation (BiFC) Assay

The CDSs of *PpSAUR43* and *PpCMB1*, excluding the stop codons, were inserted into both pSPYCE and pSPYNE using a Seamless Cloning Kit (Sangon Biotech), respectively. Different combinations of plasmids (pSPYCE-*PpSAUR43* + pSPYNE-*PpCMB1* and pSPYCE-*PpSAUR43* + pSPYNE) were transformed into *Arabidopsis* protoplasts according to the polyethylene glycol (PEG) mediation method, as reported by Yoo et al. [81]. The protoplasts were then cultured at 22.0 °C in an incubator. After 18–22 h, the yellow fluorescent protein (YFP) signal was detected by confocal microscopy (Leica TCS-SP8 SR) at a wavelength of 514 nm.

2.12. Statistical Analysis

All data are presented as the mean and standard deviation (SD) and were derived from at least three biological replicates. An ANOVA with Duncan's test was conducted using IBM SPSS Statistics 23.0 (SPSS Inc., Chicago, IL, USA). A pairwise comparison was computed using the Duncan test. A single asterisk (*) and double asterisks (**) in the figures indicate significant differences of $p < 0.05$ and $p < 0.01$, respectively. Duncan's test was used, and different letters indicate significant differences ($p < 0.05$). Origin 7.0 (OriginLab, Northampton, MA, USA) was used to prepare the figures.

3. Results

3.1. Screening and Identification of the PpSAUR Family Genes

We identified 72 *SAUR* genes from the peach genome, and all of them contained the domains (PF02519) of the *SAUR* gene family (Table 1). According to the sequence of their chromosomes or scaffold positions, they were numbered from top to bottom and named *PpSAUR1*–*PpSAUR72* (Figure S1). These members are distributed among eight chromosomes, but the number of members on different chromosomes varies widely. For example, chromosome 8 contains 40 members, while chromosome 4 contains only one member (Figure S1). These *SAUR* gene family members encode 73–206 amino acids (aa), with an isoelectric point (Theoretical Isoelectric Point, pI) of 4.53–11.12, and a molecular weight (MW) of 8.77–23.31 KDa (Table 1). The calculated MW and pI were nearly identical to those determined previously in other plant species [53–56].

Table 1. Small auxin-up RNA (SAUR) gene family in peach.

Gene Name	Gene ID	pI	MW (KDa)	Length (aa)
PpSAUR1	Prupe.1G000200	8.38	15.51	139
PpSAUR2	Prupe.1G067400	9.8	18.59	165
PpSAUR3	Prupe.1G221300	9.51	18.52	163
PpSAUR4	Prupe.1G235700	9.79	17.03	104
PpSAUR5	Prupe.1G368100	8.52	21.20	149
PpSAUR6	Prupe.1G455700	10.42	19.38	187
PpSAUR7	Prupe.2G131400	7.95	14.73	124
PpSAUR8	Prupe.2G131500	8.37	14.62	120
PpSAUR9	Prupe.2G131600	6.85	15.43	124
PpSAUR10	Prupe.2G140600	9.22	21.23	171
PpSAUR11	Prupe.2G194600	8.48	16.60	125

Table 1. *Cont.*

Gene Name	Gene ID	pI	MW (KDa)	Length (aa)
PpSAUR12	Prupe.2G211600	10.63	19.44	125
PpSAUR13	Prupe.2G317100	9.51	19.73	129
PpSAUR14	Prupe.3G023900	6.43	14.37	187
PpSAUR15	Prupe.3G024000	5.93	15.19	144
PpSAUR16	Prupe.3G024100	6.31	16.11	152
PpSAUR17	Prupe.3G035100	8.66	14.37	103
PpSAUR18	Prupe.4G136800	9.8	23.31	138
PpSAUR19	Prupe.5G076200	8.98	17.14	165
PpSAUR20	Prupe.5G147100	9.76	17.12	174
PpSAUR21	Prupe.6G108300	6.83	14.27	127
PpSAUR22	Prupe.6G108400	8.42	14.82	137
PpSAUR23	Prupe.6G108500	6.32	21.34	141
PpSAUR24	Prupe.6G232300	11.12	16.71	139
PpSAUR25	Prupe.6G234900	9.56	11.35	124
PpSAUR26	Prupe.7G048400	8.2	9.27	206
PpSAUR27	Prupe.7G048500	5.59	12.31	151
PpSAUR28	Prupe.7G048600	5.72	11.11	152
PpSAUR29	Prupe.7G104000	4.53	9.46	128
PpSAUR30	Prupe.7G120400	6.25	9.91	131
PpSAUR31	Prupe.7G192900	6.39	11.17	105
PpSAUR32	Prupe.8G072200	6.71	9.998	191
PpSAUR33	Prupe.8G072300	10	9.20	97
PpSAUR34	Prupe.8G072400	9.2	14.43	95
PpSAUR35	Prupe.8G078300	6.06	10.37	82
PpSAUR36	Prupe.8G078600	5.55	10.19	108
PpSAUR37	Prupe.8G078700	6.72	10.32	96
PpSAUR38	Prupe.8G078900	5.21	10.13	89
PpSAUR39	Prupe.8G079000	6.71	11.63	98
PpSAUR40	Prupe.8G079100	5.53	11.01	89
PpSAUR41	Prupe.8G079200	5.73	11.08	89
PpSAUR42	Prupe.8G079400	6.04	10.17	128
PpSAUR43	Prupe.8G079500	5.73	11.07	92
PpSAUR44	Prupe.8G079600	7.88	11.24	90
PpSAUR45	Prupe.8G079700	6.04	10.15	92
PpSAUR46	Prupe.8G079900	5.73	11.15	76
PpSAUR47	Prupe.8G080000	6.06	11.11	92
PpSAUR48	Prupe.8G080100	7.78	11.22	103
PpSAUR49	Prupe.8G080200	7.78	11.10	99
PpSAUR50	Prupe.8G080300	8.93	11.25	99
PpSAUR51	Prupe.8G080400	9.2	11.19	73
PpSAUR52	Prupe.8G080500	7.92	11.23	92
PpSAUR53	Prupe.8G080600	5.27	10.96	99
PpSAUR54	Prupe.8G080700	5.27	9.07	99
PpSAUR55	Prupe.8G080800	8.85	10.59	92
PpSAUR56	Prupe.8G081000	9.21	16.86	99
PpSAUR57	Prupe.8G081100	8.66	10.12	99
PpSAUR58	Prupe.8G081200	8.54	10.49	100
PpSAUR59	Prupe.8G081300	8.52	10.56	99
PpSAUR60	Prupe.8G081400	6.89	11.09	100
PpSAUR61	Prupe.8G081500	6.64	11.23	98
PpSAUR62	Prupe.8G081600	10	11.20	100
PpSAUR63	Prupe.8G081700	9.56	17.30	98
PpSAUR64	Prupe.8G081800	9.38	15.99	80
PpSAUR65	Prupe.8G081900	7.82	15.99	93
PpSAUR66	Prupe.8G082200	9.48	11.94	149
PpSAUR67	Prupe.8G157700	8.66	14.19	90
PpSAUR68	Prupe.8G157800	8.96	14.25	93
PpSAUR69	Prupe.8G157900	9.62	11.20	93
PpSAUR70	Prupe.8G158000	9.1	13.71	100
PpSAUR71	Prupe.8G158100	6.27	8.77	96
PpSAUR72	Prupe.8G158200	8.73	11.79	98

3.2. Sequence and Phylogenetic Analysis

Using the protein sequences of 72 peach SAUR family members and 72 *Arabidopsis* SAUR family members, we constructed a phylogenetic tree to better understand the evolutionary relationship between the PpSAUR and AtSAUR proteins (Figure 1). The peach

SAUR family was divided into groups (I–VIII) according to the 72 SAUR members of *Arabidopsis* (Figure 1). The results show that the number of family members in different groups varied greatly. For example, the largest group I contained 29 members, while the smallest group VI contained only 8 members. In group I, only one member was from *Arabidopsis* and the rest were from peach. PpSAUR43 and PpSAUR47 belong to paralogous genes and showed a closer relationship with AtSAUR1.

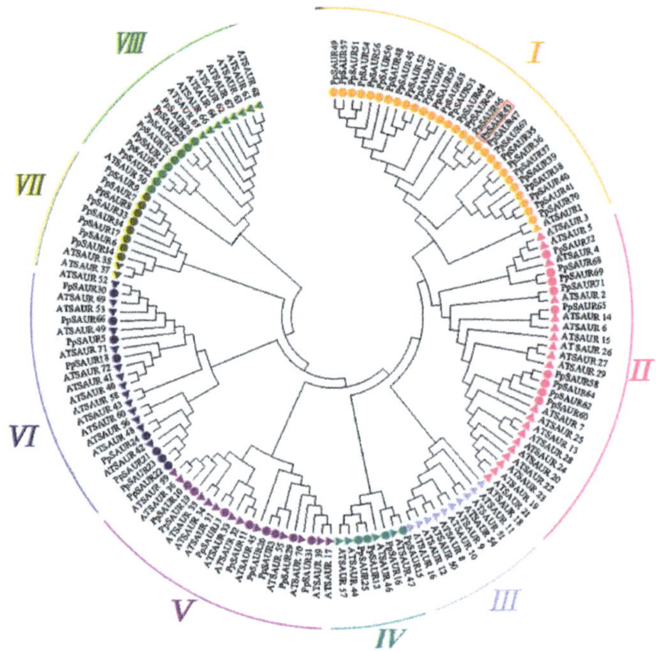

Figure 1. Phylogenetic analysis of SAUR from peaches (PpSAUR) and *Arabidopsis* (AtSAUR). The unrooted tree was generated using the MEGA7 program by the neighbor-joining method with the following parameters: A bootstrap analysis with 1000 replicates was carried out. Gene IDs corresponding to the peach SAUR proteins listed in the tree are presented in Table 1.

3.3. *Expression Patterns of PpSAUR43 in the MF and NMF Cultivars during Peach Post-Ripening*

Previous studies have shown that the expression profile of the differentially expressed gene *PpSAUR43*, as a candidate for the fruit softening rate gene, experimentally confirmed by RT-qPCR, is upregulated in low-softening-rate siblings [75]. We were interested in *PpSAUR43*, which has been predicted by previous studies during peach post-ripening. Therefore, we selected the fruit of the MF peach 'YHL' and the NMF peach 'B5' for study at the post-ripening stage. Previous studies have shown that the firmness of the NMF fruit 'B5' has a declining tendency in the post-ripening stage compared to the MF fruit 'YHL' [82]. Meanwhile, the ethylene peak for the 'YHL' fruit was higher than that found for the 'B5' fruit [82]. We detected the expression patterns of *PpSAUR43* in the fruit of the MF peach 'YHL' and the NMF peach 'B5' (Figure 2). In total, *PpSAUR43* expression was decreased during post-ripening. In addition, we observed a high expression level of *PpSAUR43* in 'B5' and a low expression level in 'YHL'. These results suggest that *PpSAUR43* may be involved in peach post-ripening.

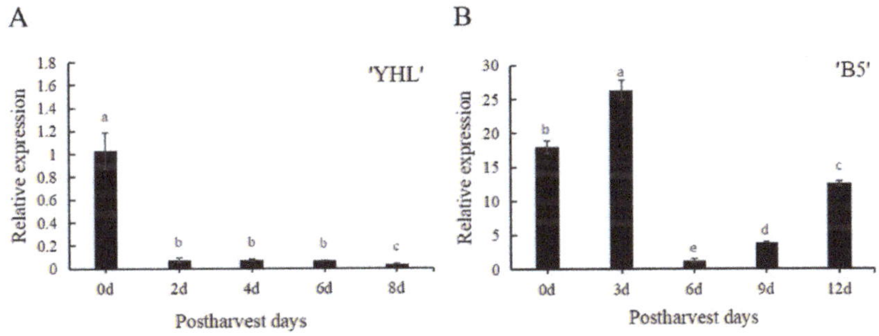

Figure 2. Expression level of *PpSAUR43* of 'YHL' and 'B5' in the post-ripening stage. (**A**) Expression level of *PpSAUR43* of 'YHL'. (**B**) Expression level of *PpSAUR43* of 'B5'. The fruits were stored at 25.0 ± 1 °C after harvesting. Three biological replicates were used and each repetition was carried out by mixing flesh samples from three fruits. The expression of *PpSAUR43* in 'YHL' at day 0 was used as the standard for normalization. Duncan's test was used with the expression of *PpSAUR43* at day 0 as a control, with different letters indicating significant differences ($p < 0.05$).

3.4. *PpSAUR43 Proteins Are Localized in the Cell Membrane and Nucleus*

To determine the subcellular localization of PpSAUR43, a 35S-PpSAUR43-GFP recombinant plasmid was generated (Figure 3A). Here, the 35S-PpSAUR43-GFP recombinant plasmid was transformed into transgenic tobacco containing a nuclear localization signal (NLS-mCherry). At the same time, empty pCAMBIA2300 was used as a control. The results show that the PpSAUR43 signal overlapped the nuclear localization signal, suggesting that PpSAUR43 was located in the nucleus (Figure 3B). At the same time, we also detected a GFP signal on the cell membrane. Therefore, PpSAUR43 may be localized in the membrane and nucleus.

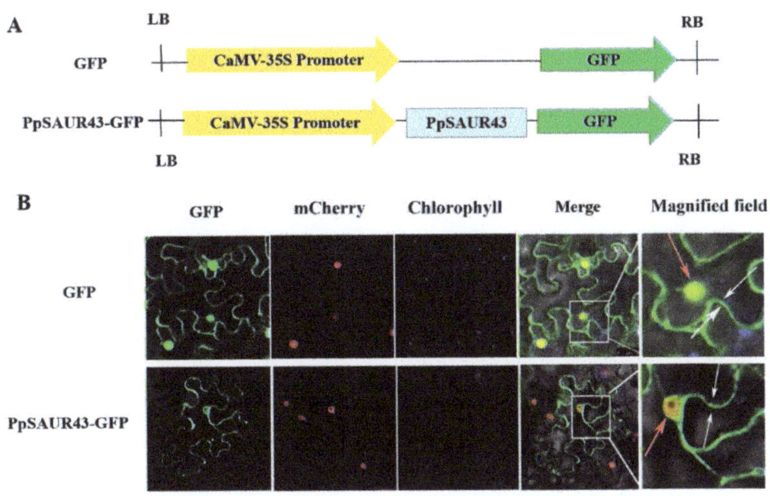

Figure 3. Subcellular localization of PpSAUR43 proteins. (**A**) Schematic diagram of GFP constructs. (**B**) Subcellular localization of proteins in *Nicotiana tabacum* leaf mesophyll cells. The nucleus was indicated by mCherry carrying a nuclear localization signal. The GFP fluorescence, chlorophyll autofluorescence, and mCherry were merged. The magnified field showed GFP signals in different parts of cells. White and red arrows indicate the presence of GFP signals in the cell membrane and nucleus, respectively.

3.5. *PpSAUR43* Impacts Peach Fruit Post-Ripening and Softening as an Inhibitor

To gain insight into the role of PpSAUR43 in peach fruit post-ripening, we investigated the effects of the transient overexpression of *PpSAUR43* on peach fruit post-ripening by performing fruit peel injection assays. As shown in Figure 4, the infested portion of the pCAMBIA2300–*PpSAUR43* infiltrated fruit surface was dark green, while the surface of the empty pCAMBIA2300 vector-infiltrated control fruit was light green or normal red (Figure 4A). The firmness of the peach tended to decrease during storage at 25.0 ± 1 °C, and the transient overexpression of *PpSAUR43* significantly reduced the rate of decline in fruit firmness after the fourth day of storage (Figure 4B). In addition, we also measured the ethylene release rate of overexpressed fruits. During the storage period, the ethylene release rate in *PpSAUR43*-overexpressing fruits and the corresponding control fruits had the same trend, both of which increased first and then decreased, reaching the peak ethylene release at four days. The ethylene release rate for the PC2300-PpSAUR43 fruit showed no significant difference when compared with the control fruit (Figure 4B). Subsequently, we selected the agroinfiltrated parts of the fruit to analyze the changes in the free-state IAA content (Figure 4B). The results show that the free-state IAA content in the PpSAUR43 overexpressing fruits was significantly lower than that of the control on days 0, 2, 4, and 6 of storage. We confirmed that *PpSAUR43* expression is effectively overexpressed at the molecular level using RT-qPCR analyses at seven days post-inoculation. The levels of *PpSAUR43* expression at the injection site of the overexpressed fruit were significantly higher than the control by approximately 10-, 14-, 2-, and 45-fold on days 0, 2, 4, and 6 of storage. At day 8 of storage, the transient overexpression effect decayed, and the expression of *PpSAUR43* at the injection site of the overexpressed and control fruits was not significantly different. This indicates that the transient overexpression of *PpSAUR43* in peaches is significant (Figure 4C). To further understand the role of *PpSAUR43* in the regulation of peach fruit post-ripening, we analyzed several ripening-related genes in the *PpSAUR43* overexpressing fruits and control fruits by RT-qPCR analyses: cell wall metabolism-related gene (polygalacturonase, *PpPG*), 1-amino cyclopropane-1-carboxylic acid synthase gene (*PpACS1*), and 1-amino cyclopropane-1-carboxylic acid oxidase gene (*PpACO1*). The transient overexpression of *PpSAUR43* led to the downregulation of the ethylene synthesis gene *PpASC1* and the cell wall degradation gene *PpPG*, and no significant difference in *PpACO1*, except that it was significantly lower than the control at four days. We additionally measured the expression levels of *PpCMB1*, a member of the MADS-box family that is closely related to fruit ripening. The transcript abundance of *PpCMB1* in the *PpSAUR43* overexpressing fruits was significantly lower than that of the control fruit on days 0, 2, and 4. These results indicate that *PpSAUR43* is essential for peach fruit post-ripening by acting as an inhibitor. To elucidate the mechanism of *PpSAUR43* overexpression on the delay of peach post-ripening and softening, we also estimated the transcript levels of a set of genes related to auxin transport and homeostasis, including the polar auxin transport gene (*PpPIN1*), the IAA-amido synthetase gene (Gretchen-Hagen 3 (*PpGH3.1*)), and the indole-3-acetic acid (IAA)-amino hydrolase gene (*PpILR1*). The expressions of *PpPIN1* and *PpGH3.1* were upregulated more in the transient overexpression of the *PpSAUR43* overexpressing fruit compared to those of the control fruit, while the expression of *PpILR1* was markedly more downregulated than in the control fruit (Figure 4C), indicating that *PpSAUR43* might promote the transport of auxin and the conversion of the free state to the conjugate state.

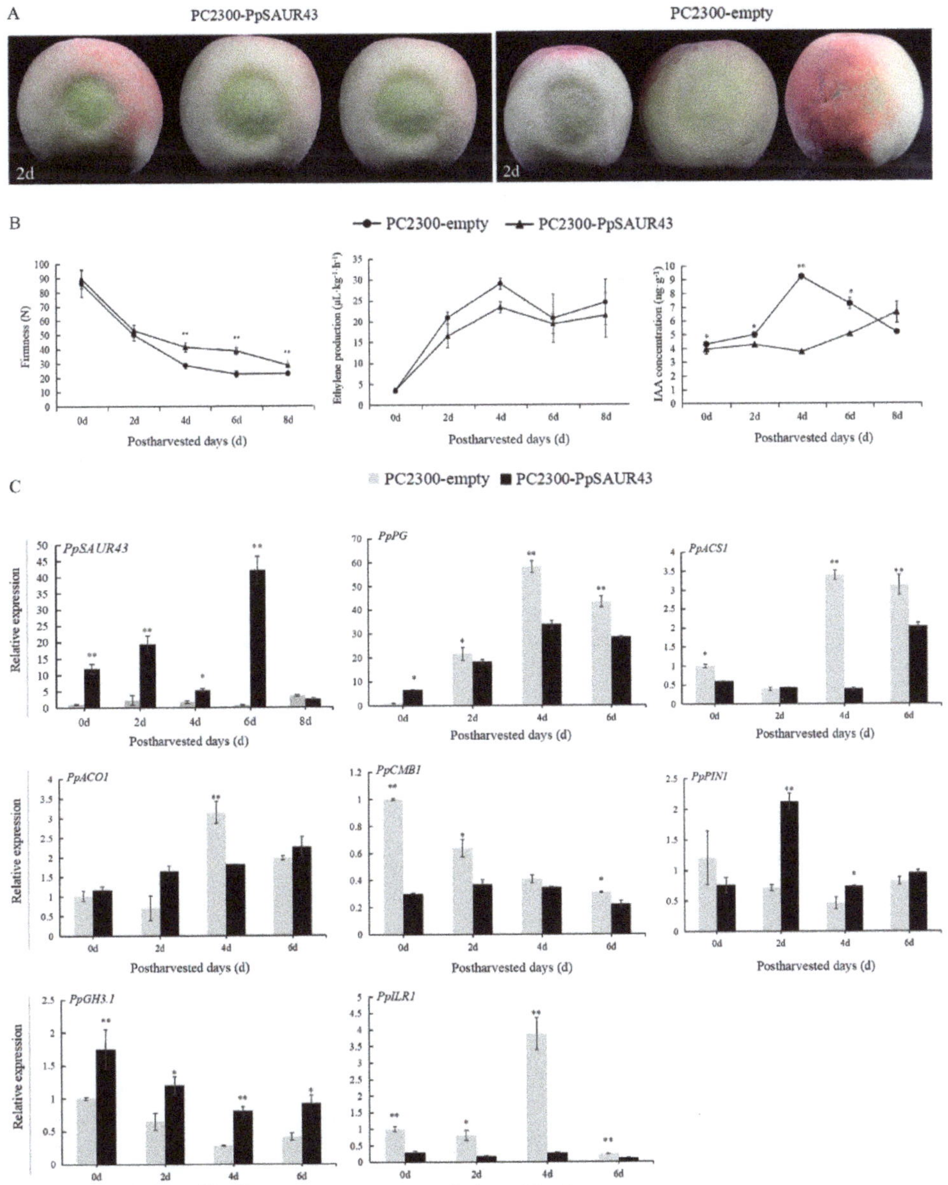

Figure 4. Transient overexpression of *PpSAUR43* in 'YHL' fruit. (**A**) Phenotypes of infiltrated fruits. Empty pCAMBIA2300 vectors (PC2300-empty) were used as a control. (**B**) Changes in fruit firmness, ethylene production, and IAA concentration during fruit storage. (**C**) Gene expression of *PpSAUR43*, *PpPG*, *PpACS1*, *PpACO1*, *PpCMB1*, *PpPIN1*, *PpGH3.1*, and *PpILR1*. Gene expression was measured relative to that of *PpTUA5*. Fruit was stored at 25.0 ± 1 °C after harvest. At the same storage days, PC2300-*PpSAUR43* was compared with controls. Error bars represent the SD of the measurements from three independent repeat experiments. Duncan's test was used to compare the differences. * $p < 0.05$; ** $p < 0.01$.

3.6. *PpSAUR43 Interacts with PpCMB1*

To identify the proteins that interact with PpSAUR43, we performed a yeast two-hybrid (Y2H) library assay using PpSAUR43 as bait. We first validated the self-activation of PpSAUR43. It turned out that neither the control nor the experimental groups showed a blue color on the SD/-Trp/-Leu/-His/x-α-gal plates, indicating that PpSAUR43 does not have transcriptional activation activity (Figure S3). Subsequently, we co-transformed the yeast strain with BD-PpSAUR43 and the library plasmid, coated the co-transformed bacteria on SD/-Ade/-His/-Leu/-Trp/x-α-gal plates, and after three days, the blue spots were subjected to PCR. The products with bands greater than 500 bp were selected for sequencing and for blast screening of the possible interacting proteins with the NCBI website (Table S2 and Figure S4). Notably, we identified PpCMB1, which has been shown to regulate fruit ripening and softening in tomatoes, as a potential PpSAUR43-interacting protein [44]. To further validate the interaction between PpSAUR43 and PpCMB1, we inserted the PpCMB1 full-length CDS into the pGADT7 vector to obtain AD-PpCMB1 and co-transformed AD-PpCMB1 and BD-PpSAUR43 into the Y2H-Gold strain (Figure 5A). As expected, the Y2H-Gold strain harboring BD-PpSAUR43/AD-PpCMB1 grew and turned blue on SD/-Leu/-Trp/-His/-Ade medium supplemented with X-α-gal (Figure 5A). The results confirm the interaction between PpSAUR43 and PpCMB1 in yeast. We further examined the subcellular localization of PpCMB1 by transforming *Nicotiana tabacum* leaves with *Agrobacterium tumefaciens*-containing 35S-PpCMB1-GFP vectors. The results show that PpCMB1 was localized in the cell membrane and nucleus. (Figure S5), which was the same for PpSAUR43. Furthermore, a BIFC assay confirmed this interaction in *Arabidopsis* protoplasts. Different combinations, including pSPYCE-PpSAUR43/pSPYNE-PpCMB1 and pSPYCE-PpSAUR43/pSPYNE, were co-transformed into *Arabidopsis* protoplasts (Figure 5B). The YFP signal that was detected in the nucleus could be detected only when pSPYCE-PpSAUR43 was co-transformed with pSPYNE-PpCMB1. These results demonstrate that PpSAUR43 interacts physically with PpCMB1 in the nucleus (Figure 5C).

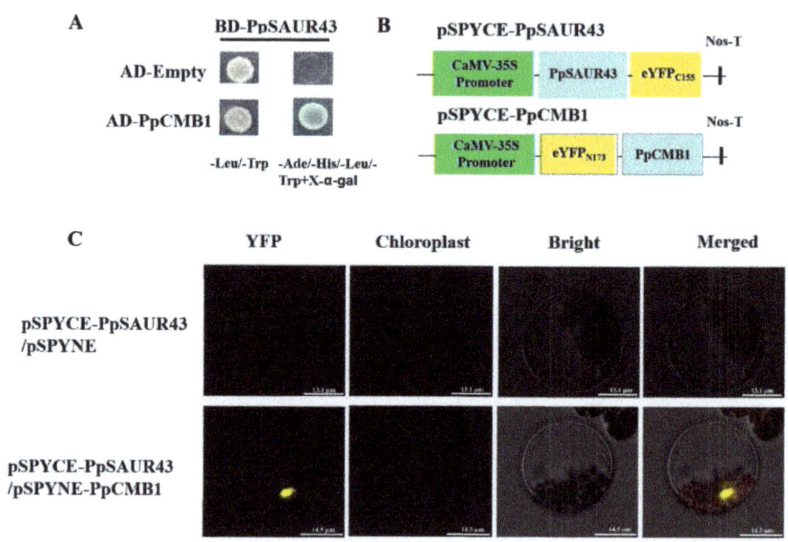

Figure 5. Analysis of the interaction between PpSAUR43 and PpCMB1. (**A**) The Y2H assay confirmed the interaction between PpSAUR43 and PpCMB1 in yeast. (**B**) Schematic diagrams of the vectors used in the BIFC assay. (**C**) The BIFC assay confirmed the interaction between PpSAUR43 and PpCMB1 *Arabidopsis* protoplasts.

4. Discussion

Fruit ripening is a complex process involving metabolic and physiological changes that depend on the coordinated role of multiple hormone signals, microRNAs, epigenetic maintenance, epigenetic modifying genes, and other major regulatory factors [83–85]. As is already known, ethylene is regarded as the major regulator of climacteric fruit ripening [84,86]. However, an increasing body of evidence has pointed toward a more complex role of auxin in climacteric fruit ripening [28]. Primary auxin-responsive genes, the SAUR gene family, are present as large gene families in diverse plant species and have been well established in plant growth and development, such as cell elongation [87] and leaf senescence [65,66].

In this study, we identified 72 *SAUR* genes in peaches and named them according to their positions on the chromosome site. All of the members were encoded with the SAUR domains. The number of SAUR family members in peaches is almost the same as that of *Arabidopsis* (81) [53], rice (58) [54], and tomatoes (99) [55]. Based on their chromosomal distribution, it was found that the distribution of SAUR family members is often concentrated in tandem clusters on parts of the chromosome (Figure S1). *SAUR* gene clusters are also commonly found in other species. For example, four *SAUR* gene clusters have been identified in *Arabidopsis*, located on chromosomes 1, 3, 4, and 5, while five *SAUR* gene clusters have been identified in rice, located on chromosomes 1, 2, 3, 9, and 12 [53,54]. There were two *SAUR* genes clusters on chromosome 8 in the peach genome. We deduced that this may be due to gene duplication in the family. Gene duplication is a major way of increasing the number of gene families in which genes are usually similar, such as unequal exchange, retro-transposition, or whole-gene duplication, to produce genes or base sequences similar to the original gene [88]. It was reported that rapid gene amplification is prevalent in gene families associated with morphogenesis and stress responses, and gene duplication plays an important role in plant evolution [89]. To further investigate the evolutionary relationships of *SAUR* family genes, we constructed a phylogenetic tree of PpSAUR and divided it into eight subfamilies. We found a high degree of sequence similarity among the gene cluster members. The gene cluster members located on chromosome 8 are part of the first subgroup and are genetically close to *AtSAUR1* (AT4G34770) and are presumed to have similar functions and to be important in hormone crosstalk [90]. Furthermore, we analyzed the gene structure of the SAUR family members (Figure S2). We found that 67 of the 72 SAUR families lacked intron structures, a feature also found in SAUR gene families of other species, such as rice and tomatoes [90,91]. This may be due to the fact that the genes that respond quickly to auxin tend to be more likely to have genes without introns [92].

Previous studies developed by Carrasco-Valenzuela et al. [75] used quantitative trait locus (QTL) and expression QTL (eQTL) analysis to screen the candidate gene, indole-3-acetic acid-induced protein (*PpSAUR43*), which is related to the fruit softening rate. Moreover, *PpSAUR43* was identified to be differentially expressed in peaches with different softening rates by RNA-seq and RT-qPCR analysis, so we put the spotlight on *PpSAUR43* (the RNA-seq data are available at the NCBI's Sequence Read Achieve (available at: http://www.ncbi.nlm.nih.gov/sra (accessed on 22 June 2020)) with SRA accession number SRP186384). First, we analyzed the expression pattern of *PpSAUR43* in different textured peach cultivars during post-ripening. Interestingly, the expression of *PpSAUR43* was negatively correlated with the post-ripening process in NMF and MF fruits, and the expression of *PpSAUR43* was high in the NMF varieties and low in the MF varieties. These results are concordant with the previous findings that the expression of *PpSAUR43* is higher in the fruits of low-softening-rate cultivars than in high-softening-rate cultivars [75]. It is speculated that *PpSAUR43* plays a negative role in peach fruit post-ripening and softening.

Some studies have shown that SAUR proteins may be localized in subcellular structures such as the nucleus, cytoplasm, and plasma membrane (PM), which may imply that these proteins have multiple functions in plants [50,93]. Our data show that PpSAUR43 is located in the nucleus and cell membrane. This dual localization hints at its function in transmitting auxin signals between the cell membrane and the nucleus. To further validate the role of *SAUR* in peach fruit post-ripening and softening, we characterized the biological function of *PpSAUR43* in peaches by transient gene overexpression. The overexpression of *PpSAUR43* significantly inhibits peach post-ripening, as evidenced by noticeable increases in the fruit firmness and delays in the fruit softening. A more noticeable metabolic change in fruit ripening is the softening caused by changes in the structure of the cell wall [94]. The main component of the gel layer in the cell wall is pectin, which plays an essential role in intercellular adhesion [95,96]. The degradation of pectin causes a rapid reduction in cell viscosity and leads to a reduction in fruit hardness. As polygalacturonase is the predominant enzyme that degrades pectin in the cell wall (PG), we measured the expression of the *PpPG* gene, a key gene that regulates fruit softening, to assess the degree of cell wall degradation between the overexpression of *PpSAUR43* and the control fruit [97]. We found that although there was no significant difference between the experimental and the control groups on day 0 of storage, the overexpression of *PpSAUR43* led to a decrease in the transcript level of *PpPG* in the fruit as the storage days increased. Ethylene is known to play a key role in fruit ripening. Hence, we also measured the amount of ethylene produced. Unfortunately, no significant differences were found between the overexpressed and control fruit. We speculate that the reason for this may be that we used whole peaches to measure the ethylene content and the small area of the injection site did not result in a significant difference. Subsequently, we further determined the expression of *PpACS1*, which is related to ethylene synthesis in the fruit, and found that the overexpression of *PpSAUR43* resulted in a significantly lower expression of *PpACS1* than the control. Recent studies have shown that a low level of auxin in peaches inhibits the expression of the *PpACS1* gene [17–19]. Furthermore, we measured the transcript levels of *PpACO1*, a key ACC oxidase gene involved in peach ethylene synthesis, and found no significant change in *PpACO1* expression after the overexpression of *PpSAUR43*, except for days. Based on the above data, we speculate that *PpSAUR43* affects the transcription of *PpACS1*, probably due to reduced auxin levels in the fruit. The ability of SAUR proteins to alter auxin levels in vivo has been reported in other species for a long time. In *Arabidopsis*, *AtSAUR19* overexpression in seedling hypocotyls increases translocation to auxin [68]. Similar findings have been obtained with *OsSAUR39* in rice [98]. We measured the content of free-state IAA in the *PpSAUR43* overexpressing fruit and control fruits and found that the overexpression of *PpSAUR43* significantly reduced the level of IAA in vivo. Reducing auxin levels can be achieved through auxin transport and auxin conjugation or degradation mechanisms. The PIN-FORMED (PIN) gene family of transmembrane proteins as auxin efflux carriers plays a crucial role in polar auxin transport [99]. It has been suggested that PpPIN1 may regulate the distribution of auxin involved in peach ripening [22]. The expression of *PpPIN1* was significantly upregulated in response to the overexpression of *PpSAUR43*, suggesting that *PpSAUR43* facilitates the transport of auxin. GH3 genes, which catalyze the conversion of free-form auxin to the conjugated state, play a role in the maintenance of cellular auxin levels in a range of plant species [100]. Our results show that the expression of *PpGH3.1* was elevated in comparison to the control. Interestingly, we then analyzed the transcript levels of the indole-3-acetic acid (IAA)-amino hydrolase *PpILR1* and showed that the expression of *PpILR1* was significantly lower when overexpressed by *PpSAUR43*. As previously reported, *PpILR1* acts as a key gene for IAA-amino hydrolase and is positively involved in the regulation of peach fruit ripening [101]. This supports our speculation that *PpSAUR43* is associated with auxin homeostasis and reduces the level of free-state auxin in vivo by promoting the formation of the auxin-conjugated state.

To further investigate the regulatory mechanism of *PpSAUR43* on fruit softening, we demonstrated the specific interaction of PpSAUR43 with PpCMB1, a MADS-box protein of the SEP class, by Y2H and BIFC analysis. In our study, PpSAUR43 and PpCMB1 were both localized and interacted in the cell nucleus and membrane, which provided the possibility for their physical interaction. Some research indicates that SlCMB1, which shares high homology with RIN, SlMADS1, and SlMBP21, is a new component of the current model of the regulatory network that regulates ethylene biosynthesis during tomato fruit ripening [44]. Then, RT-qPCR analysis showed that *PpCMB1* was downregulated during post-ripening in *PpSAUR43* overexpressed fruits. These results imply that there is an antagonistic relationship between PpSAUR43 and PpCMB1 in the regulation of peach post-ripening and that PpSAUR43 may act as a negative regulator, while PpCMB1 is a positive regulator. This coincides with previous studies on tomatoes that have shown that the suppression of *SlCMB1* results in delayed fruit ripening by affecting ethylene biosynthesis and signal transduction [44]. Notably, promoter analysis by website prediction showed that the *PpCMB1* and *PpSAUR43* gene promoter region contained the auxin-responsive element AuxRR-Core (GGTCCAT) and the ethylene-responsive element ERE motif (ATTTTAAA/ATTTCATA) (Table S3) [102,103]. These results suggest that PpSAUR43 and PpCMB1 act as modulators of the optimal balance between auxin and ethylene actions. Hence, on the one hand, overexpressing *PpSAUR43* inhibits the transcriptional activation of *PpCMB1*, thereby suppressing the expression of downstream ethylene synthesis and signal transduction genes. On the other hand, *PpSAUR43* could work in a regulation loop to decrease the free auxin level, which, in turn, would decrease the expression of *PpACS1* via the auxin signal transduction pathway.

5. Conclusions

In conclusion, our study reports 72 SAUR family members in peaches after we performed bioinformatics analysis. The functions of peach *PpSAUR43* were then identified by transient overexpression in peaches, and a new mechanism was found for the interaction between PpSAUR43 and PpCMB1 regulating peach post-ripening and softening. The findings of this study will help to develop methods to precisely control peach post-ripening and extend shelf life.

Supplementary Materials: The following supporting information can be downloaded at https://www.mdpi.com/article/10.3390/horticulturae8050379/s1, Figure S1: Chromosomal localization of SAUR family members; Figure S2: Gene structure of SAUR family members; Figure S3: Transcriptional activation identification of the full length of PpSAUR43; Figure S4: The interacting confirmation between PpSAUR43 and PpCMB1, PpMY66, PpWHY1, and PpPP2C5 in yeast two-hybrid system; Figure S5: Subcellular localization of GFP constructs of PpCMB1 proteins; Table S1: Primers for cloning and RT-qPCR in this study; Table S2: Result of partial candidate positive clones by yeast two-hybrid (Y2H) library assay in peach; Table S3: Cis-acting regulatory elements were predicted in the promoter regions of *PpSAUR43*, *PpCAMB1* related to fruit development and ripening in peach.

Author Contributions: C.Z. conceived and designed the research. J.W., W.S., K.L. and Z.X. conducted the experiments. J.M., D.Z. and C.Z. contributed reagents and gave advice. J.W. and Y.H. wrote the manuscript. Y.H. and K.S. edited the manuscript. All authors have read and agreed to the published version of the manuscript.

Funding: This research and the APC were funded by the National Key R&D Program of China (Grant No. 2019YFD1000203).

Institutional Review Board Statement: This research content of the manuscript does not involve ethical issues.

Informed Consent Statement: This research content of the manuscript does not involve humans.

Data Availability Statement: All results are included with in the article.

Acknowledgments: We thank Jing Zhang (Horticulture Science Research Center, Northwest A&F University, Xianyang, China) for providing professional technical assistance with LC-MS/MS analysis.

Conflicts of Interest: The authors declare no conflict of interest.

References

1. Grierson, D.; Kader, A.A. Fruit ripening and quality. In *The Tomato Crop*; Rudich, J., Ed.; Springer: London, UK, 1986; pp. 241–280.

2. Li, Y.; Wang, L. Genetic Resources, Breeding Programs in China, and Gene Mining of Peach: A Review. *Hortic. Plant J.* **2020**, *6*, 205–215. [CrossRef]
3. Food and Agricultural Organization. Available online: www.fao.org (accessed on 18 December 2021).
4. Yoshida, M. Genetical studies on the fruit quality of peach varieties. III. Texture and keeping quality. *Bull. Fruit Tree Res. Stn.* **1976**, *3*, 1–16.
5. Given, N.K.; Venis, M.A.; Gierson, D. Hormonal regulation of ripening in the strawberry, a non-climacteric fruit. *Planta* **1988**, *174*, 402–406. [CrossRef]
6. Fuentes, L.; Figueroa, C.R.; Valdenegro, M. Recent Advances in Hormonal Regulation and Cross-Talk during Non-Climacteric Fruit Development and Ripening. *Horticulturae* **2019**, *5*, 45. [CrossRef]
7. Obroucheva, N.V. Hormonal regulation during plant fruit development. *Russ. J. Dev. Biol.* **2014**, *45*, 14. [CrossRef]
8. Wang, P.; Lu, S.; Zhang, X.; Hyden, B.; Qin, L.; Liu, L.; Bai, Y.; Han, Y.; Wen, Z.; Xu, J.; et al. Double NCED isozymes control ABA biosynthesis for ripening and senescent regulation in peach fruits. *Plant Sci.* **2021**, *304*, 110739. [CrossRef]
9. Tan, B.; Lian, X.; Cheng, J.; Zeng, W.; Zheng, X.; Wang, W.; Ye, X.; Li, J.; Li, Z.; Zhang, L.; et al. Genome-wide identification and transcriptome profiling reveal that E3 ubiquitin ligase genes relevant to ethylene, auxin and abscisic acid are differentially expressed in the fruits of melting flesh and stony hard peach varieties. *BMC Genom.* **2019**, *20*, 892. [CrossRef]
10. Burg, S.P. Ethylene action and the ripening of fruits. *Science* **1965**, *148*, 1190–1196. [CrossRef]
11. Alexander, L.; Grierson, D. Ethylene biosynthesis and action in tomato: A model for climacteric fruit ripening. *J. Exp. Bot.* **2002**, *377*, 2039–2055. [CrossRef]
12. Lelièvre, J.M.; Latchè, A.; Jones, B.; Hall, M.A. Ethylene and fruit ripening. *Physiol. Plant.* **1997**, *101*, 727–739. [CrossRef]
13. Zarembinski, T.I.; Theologis, A. Ethylene biosynthesis and action: A case of conservation. *Plant Mol. Biol.* **1994**, *26*, 1579–1597. [CrossRef] [PubMed]
14. Jerie, P.H.; Hall, M.A.; Jones, B. Aspects of the role of ethylene in fruit ripening. *Acta Hortic.* **1978**, *80*, 325–332. [CrossRef]
15. Haji, T.; Yaegaki, H.; Yamaguchi, M. Inheritance and expression of fruit texture melting, non-melting and stony hard in peach. *Sci. Hortic.* **2005**, *105*, 241–248. [CrossRef]
16. Bailey, J.S.; French, A.P. The inheritance of certain characters in the peach. *Proc. Am. Soc. Hortic. Sci.* **1932**, *29*, 127–130.
17. Tatsuki, M.; Yamaguchi, H.M. The involvement of 1-aminocyclopropane-1-carboxylic acid synthase isogene, *PpACS1*, in peach fruit softening. *J. Exp. Bot.* **2006**, *57*, 1281–1289. [CrossRef]
18. Tatsuki, M.; Haji, T.; Yamaguchi, M. The peach 1-aminocyclopropane-1-carboxylic acid synthase isogene, *Pp-ACS1*, is required for fruit softening. In *Advances in Plant Ethylene Research*; Springer: Dordrecht, Germany, 2007; pp. 227–228. [CrossRef]
19. Tatsuki, M.; Nakajima, N.; Fujii, H.; Shimada, T.; Nakano, M.; Hayashi, K.-I.; Hayama, H.; Yoshioka, H.; Nakamura, Y. Increased levels of IAA are required for system 2 ethylene synthesis causing fruit softening in peach (*Prunus persica* L. Batsch). *J. Exp. Bot.* **2013**, *64*, 1049–1059. [CrossRef]
20. Pan, L.; Zeng, W.; Liang, N. *PpYUC11*, a strong candidate gene for the stony hard phenotype in peach (*Prunus persica* L. Batsch), participates in IAA biosynthesis during fruit ripening. *J. Exp. Bot.* **2015**, *22*, 7031–7044. [CrossRef]
21. Zeng, W.; Ding, Y.; Pan, L.; Wang, X.; Niu, L.; Lu, Z.; Cui, G.; Wang, Z. A CACTA transposable element in a *PpYUC11* gene promoter is associated with the stony hard phenotype in peach. *J. Fruit Sci.* **2017**, *4*, 1239–1248. [CrossRef]
22. Tadiello, A.; Ziosi, V.; Negri, A.S.; Noferini, M.; Fiori, G.; Busatto, N.; Espen, L.; Costa, G.; Trainotti, L. On the role of ethylene, auxin and a GOLVEN-like peptide hormone in the regulation of peach ripening. *BMC Plant Biol.* **2016**, *16*, 44. [CrossRef]
23. Busatto, N.; Tadiello, A.; Trainotti, L.; Costa, F. Climacteric ripening of apple fruit is regulated by transcriptional circuits stimulated by cross-talks between ethylene and auxin. *Plant Signal. Behav.* **2016**, *12*, e1268312. [CrossRef]
24. Khaksar, G.; Sirikantaramas, S. Auxin Response Factor 2A Is Part of the Regulatory Network Mediating Fruit Ripening Through Auxin-Ethylene Crosstalk in Durian. *Front. Plant Sci.* **2020**, *11*, 543747. [CrossRef] [PubMed]
25. Zhang, T.; Li, W.; Xie, R.; Xu, L.; Zhou, Y.; Li, H.; Yuan, C.; Zheng, X.; Xiao, L.; Liu, K. CpARF2 and CpEIL1 interact to mediate auxin–ethylene interaction and regulate fruit ripening in papaya. *Plant J.* **2020**, *103*, 1318–1337. [CrossRef]
26. Hao, Y.; Hu, G.; Breitel, D.; Liu, M.; Mila, I.; Frasse, P.; Fu, Y.; Aharoni, A.; Bouzayen, M.; Zouine, M. Auxin Response Factor SlARF2 Is an Essential Component of the Regulatory Mechanism Controlling Fruit Ripening in Tomato. *PLoS Genet.* **2015**, *11*, e1005649. [CrossRef] [PubMed]
27. Zhang, L.; Chen, L.; Pang, S.; Zheng, Q.; Quan, S.; Liu, Y.; Xu, T.; Liu, Y.; Qi, M. Function Analysis of the ERF and DREB Subfamilies in Tomato Fruit Development and Ripening. *Front. Plant Sci.* **2022**, *13*, 849048. [CrossRef] [PubMed]
28. Trainotti, L.; Tadiello, A.; Casadoro, G. The involvement of auxin in the ripening of climacteric fruits comes of age: The hormone plays a role of its own and has an intense interplay with ethylene in ripening peaches. *J. Exp. Bot.* **2007**, *58*, 3299–3308. [CrossRef] [PubMed]
29. Gu, T.; Jia, S.; Huang, X.; Wang, L.; Fu, W.; Huo, G.; Gan, L.; Ding, J.; Li, Y. Transcriptome and hormone analyses provide insights into hormonal regulation in strawberry ripening. *Planta* **2019**, *250*, 145–162. [CrossRef]
30. Liu, G.S.; Li, H.L.; Grierson, D.; Fu, D.Q. NAC transcription factor family regulation of fruit ripening and quality: A Review. *Cells* **2022**, *11*, 525. [CrossRef]
31. Xie, F.; Hua, Q.; Chen, C.; Zhang, Z.; Zhang, R.; Zhao, J.; Hu, G.; Chen, J.; Qin, Y. Genome-Wide Characterization of R2R3-MYB Transcription Factors in Pitaya Reveals a R2R3-MYB Repressor *HuMYB1* Involved in Fruit Ripening through Regulation of Betalain Biosynthesis by Repressing Betalain Biosynthesis-Related Genes. *Cells* **2021**, *10*, 1949. [CrossRef]

32. Jiang, G.; Zeng, J.; Li, Z.; Song, Y.; Yan, H.; He, J.; Jiang, Y.; Duan, X. Redox Regulation of the NOR Transcription Factor Is Involved in the Regulation of Fruit Ripening in Tomato. *Plant Physiol.* **2020**, *183*, 671–685. [CrossRef]
33. Deng, H.; Chen, Y.; Liu, Z.; Liu, Z.Q.; Shu, P.; Wang, R.; Hao, Y.; Su, D.; Pirrello, J.; Liu, Y.; et al. SlERF.F12 modulates the transition to ripening in tomato fruit by recruiting the co-repressor TOPLESS and histone deacetylases to repress key ripening genes. *Plant Cell.* **2022**, *4*, 1250–1272. [CrossRef]
34. Wang, X.; Pan, L.; Wang, Y.; Meng, J.; Deng, L.; Niu, L.; Liu, H.; Ding, Y.; Yao, J.-L.; Nieuwenhuizen, N.J.; et al. *PpIAA1* and *PpERF4* form a positive feedback loop to regulate peach fruit ripening by integrating auxin and ethylene signals. *Plant Sci.* **2021**, *313*, 111084. [CrossRef] [PubMed]
35. Li, S.; Chen, K.; Grierson, D. A critical evaluation of the role of ethylene and MADS transcription factors in the network controlling fleshy fruit ripening. *New Phytol.* **2019**, *221*, 1724–1741. [CrossRef]
36. Vrebalov, J.; Ruezinsky, D.; Padmanabhan, V.; White, R.; Medrano, D.; Drake, R.; Schuch, W.; Giovannoni, J. A MADS-Box Gene Necessary for Fruit Ripening at the Tomato *Ripening-Inhibitor* (*Rin*) Locus. *Science* **2002**, *296*, 343–346. [CrossRef] [PubMed]
37. Schaffer, R.J.; Ireland, H.S.; Ross, J.J.; Ling, T.J. SEPALLATA1/2-suppressed mature apples have low ethylene, high auxin and reduced transcription of ripening-related genes. *AOB Plants* **2013**, *5*, 047. [CrossRef] [PubMed]
38. Elitzur, T.; Yakir, E.; Quansah, L.; Zhangjun, F.; Vrebalov, J.; Khayat, E.; Giovannoni, J.J.; Friedman, H. Banana *MaMADS* Transcription Factors Are Necessary for Fruit Ripening and Molecular Tools to Promote Shelf-Life and Food Security. *Plant Physiol.* **2016**, *171*, 380–391. [CrossRef]
39. Li, J.; Li, F.; Qian, M.; Han, M.; Liu, H.; Zhang, D.; Ma, J.; Zhao, C. Characteristics and regulatory pathway of the PrupeSEP1 SEPALLATA gene during ripening and softening in peach fruits. *Plant Sci.* **2017**, *257*, 63–73. [CrossRef]
40. Pi, M.; Hu, S.; Cheng, L.; Zhong, R.; Cai, Z.; Liu, Z.; Yao, J.-L.; Kang, C. The MADS-box gene FveSEP3 plays essential roles in flower organogenesis and fruit development in woodland strawberry. *Hortic. Res.* **2021**, *8*, 247. [CrossRef]
41. Lu, S.; Ye, J.; Zhu, K.; Zhang, Y.; Zhang, M.; Xu, Q.; Deng, X. A fruit ripening-associated transcription factor CsMADS5 positively regulates carotenoid biosynthesis in citrus. *J. Exp. Bot.* **2021**, *72*, 3028–3043. [CrossRef]
42. Slugina, M.A. Transcription Factor Ripening Inhibitor and Its Homologs in Regulation of Fleshy Fruit Ripening of Various Plant Species. *Russ. J. Plant Physiol.* **2021**, *68*, 783–799. [CrossRef]
43. Xu, Y.; Lin, Z.; Hua, X.; Zhang, Y.Q.; Oliveira, M.M. Expression analysis and genetic mapping of three SEPALLATA-like genes from peach (*Prunus persica* (L.) Batsch). *Tree Genet. Genomes* **2008**, *4*, 693–703. [CrossRef]
44. Zhang, J.; Hu, Z.; Yao, Q.; Guo, X.; Nguyen, V.; Li, F.; Chen, G. A tomato MADS-box protein, SlCMB1, regulates ethylene biosynthesis and carotenoid accumulation during fruit ripening. *Sci. Rep.* **2018**, *8*, 3413. [CrossRef]
45. Li, Q.; Wang, T.; Xu, C.; Li, M.; Tian, J.; Wang, Y.; Zhang, X.; Xu, X.; Han, Z.; Wu, T. *MdMADS6* Recruits Histone Deacetylase *MdHDA19* to Repress the Expression of the Carotenoid Synthesis-Related Gene *MdCCD1* during Fruit Ripening. *Plants* **2022**, *11*, 668. [CrossRef]
46. Qi, X.; Liu, C.; Song, L.; Li, M. *PaMADS7*, a MADS-box transcription factor, regulates sweet cherry fruit ripening and softening. *Plant Sci.* **2020**, *301*, 110634. [CrossRef] [PubMed]
47. Tani, E.; Polidoros, A.; Flemetakis, E.; Stedel, C.; Kalloniati, C.; Demetriou, K.; Katinakis, P.; Tsaftaris, A.S. Characterization and expression analysis of AGAMOUS-like, SEEDSTICK-like, and SEPALLATA-like MADS-box genes in peach (*Prunus persica*) fruit. *Plant Physiol. Biochem.* **2009**, *47*, 690–700. [CrossRef]
48. Li, M.; Galimba, K.; Xiao, Y.; Dardick, C.; Mount, S.M.; Callahan, A.; Liu, Z. Comparative transcriptomic analysis of apple and peach fruits: Insights into fruit type specification. *Plant J.* **2022**, *109*, 1614–1629. [CrossRef]
49. Li, C.; Lei, C.; Wang, K.; Tan, M.; Xu, F.; Wang, J.; Zheng, Y. A novel MADS-box gene regulated a priming defence in postharvest peach through SA- and ABA-signaling collaboration. *J. Exp. Bot.* **2022**, erac099, accepted manuscript. [CrossRef] [PubMed]
50. Abel, S.; Oeller, P.W.; Theologis, A. Early auxin-induced genes encode short-lived nuclear proteins. *Proc. Natl. Acad. Sci. USA* **1994**, *91*, 326–330. [CrossRef] [PubMed]
51. Stortenbeker, N.; Bemer, M. The SAUR gene family: The plant's toolbox for adaptation of growth and development. *J. Exp. Bot.* **2018**, *70*, 17–27. [CrossRef]
52. Mcclure, B.A.; Guilfoyle, T. Characterization of a class of small auxin-inducible soybean polyadenylated RNAs. *Plant Mol. Biol.* **1987**, *9*, 611–623. [CrossRef] [PubMed]
53. Hagen, G.; Guilfoyle, T. Auxin-responsive gene expression: Genes, promoters and regulatory factors. *Plant Mol. Biol.* **2002**, *49*, 373–385. [CrossRef]
54. Jain, M.; Tyagi, A.K.; Khurana, J.P. Genome-wide analysis, evolutionary expansion, and expression of early auxin-responsive SAUR gene family in rice (*Oryza sativa*). *Genomics* **2006**, *88*, 360–371. [CrossRef]
55. Wu, J.; Liu, S.; He, Y.; Guan, X.; Zhu, X.; Cheng, L.; Wang, J.; Lu, G. Genome-wide analysis of SAUR gene family in Solanaceae species. *Gene* **2012**, *509*, 38–50. [CrossRef]
56. Wang, P.; Lu, S.; Xie, M.; Wu, M.; Ding, S.; Khaliq, A.; Ma, Z.; Mao, J.; Chen, B. Identification and expression analysis of the small auxin-up RNA (SAUR) gene family in apple by inducing of auxin. *Gene* **2020**, *750*, 144725. [CrossRef]
57. Park, J.E.; Kim, Y.S.; Yoon, H.K.; Park, C.M. Functional characterization of a small auxin-up RNA gene in apical hook development in Arabidopsis. *Plant Sci.* **2007**, *172*, 150–157. [CrossRef]
58. Hong, R.; Gray, W.M. SAUR proteins as effectors of hormonal and environmental signals in plant growth. *Mol. Plant* **2015**, *8*, 1153–1164. [CrossRef]

59. Kathare, P.K.; Dharmasiri, S.; Arellano, I.; Dharmasiri, N. Interaction of SAUR53 and its close homologs with calmodulin may play a role in early development in Arabidopsis. *Plant Mol. Biol. Rep.* **2020**, *38*, 343–351. [CrossRef]
60. Yang, T.; Poovaiah, B.W. Molecular and biochemical evidence for the involvement of calcium/calmodulin in auxin action. *J. Biol. Chem.* **2000**, *275*, 3137. [CrossRef]
61. Kathare, P.K.; Dharmasiri, S.; Dharmasiri, N. *SAUR53* regulates organ elongation and apical hook development in *Arabidopsis*. *Plant Signal. Behav.* **2018**, *13*, 1–7. [CrossRef] [PubMed]
62. Wang, J.; Sun, N.; Zhang, F.; Yu, R. Differential regulation of *Arabidopsis* PP2C-D1 by *SAUR17* and *SAUR50* in apical hook development and cotyledon opening. *Plant Cell* **2020**, *32*, 3792–3811. [CrossRef]
63. Guo, Y.; Jiang, Q.; Hu, Z.; Sun, X.; Fan, S.; Zhang, H. Function of the auxin-responsive gene *TaSAUR75* under salt and drought stress. *Crop. J.* **2018**, *6*, 181–190. [CrossRef]
64. Gastaldi, V.; Lucero, L.E.; Ariel, F.D.; Gonzalez, D.H. Class-I TCP transcription factors activate the *SAUR63* gene subfamily in gibberellin-dependent stamen filament elongation. *Plant Physiol.* **2020**, *182*, 01501. [CrossRef] [PubMed]
65. Zhou, J.; Wen, Z.W.; Mei, Y.Y.; Gonzalez, D.H. The mechanism underlying the role of *SAUR72* in Arabidopsis leaf senescence regulation. *Plant Physiol. J.* **2018**, *54*, 379–385. [CrossRef]
66. Wen, Z.; Mei, Y.; Zhou, J.; Cui, Y.; Wang, D.; Wang, N.N. *SAUR49* Can Positively Regulate Leaf Senescence by Suppressing SSPP in *Arabidopsis*. *Plant Cell Physiol.* **2019**, *61*, 644–658. [CrossRef]
67. Wong, J.H.; Spartz, A.K.; Park, M.Y.; Gonzalez, D.H. Mutation of a conserved motif of PP2C.D phosphatases confers SAUR immunity and constitutive activity. *Plant Physiol.* **2019**, *181*, 353–366. [CrossRef] [PubMed]
68. Strader, L.; Ak, S.; Sang, H.L.; Wenger, J.P. Faculty Opinions recommendation of The SAUR19 subfamily of SMALL AUXIN UP RNA genes promote cell expansion. *Plant J.* **2012**, *70*, 978–990. [CrossRef]
69. Kant, S.; Rothstein, S. Auxin-responsiveSAUR39gene modulates auxin level in rice. *Plant Signal. Behav.* **2009**, *4*, 1174–1175. [CrossRef]
70. He, Y.; Liu, Y.; Li, M.; Lamin-Samu, A.T.; Yang, D.; Yu, X.; Izhar, M.; Jan, I.; Ali, M.; Lu, G. The Arabidopsis SMALL AUXIN UP RNA32 Protein Regulates ABA-Mediated Responses to Drought Stress. *Front. Plant Sci.* **2021**, *12*, 625493. [CrossRef]
71. Goda, H.; Sawa, S.; Asami, T.; Fujioka, S.; Shimada, Y.; Yoshida, S. Comprehensive Comparison of Auxin-Regulated and Brassinosteroid-Regulated Genes in Arabidopsis. *Plant Physiol.* **2004**, *134*, 1555–1573. [CrossRef]
72. Li, Z.G.; Chen, H.W.; Li, Q.T.; Shimada, Y. Three SAUR proteins SAUR76, SAUR77 and SAUR78 promote plant growth in Arabidopsis. *Sci. Rep.* **2015**, *5*, 12477. [CrossRef]
73. Zhai, Y. Functional analysis of ERF19 and the ERF19 downstream gene SAUR32 in Arabidopsis. Ph.D. Thesis, Institute of Biotechnology, National Chung Hsing University, Taiwan, China, 2016.
74. Shin, J.Y.; Mila, I.; Liu, M.; Rodrigues, M.A. The RIN-regulated small auxin-up RNA SAUR69 is involved in the unripe-to-ripe phase transition of tomato fruit via enhancement of the sensitivity to ethylene. *New Phytol.* **2019**, *222*, 820–836. [CrossRef]
75. Carrasco-Valenzuela, T.; Muñoz-Espinoza, C.; Riveros, A.; Pedreschi, R.; Arús, P.; Campos-Vargas, R.; Meneses, C. Expression QTL (eQTLs) Analyses Reveal Candidate Genes Associated with Fruit Flesh Softening Rate in Peach [*Prunus persica* (L.) Batsch]. *Front. Plant Sci.* **2019**, *10*, 1581. [CrossRef] [PubMed]
76. Müller, M.; Munné-Bosch, S. Rapid and sensitive hormonal profiling of complex plant samples by liquid chromatography coupled to electrospray ionization tandem mass spectrometry. *Plant Methods* **2011**, *7*, 37. [CrossRef] [PubMed]
77. Xing, L.; Zhang, D.; Li, Y.; Zhao, C.; Zhang, S.; Shen, Y.; An, N.; Han, M. Genome-wide identification of vegetative phase transition-associated microRNAs and target predictions using degradome sequencing in *Malus hupehensis*. *BMC Genom.* **2014**, *15*, 1125. [CrossRef]
78. Livak, K.J.; Schmittgen, T. Analysis of relative gene expression data using real-time quantitative PCR and the $2^{-\Delta\Delta CT}$ method. *Methods* **2001**, *25*, 402–408. [CrossRef]
79. Yang, Q.; Niu, Q.; Li, J.; Zheng, X. PpHB22, a member of HD-Zip proteins, activates PpDAM1 to regulate bud dormancy transition in 'suli' pear (*Pyrus pyrifolia* White Pear Group). *Plant Physiol. Biochem.* **2018**, *127*, 355–365. [CrossRef]
80. Jia, H.-F.; Chai, Y.-M.; Li, C.-L.; Qin, L.; Shen, Y.-Y. Cloning and Characterization of the H Subunit of a Magnesium Chelatase Gene (PpCHLH) in Peach. *J. Plant Growth Regul.* **2011**, *30*, 445–455. [CrossRef]
81. Yoo, S.D.; Cho, Y.H.; Sheen, J. Arabidopsis mesophyll protoplasts: A versatile cell system for transient gene expression analysis. *Nat. Protoc.* **2007**, *2*, 1565–1572. [CrossRef]
82. Qian, M.; Xu, Z.; Zhang, Z.; Li, Q.; Yan, X.; Liu, H.; Han, M.; Li, F.; Zheng, J.; Zhang, D.; et al. The downregulation of PpPG21 and PpPG22 influences peach fruit texture and softening. *Planta* **2021**, *254*, 22. [CrossRef]
83. Gray, J.; Picton, S.; Shabbeer, J.; Schuch, W.; Grierson, D. Molecular biology of fruit ripening and its manipulation with antisense genes. *Plant Mol. Biol.* **1992**, *19*, 69–87. [CrossRef]
84. Liu, M.; Pirrello, J.; Chervin, C.; Roustan, J.-P.; Bouzayen, M. Ethylene control of fruit ripening: Revisiting the complex network of transcriptional regulation. *Plant Physiol.* **2015**, *169*, 2380–2390. [CrossRef]
85. Adams-Phillips, L.; Barry, C.; Giovannoni, J. Signal transduction systems regulating fruit ripening. *Trends Plant Sci.* **2004**, *9*, 331–338. [CrossRef] [PubMed]
86. Iqbal, N.; Khan, N.A.; Ferrante, A.; Trivellini, A.; Francini, A.; Khan, M.I.R. Ethylene Role in Plant Growth, Development and Senescence: Interaction with Other Phytohormones. *Front. Plant Sci.* **2017**, *8*, 475. [CrossRef] [PubMed]

87. Du, M.; Spalding, E.P.; Gray, W.M. Rapid auxin-mediated cell expansion. *Annu. Rev. Plant Biol.* **2020**, *71*, 1–24. [CrossRef] [PubMed]
88. Zhang, J. Evolution by gene duplication: An update. *Trends Ecol. Evol.* **2003**, *18*, 292–298. [CrossRef]
89. Hanada, K.; Zou, C.; Lehti-Shiu, M.D.; Shinozaki, K.; Shiu, S.-H. Importance of Lineage-Specific Expansion of Plant Tandem Duplicates in the Adaptive Response to Environmental Stimuli. *Plant Physiol.* **2008**, *148*, 993–1003. [CrossRef]
90. Nemhauser, J.L.; Hong, F.; Chory, J. Different plant hormones regulate similar processes through largely nonoverlapping transcriptional responses. *Cell* **2006**, *126*, 467–475. [CrossRef]
91. Boivin, S.; Fonouni-Farde, C.; Frugier, F. How Auxin and Cytokinin Phytohormones Modulate Root Microbe Interactions. *Front. Plant Sci.* **2016**, *7*, 1240. [CrossRef]
92. Jeffares, D.C.; Penkett, C.J.; Jürg, B. Rapidly regulated genes are intron poor. *Trends Genet.* **2008**, *24*, 375–378. [CrossRef]
93. Guo, Y.; Xu, C.-B.; Sun, X.-J.; Hu, Z.; Fan, S.-J.; Jiang, Q.-Y.; Zhang, H. TaSAUR78 enhances multiple abiotic stress tolerance by regulating the interacting gene TaVDAC1. *J. Integr. Agric.* **2019**, *18*, 2682–2690. [CrossRef]
94. Bank, A.D. Cell wall disassembly in ripening fruit. *Funct. Plant Biol.* **2006**, *33*, 103–119. [CrossRef]
95. Orfila, C.; Huisman, M.M.; Willats, W.G.; van Alebeek, G.-J.W.; Schols, H.A.; Seymour, G.B.; Knox, P.J. Altered cell wall disassembly during ripening of Cnr tomato fruit: Implications for cell adhesion and fruit softening. *Planta* **2002**, *215*, 440–447. [CrossRef] [PubMed]
96. Payasi, A.; Mishra, N.N.; Chaves, A.L.S.; Singh, R. Biochemistry of fruit softening: An overview. *Physiol. Mol. Biol. Plants* **2009**, *15*, 103–113. [CrossRef] [PubMed]
97. Qian, M.; Zhang, Y.; Yan, X.; Han, M.; Li, J.; Li, F.; Li, F.; Zhang, D.; Zhao, C. Identification and Expression Analysis of Polygalacturonase Family Members during Peach Fruit Softening. *Int. J. Mol. Sci.* **2016**, *17*, 1933. [CrossRef]
98. Kant, S.; Bi, Y.M.; Tong, Z. SAUR39, a small auxin-up RNA gene, acts as a negative regulator of auxin synthesis and transport in rice. *Plant Signal. Behav.* **2009**, *151*, 691–701. [CrossRef] [PubMed]
99. Palme, K.; Gälweiler, L. PIN-pointing the molecular basis of auxin transport. *Curr. Opin. Plant Biol.* **1999**, *2*, 375–381. [CrossRef]
100. Zeng, W.F.; Pan, L.; Niu, L.; Lu, Z.H. Bioinformatics analysis and expression of the nectarine indole-3-aceticacid-amido synthase (GH3) gene family during fruit development. *Acta Hortic. Sin.* **2015**, *42*, 833–842. [CrossRef]
101. Wang, X.; Meng, J.; Deng, L.; Wang, Y.; Liu, H.; Yao, J.-L.; Nieuwenhuizen, N.J.; Wang, Z.; Zeng, W. Diverse Functions of IAA-Leucine Resistant PpILR1 Provide a Genic Basis for Auxin-Ethylene Crosstalk During Peach Fruit Ripening. *Front. Plant Sci.* **2021**, *12*, 655758. [CrossRef]
102. Guilfoyle, T.J.; Hagen, G. Auxin response factors. *Curr. Opin. Plant Biol.* **2007**, *10*, 453–460. [CrossRef]
103. Shinshi, O.T. Ethylene-inducible DNA binding proteins that interact with an ethylene-responsive element. *Plant Cell* **1995**, *7*, 173–182. [CrossRef]

horticulturae

Article

Rapid Identification of Apple Maturity Based on Multispectral Sensor Combined with Spectral Shape Features

Mengsheng Zhang [1,2,3,†], Maosheng Shen [1,2,†], Yuge Pu [1,2], Hao Li [1,2], Bo Zhang [1,2], Zhongxiong Zhang [1,2], Xiaolin Ren [4] and Juan Zhao [1,2,3,*]

1 College of Mechanical and Electronic Engineering, Northwest A&F University, Xianyang 712100, China; mszhang@nwsuaf.edu.cn (M.Z.); shenms@nwafu.edu.cn (M.S.); puyuge@nwafu.edu.cn (Y.P.); oriental@nwafu.edu.cn (H.L.); zhangbo1996@nwafu.edu.cn (B.Z.); zzx9519@nwafu.edu.cn (Z.Z.)
2 Key Laboratory of Agricultural Internet of Things, Ministry of Agriculture and Rural Affairs, Xianyang 712100, China
3 Shaanxi Key Laboratory of Agricultural Information Perception and Intelligent Service, Xianyang 712100, China
4 College of Horticulture, Northwest A&F University, Xianyang 712100, China; renxl@nwsuaf.edu.cn
* Correspondence: zhaojuan@nwsuaf.edu.cn
† These authors contributed equally to this work.

Abstract: The rapid and convenient detection of maturity is of great significance to determine the harvest time and postharvest storage conditions of apples. In this study, a portable visible and near-infrared (VIS/NIR) analysis device prototype was developed based on a multispectral sensor and applied to 'Fuji' apple maturity detection. The multispectral data of apples with maturity variation was measured, and the prediction model was established by a least-square support vector machine and linear discriminant analysis. Due to the low resolution of the multispectral data, regular preprocessing methods cannot improve the prediction accuracy. Instead, the spectral shape features (spectral ratio, spectral difference, and normalized spectral intensity difference) were used for preprocessing and model establishment, and the combination of the three features effectively improved the model performance with a prediction accuracy of 88.46%. In addition, the validation accuracy of the optimal model was 84.72%, and the area under curve (AUC) value of each maturity level was higher than 0.8972. The results show that the multispectral sensor is an appliable choice for the development of the portable detection device of apple maturity, and the data processing method proposed in this study provides a potential solution to improve the detection accuracy for multispectral sensors.

Keywords: 'Fuji' apple; maturity; visible and near-infrared spectroscopy; multispectral sensor; spectral shape feature

Citation: Zhang, M.; Shen, M.; Pu, Y.; Li, H.; Zhang, B.; Zhang, Z.; Ren, X.; Zhao, J. Rapid Identification of Apple Maturity Based on Multispectral Sensor Combined with Spectral Shape Features. *Horticulturae* **2022**, *8*, 361. https://doi.org/10.3390/horticulturae8050361

Academic Editors: Jianwei Qin and Luigi De Bellis

Received: 15 December 2021
Accepted: 17 April 2022
Published: 21 April 2022

1. Introduction

Apples are one of the most important agricultural products in the global market, which are nutritious and crisp and are deeply loved by consumers [1]. Maturity is closely related to the harvest time and postharvest quality of apples, making it a reliable index to scientifically manage the harvest and storage of apples, thereby prolonging their shelf life and ensuring their final quality [2,3]. Maturity indicators are determined by destructive measurements, such as starch pattern index (SPI), firmness, soluble solids content (SSC), Streif index, etc. [4]. In recent years, investigations have been done to fulfill the non-destructive detection need of the fruit industry. Visible and near-infrared (VIS/NIR) spectroscopy is a promising solution for the quality characterization of fruits and other agricultural products [5,6] due to its fast, non-destructive measurement for simultaneous analysis of multiple components without sample preparation.

VIS/NIR spectroscopy has been applied to the non-destructive detection of the maturity of apples, pears, and other fruits [7]. Peirs et al. [8,9] used VIS/NIR spectroscopy to study the prediction of the Streif index of different apple varieties and analyzed the effect of natural variability on the prediction. Zhang et al. [10] determined the maturity according to starch dyeing and accurately predicted the apple maturity level by VIS/NIR spectroscopy. Pourdarbani et al. [11,12] used color and spectral data to make non-destructive discrimination of the four ripening stages of an apple.

In the past decade, research on portable VIS/NIR analysis devices has gradually sprung up. The devices are of small size and low manufacturing cost and can realize the on-site detection of fruit quality [13]. Fan, et al. [14] developed a VIS/NIR analysis device prototype using a commercial spectrometer (USB2000+, Ocean Optics Inc., Dunedin, FL, USA) combined with partial least squares regression to achieve non-destructive testing of apple SSC. Guo, et al. [15] developed a hand-held fruit SSC detector using an STS micro-spectrometer (Ocean Optics Inc., Dunedin, FL, USA), which was combined with partial least squares regression to realize the SSC non-destructive testing of kiwifruit, nectarine, and apricot. Because of the high price of high-resolution spectrometers, some researchers have explored the potential of some low-cost optical sensors as the collector for spectra of multiple wavelengths instead. In this case, two common strategies have been applied. The first one is using photodiodes to collect the spectrum of characteristic LEDs. Zhao, et al. [16] realized the characteristic spectrum collection of apple quality by using characteristic LED light sources combined with a photodiode and established a related quantitative analysis model. Abasi, et al. [17] used six characteristic LED light sources to form an array, combined with photodiodes to develop a portable fruit quality detector. The other solution is to use multispectral sensors, which can detect the spectral information of multiple wavelengths at the same time and has the advantage of low cost. Li, et al. [18] used a 7-wavelength spectral sensor combined with a tungsten halogen lamp to achieve the non-destructive detection of moldy apple cores. Yang, et al. [19] designed a VIS/NIR analysis device using an 18-wavelength spectral sensor for measuring the composition of milk.

Multispectral sensors have low spectral resolution and can only collect a few wavelengths of spectral information. The conventional preprocessing methods for the whole spectrum are no longer applicable for multispectral data. Spectral shape features can be described by some spectral indices to enhance spectral information according to previous VIS/NIR studies. Ma, et al. [20] used the spectral absorption index as a pretreatment method to extract morphological characteristic information from the absorption spectrum and reflection spectrum to detect the water content of pork. Li, et al. [21] used the normalized spectral ratio to correct the light scattering effect in the original spectrum and improve the prediction accuracy of SSC and dry matter content of apples. Combined with the empirical threshold, the spectral indices can also be used to detect fruit diseases [22], defects [23], and maturity [24]. Spectral shape features could enhance spectral information and help to improve the prediction accuracy of multispectral sensors.

This study aims to achieve low-cost and convenient non-destructive testing of apple ripeness by combining multispectral sensors with spectral shape features. A portable VIS/NIR analyzer based on a multispectral sensor was used to collect the spectral information of apples of different maturity. A least-square support vector machine (LSSVM) and linear discriminant analysis (LDA) were used to build predictive models, and the performance of the optimal model was validated using apples in different seasons. The prediction results of multi-spectral data processed by conventional spectral data processing methods and spectral shape features were analyzed and compared. The results of the comparison proved that the spectral shape feature is a more effective method for processing multispectral data, and the multispectral sensor combined with the spectral shape features can predict apple ripeness non-destructively.

2. Materials and Methods

2.1. Experimental Samples

The experimental samples were obtained from a commercial orchard in Fufeng, Shaanxi Province, China. Twelve 'Fuji' apple trees with similar growth status were randomly selected as fixed sampling points in the orchard. In 2019, 836 apples (Set-1) at different maturity stages were collected to develop a calibration model for the device. In 2020, 360 apples (Set-2) of different maturity stages were collected in the same orchard to verify the performance of the model. All apples picked from the trees were washed and numbered after being transported back to the lab on the same day. Before spectral measurements, all samples were stored in the laboratory for 24 h, allowing the samples to reach room temperature to avoid the influence of temperature.

2.2. Spectral Data Acquisition

Spectral data acquisition was completed by a Vis/NIR analysis device based on multispectral sensors. The device was mainly composed of a microcontroller, multispectral sensor, halogen tungsten lamp, display, and power supply unit (Figure 1). The multispectral sensor (AS7265x, Austria Mikro Systeme, Styria, Austria) can perform spectral detection at 18 wavelengths (410, 435, 460, 485, 510, 535, 560, 585, 610, 645, 680, 705, 730, 760, 810, 860, 900 and 940 nm) from 410 nm to 940 nm. Four 4.5 W halogen tungsten lamps (VIVO-B, Ocean Optics Inc., USA) were selected as the light source, which have a spectral range of 360–2000 nm and provide stable illumination. The device was powered by a lithium battery. The structure of the device was designed using PTC Creo Parametric 5.0 (PTC Inc., Boston, MA, USA) and produced using a 3D printer. The appearance of the device is square. The OLED display and the detection probe were placed in the upper part of the device, the lithium battery and microcontroller were placed in the lower part of the device, and insulated cotton was added between the battery and the microcontroller. Ventilation holes were added to the back of the device to enhance heat dissipation.

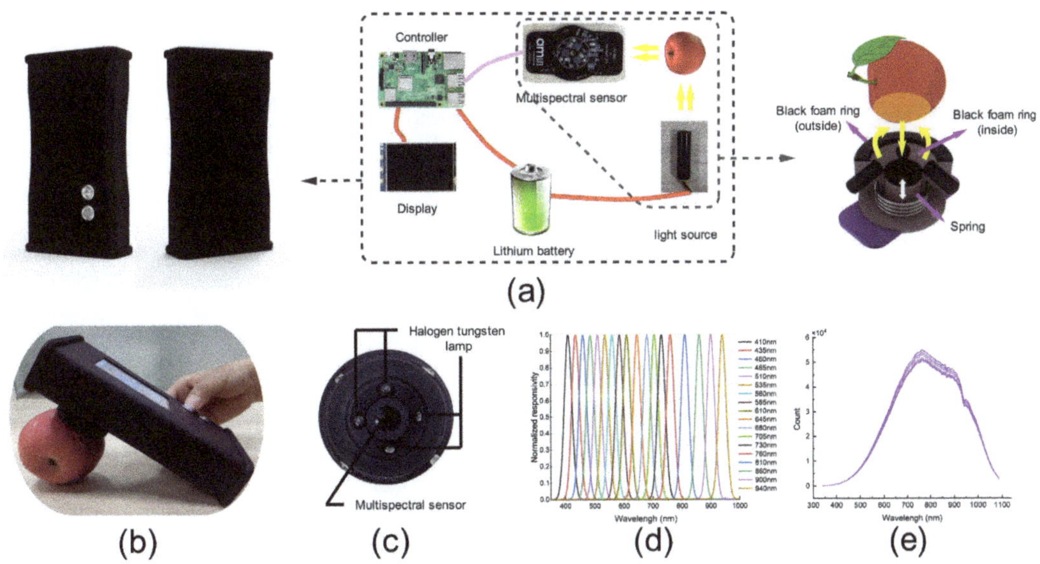

Figure 1. The schematic (**a**), the prototype (**b**), and the detection probe (**c**) of the Vis/NIR device based on the multispectral sensor, the spectral response of the sensor (**d**) and the light source (**e**).

Previous studies have shown that the interaction mode can obtain more internal optical information in fruits [25]. Maturity is a comprehensive attribute of apples, and more internal optical information could contribute to better maturity detection. Therefore, a detection probe based on interaction mode was designed to collect spectra (Figure 1a). The sensor and the light source were integrated into the detection probe to decrease the size and avoid a complex optical path design. The collimating light source with 45° incidence was arranged around the detection channel in a ring. When the device is working, the light emitted by the light source will reach the apple surface directly and integrate with fruit tissue by reflection, scattering and absorption. The sensor will then receive the reflected and scattered light signal. The diameter of the detection probe was only 50 mm. A soft black foam ring with an outer diameter of 50 mm was placed on the probe to avoid the interference of stray ambient light. A soft black foam ring with an outer diameter of 20 mm was placed around the detection channel to support the fruit and separate the collection area from the lighting area, ensuring that the measured light signal is from the fruit tissue rather than the light source. In addition, considering the variation size (diameter 70–120 mm) and shape of apples, an adaptive structure was designed to ensure that the distance between the detector and the fruit was approximately fixed. The adaptive structure was realized by a soft spring (Figure 1a). When testing, pressing the apple can make the spring stretch and the sensor move down; therefore, the foam ring was tightly attached to the surface of the apple. After the testing was completed, the spring returned to its original state.

The software of the device was developed by Python, which was used to realize the control of the device, the acquisition and analysis of spectral data, and the display of detection results. The integration time was set to 200 ms. The average value obtained by ten consecutive scans at the same position was recorded as the spectrum of each sample. The measured spectral data were automatically named and stored using predefined file name prefixes and save paths. It takes less than three seconds to complete the analysis of each fruit spectrum. To improve the data stability, we calibrated the light source intensity using a standard whiteboard (WR-D97, material PTFE) so that the light source intensity was adjusted to the standard spectrum every time the device was turned on. The dark spectrum was obtained by turning off the light source, and the original spectrum was corrected according to Equation (1) [10]:

$$R = \frac{R_{raw} - R_{dark}}{R_{white} - R_{dark}} \times 100\% \tag{1}$$

where R_{raw} represents the original spectrum, R_{dark} represents the dark spectrum, R_{white} represents the standard spectrum, and R represents the corrected spectrum.

For the spectral detection system, the signal-to-noise ratio (SNR) is an important index to evaluate the performance of the system. Walsh, et al. [26] defined SNR as the average count (intensity) of each selected wavelength divided by the standard deviation. In addition, the area change rate (ACR) was used as the indicator for the system stability. ACR describes the variation in spectra measured at the same position of a sample by calculating the difference in the spectral region area. Before ACR analysis, the spectrum data is normalized to the range of 0–1 to reduce the influence of different light source settings in the experiment. ACR is approximately the root-mean-square deviation (RMSD) of all spectral areas and can be calculated by Equation (2) [27]:

$$\text{RMSD} = \sqrt{\frac{1}{N}\sum_{i=1}^{N}(Y_i - Y_{mean})^2} \tag{2}$$

where N represents the number of the spectra, Y represents the area of the ith spectrum, and Y_{mean} represents the mean area of all spectra.

2.3. Data Measurement

After the spectra collection, the destructive tests were carried out to measure the maturity of the apples. The starch pattern index (SPI) was measured by cutting the fruit in half along the equator, soaking half of the fruit in I2-KI solution, and comparing it with the Cornell general SPI chart [28]. In addition, other qualities (texture, SSC, acidity) were measured to observe the quality changes of apples with different maturity. The texture was characterized by a texture analyzer (TA. XT Express, Stable Micro Systems, Godalming, UK). The type of probe was P/2 and the puncture depth was 10 mm. According to texture analysis, the phenotypic texture parameters including pulp firmness, peel firmness, pericarp elasticity, pulp elasticity, and fruit brittleness were determined [29,30]. The juice of the remaining half of the fruit was extracted, and the SSC and acidity were measured by a digital refractometer (PAL-BX/ACID5, Atago, Tokyo, Japan) and a fruit acidity meter (GMK-835F (apple), G-WON, Seoul, Korea).

According to the SPI, the maturity of apples was characterized by three levels: immature, harvest mature, and eatable mature [10]. The number of apples obtained in 2019 was 270, 320, and 246 for each maturity level. For apples in 2020, the number was 103, 135, and 122, respectively. The number of the harvest mature apples (326 apples) and immature apples (274 apples) harvested in 2019 was significantly higher than that of eatable mature apples (246 apples). To avoid the over-fitting problem caused by sample imbalance, we selected the same number of samples (246 apples) from immature apples and harvested mature apples. On this basis, the fruits of each level were divided into calibration set and prediction set at a ratio of 3:1 by the Kennard–Stone (KS) method. The samples from 2020 were all used to verify the model performance.

2.4. Spectra Preprocessing

Spectra preprocessing can decrease the noise of raw data and improve prediction accuracy. Common preprocessing methods include Savitzky–Golay smoothing (SGS), multivariate scattering correction (MSC), standard normal variable transformation (SNV), and so on. However, those methods can hardly apply to the data of this study since the spectra measured only contained signals of 18 wavelengths, which is far less than the data of regular spectrometers. In this study, three spectral shape features (spectral ratio (SR), spectral difference (SD), and normalized spectral intensity difference (NSID)) were used to preprocess the spectra. SR and SD were considered effective parameters to evaluate specific components or properties of fruits, like fruit maturity and surface damage. Lleó, et al. [24] calculated the spectral index related to SR and SD by three wavelengths ($\text{Index}_1 = R_{\lambda=720} + R_{\lambda=634} - 2R_{\lambda=674}$, $\text{Index}_2 = 2R_{\lambda=674}/(R_{\lambda=720} + R_{\lambda=634})$) for peach maturity detection. NSID is a standardized index, also known as the normalized vegetation index in remote sensing, which is used to generate images showing the amount of vegetation (relative biomass). Jie, et al. [31] used SR and NSID ($\text{Index}_1 = R_{\lambda=730}/R_{\lambda=803}$, $\text{Index}_2 = (R_{\lambda=730} - R_{\lambda=803})/(R_{\lambda=730} + R_{\lambda=803})$) to assess watermelon maturity. Although these spectral indices are calculated by only three wavelengths, they effectively revealed the spectral shape features highly related to fruit maturity. The equations for spectral shape features are as follows [18,31]:

$$\text{SR} = \frac{R_{\lambda=i}}{R_{\lambda=j}} \quad (3)$$

$$\text{SD} = R_{\lambda=i} - R_{\lambda=j} \quad (4)$$

$$\text{NSID} = \frac{R_{\lambda=i} - R_{\lambda=j}}{R_{\lambda=i} + \text{R}_{\lambda=j}} \quad (5)$$

where $R_{\lambda=i}$ and $R_{\lambda=j}$ represent the spectral reflection intensity of the *i*th and *j*th wavelength in a spectral curve, respectively.

In previous VIS/NIR studies, a single spectral shape feature (spectral index) has also been successfully applied to detect the firmness and maturity of peaches, strawberries, and other fruits [24,31,32]. In this study, four commonly used spectral indices were compared with the multivariate analysis of spectral shape features. Specifically, 680 nm is an important wavelength related to chlorophyll content, which is suitable for checking the maturity and ripening process of apples [33]. The peaks (645 nm and 730 nm) of the spectral curve on both sides of 680 nm were both selected with 680 nm to establish the spectral index to identify apple maturity. The equations for spectral indices are as follows [10]:

$$\text{Index}_1 = R_{\lambda=730} + R_{\lambda=645} - 2R_{\lambda=680} \tag{6}$$

$$\text{Index}_2 = \frac{R_{\lambda=730} + R_{\lambda=645}}{R_{\lambda=680}} \tag{7}$$

$$\text{Index}_3 = \frac{R_{\lambda=730} - R_{\lambda=680}}{R_{\lambda=720} + R_{\lambda=680}} \tag{8}$$

$$\text{Index}_4 = log_{10}\frac{R_{\lambda=730}}{R_{\lambda=680}} \tag{9}$$

2.5. Model Establishment and Evaluation

The least-square support vector machine (LSSVM) and linear discriminant analysis (LDA) were used to establish the calibration model for maturity prediction. LSSVM is an improved form of support vector machine, which simplifies the solution of the problem by transforming the solution of the quadratic optimization problem into the solution of a system of linear equations. LSSVM can handle linear and nonlinear multivariate analysis. Before training the LSSVM model, the optimal combination of the regularization parameter of the model and the kernel function parameter of the radial basis function is determined by using the two-dimensional grid search method [34]. LDA projects sample spectral variables to the best discriminant vector space to ensure that the same category of data after projection is as close as possible, and different categories of data are separated as far as possible [35,36].

For the spectral index, it is necessary to determine an optimal threshold for sample maturity assessment. The Otsu method is an effective algorithm to determine the threshold of image binary segmentation in image analysis [37]. Variance is a measure of the uniformity of gray distribution. The larger the between-class variance between the background and the target in the image, the greater the difference between the two parts. Misclassification will make the between-class variance lower. Therefore, the segmentation that maximizes the variance between classes could minimize the probability of misclassification. For the spectral index, after arranging the spectral index in descending order, the method for maximizing between-class variance can be used to determine the optimal threshold between immaturity and harvest maturity, as well as between harvest maturity and eatable maturity.

The accuracy was used to evaluate the overall accuracy of the classifier. The confusion matrix, recall, precision, and F1-Score were used to further analyze the prediction results. The receiver operating characteristic curve (ROC) can demonstrate the prediction ability of the classifier. In the ROC graph, the closer the curve is to the upper-left edge of the graph, the better the performance of the classifier. The equations for each evaluation index are as follows [10]:

$$\text{accuracy} = \frac{TP + TN}{TP + TN + FP + FN} \tag{10}$$

$$\text{recall} = \frac{TP}{TP + FN} \tag{11}$$

$$\text{precision} = \frac{TP}{TP + FP} \tag{12}$$

$$\text{F1-Score} = \frac{2 \times \text{precision} \times \text{recall}}{\text{precision} + \text{recall}} \tag{13}$$

where *TP* represents the correctly classified positive sample; *TN* represents correctly classified negative samples; *FP* represents the positive sample of misclassification; and *FN* represents the negative sample of misclassification.

3. Results and Discussion

3.1. The Sensor Stability Test

The white ball made of polytetrafluoroethylene was used as a reference to test the stability of the spectrum collected by the multispectral sensor. In order to observe the effect of preheating time on the stability of the sensor, the spectrum was collected every 2 min in the first 10 min. Then, spectrum was collected every 10 min, and a total of 200 spectra were collected. The SNR and ACR were calculated to reflect the stability of the sensor. Figure 2a shows the SNR at different wavelengths. The SNR of the sensor at different wavelengths was 20.840–47.308. Figure 2b shows the ACR values of each measurement. Since the device was not fully preheated at the beginning, the collected spectrum fluctuated greatly, and the maximum ACR value was 7.094. After 10 min, the fluctuation of ACR decreased. The dotted line represents that the average ACR value obtained after 10 min was 3.787. The SNR and ACR values of the sensor in this study were close to the previous research on the dynamic transmission spectrum detection system [1,38], indicating that this sensor has good stability.

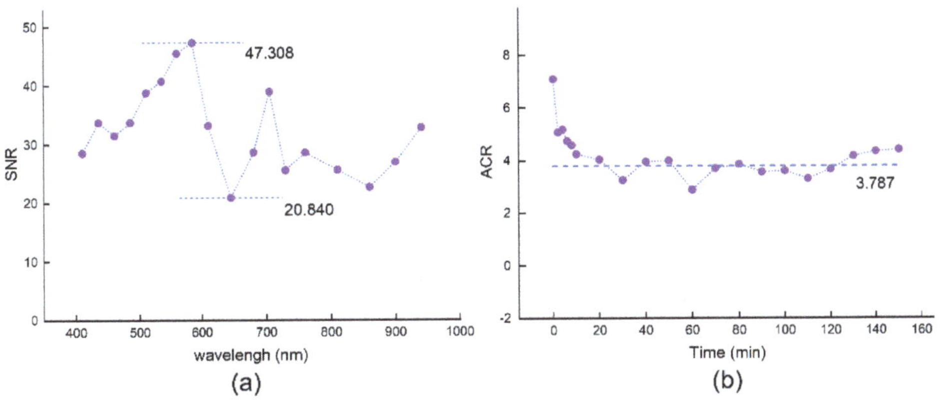

Figure 2. The device stability test results. (**a**) Signal-to-noise ratio (SNR) values at different wavelengths, and (**b**) area change rate (ACR) values at different times.

3.2. Apple Quality at Different Maturity Levels

The quality changes of 'Fuji' apples during ripening are shown in Figure 3. During apple ripening, the color gradually turned red (Figure 3a). Due to the increase in the activities of amylase, invertase, and sucrose synthase in the fruit, the starch was gradually hydrolyzed, the starch staining area decreased (Figure 3a), SSC increased (Figure 3c), and the acidity decreased (Figure 3d). The texture changed due to the changes in pectin content and cell wall composition during ripening. There were differences in firmness (Figure 3e), pericarp hardness (Figure 3f), pericarp resilience (Figure 3g), flesh resilience (Figure 3h), and fruit brittleness (Figure 3i) among apples with different maturity levels. The above quality changes will affect the measured spectral data, which provides a basis for using VIS/NIR spectroscopy to detect apple maturity.

3.3. Spectral Analysis

The spectral data before 600 nm was affected by the content of anthocyanins in apples, so it is not suitable for the maturity detection of bagged apples [39,40]. Ten wavelengths in the wavelength range of 600–940 nm were selected for analysis. The spectral curves of different maturity and the spectral intensity distribution at different wavelengths are shown in Figure 4. At a given wavelength, the spectral intensity of high-maturity apples was lower than that of low-maturity apples, and there was a similar trend. The chlorophyll absorption peak at 680 nm and the downward trend close to the water absorption peak at 940 nm can be observed. A similar phenomenon was also observed by Zhang, et al. [10]. After the Kruskal–Wallis test, the spectral intensity of apples with a different maturity was significantly different at the same wavelength, which shows that the spectral information obtained based on a multispectral sensor has the potential to effectively detect apple maturity.

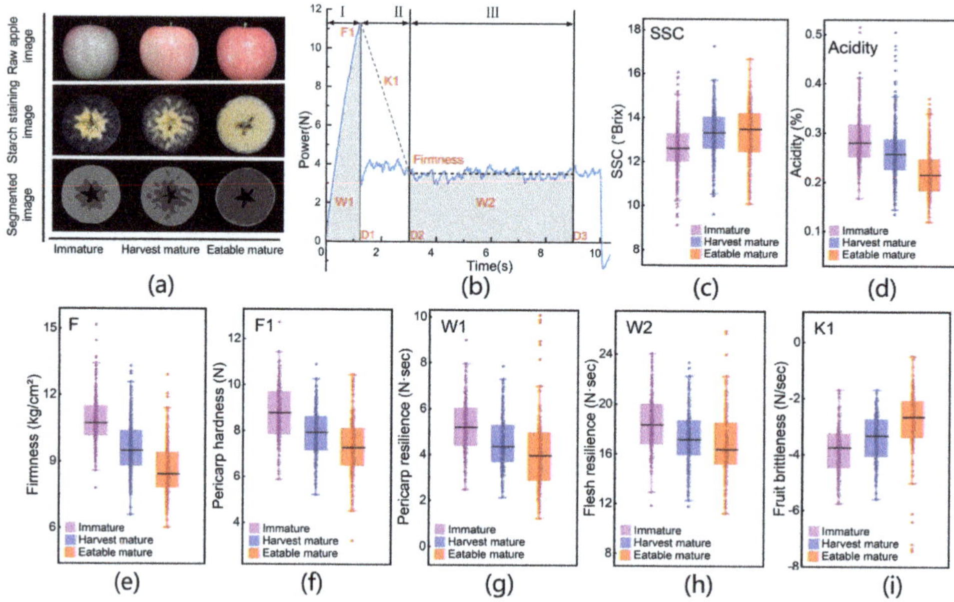

Figure 3. 'Fuji' apple quality changes at different maturity levels: (**a**) apple fruit and starch staining image, (**b**) apple-pulp puncture force-displacement curve, and (**c–i**) the distribution of SSC, acidity, firmness, pericarp firmness, pericarp resilience, flesh resilience, and fruit brittleness, respectively.

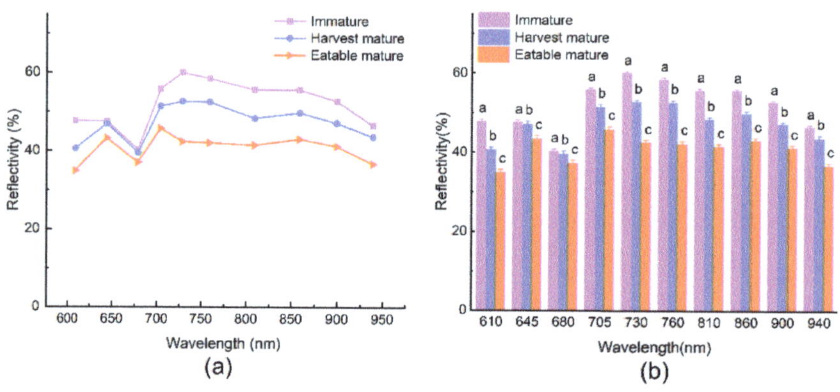

Figure 4. Spectra of apples with different maturity levels: (**a**) spectral curve, (**b**) spectral intensity distribution at different wavelengths. Mean values in the same wavelength with different letters (a, b, c) are significantly different ($p \leq 0.05$).

3.4. Modeling Based on Traditional Methods

The calibration models were developed using LSSVM and LDA. Several common pretreatment methods (SGS, MSC, and SNV) were used to compare with the original spectrum. Table 1 shows the prediction results of the developed models. The prediction performances of the LSSVM models were better than that of the LDA models because LSSVM can deal with the potential nonlinear relation between spectral data and fruit maturity [41]. The accuracy of the LSSVM model and LDA model based on the raw spectra was 84.70% and 82.51%, respectively. Compared with the raw spectra, the models based on the pretreated spectrum did not achieve better prediction accuracy. The accuracy of the calibration set was 80.36–89.01%, and that of the prediction set was 71.58–81.97%. Compared with the previous study, the prediction accuracy of multispectral sensors in predicting apple ripeness was significantly lower than that of the Vis/NIR spectrometer [10]. This result could be attributed to the fact that the pretreatment methods effective for regular spectra with high resolution are hardly applicable for multispectral data.

Table 1. Prediction results of the model based on different pretreatment methods.

Model	Pretreatment	Calibration Set			Prediction Set		
		Sample	Correct	Accuracy/%	Sample	Correct	Accuracy/%
LSSVM	RAW	555	485	87.39	183	155	84.70
	SGS	555	470	84.68	183	142	77.60
	MSC	555	494	89.01	183	150	81.97
	SNV	555	492	88.65	183	147	80.33
LDA	RAW	555	483	87.03	183	151	82.51
	SGS	555	446	80.36	183	131	71.58
	MSC	555	466	83.96	183	144	78.69
	SNV	555	474	85.41	183	146	79.78

The prediction results of the model developed based on the raw spectra and LSSVM is shown in Table 2. The recall rate of the harvested mature samples (88.52%) was higher than that of the immature samples (83.61%) and the eatable mature samples (81.90%), and the precision (77.14%) was significantly lower than that of the immature samples (87.93%) and the eatable mature samples (90.91%). According to the confusion matrix, this is mainly because more immature and eatable mature samples were misclassified as being of harvest maturity, resulting in poor credibility of the developed model for the prediction of harvest maturity. Therefore, it is necessary to further improve the prediction accuracy of the model for immature and eatable mature samples.

Table 2. Confusion matrix, recall, precision, and F-score based on the Raw-LSSVM model.

Model	Maturity Category	1	2	3	No.	Recall/%	Precision/%	F-Score/%	Accuracy/%
Raw-LSSVM	1	51	7	3	61	83.61	87.93	85.71	84.70%
	2	5	54	2	61	88.52	77.14	82.44	
	3	2	9	50	61	81.90	90.91	86.21	

3.5. Modeling Based on Spectral Shape Features

3.5.1. Spectral Index

The spectral indices can reflect the shape features of the spectral curves. Specifically, 680 nm is an important wavelength related to chlorophyll content. The peaks (645 nm and 730 nm) of the spectral curve on both sides of the 680 nm were both selected with 680 nm to establish a spectral index to identify the maturity level. The Otsu method was used to determine the threshold in the calibration set, and the prediction set was used to verify the effectiveness of the threshold. The distribution and threshold determination process for $Index_1$ is shown in Figure 5a,b. In the process of determining the threshold, the between-class variance increases gradually in the initial iterative stage and then decreases gradually. When the between-class variance was at its maximum, the optimal threshold could be determined. Table 3 shows the threshold of four spectral indices.

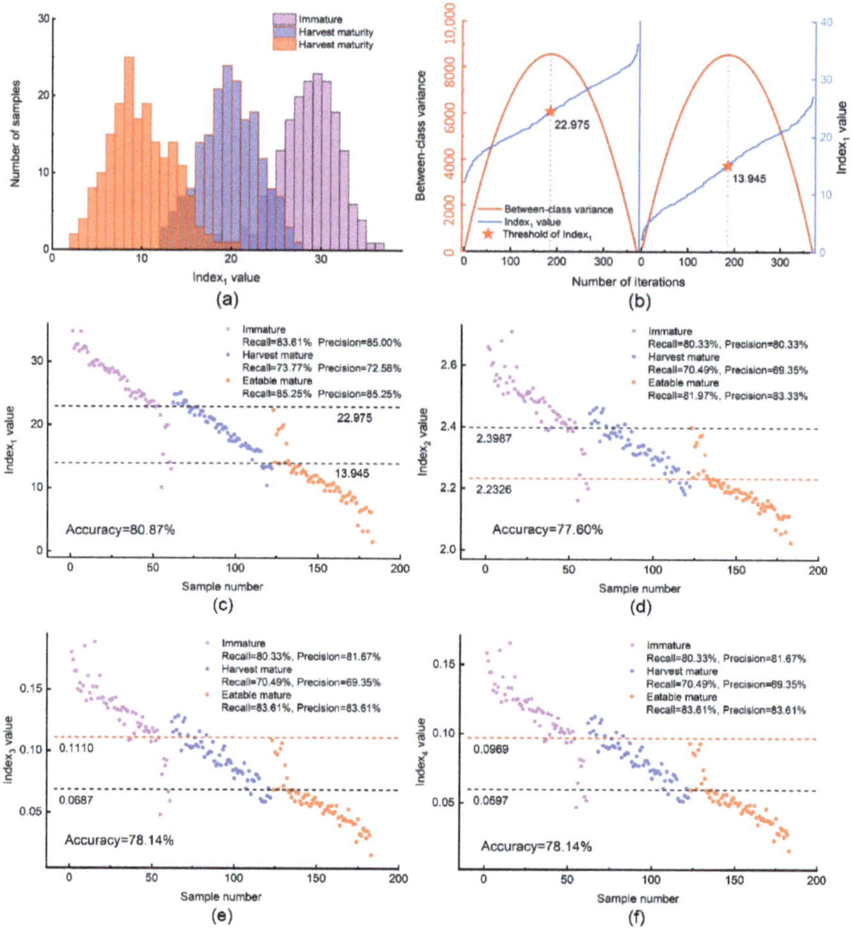

Figure 5. The prediction results based on the spectral index: (**a**) the distribution of $Index_1$ in the calibration set, (**b**) the process of determining the threshold of $Index_1$, and the distribution of (**c**) $Index_1$, (**d**) $Index_2$, (**e**) $Index_3$, and (**f**) $Index_4$ in the prediction set.

Table 3. The threshold of spectral indices.

Threshold	Spectral Index			
	$Index_1$	**$Index_2$**	**$Index_3$**	**$Index_4$**
Threshold-1	22.975	2.3987	0.1110	0.0969
Threshold-2	13.945	2.2326	0.0687	0.0567

After determining the optimal threshold, the reliability of the threshold was verified on the prediction set. The distribution of the spectral index of the prediction set is shown in Figure 5d–f, where the dotted line represents the threshold. The prediction accuracy of the spectral indices was 77.60–80.87%. $Index_1$ has the best prediction performance, with an accuracy of 80.87%. According to the results, the maturity level of apples can be easily identified by the spectral index combined with a threshold. However, because the selected wavelength (645 nm, 680 nm, and 730 nm) was mainly related to the absorption peak of chlorophyll, the spectral index contains less internal quality information of apples, and the prediction accuracy was low, so it should not be directly applied.

3.5.2. Preprocessing Based on Spectral Shape Features

Table 4 shows the prediction results based on different spectral shape features. Similarly, the prediction results of the LSSVM model were better than that of the LDA model. Compared with preprocessing methods such as MSC, better prediction results were obtained using spectral shape features, and the prediction results of SR and NSID were better than the raw spectra, in which SR-LSSVM has the highest prediction accuracy, and the accuracy of the calibration set and prediction set was 89.73% and 87.43%, respectively. The prediction result of the combination of the three features was higher than that of every single feature. The prediction performance of the LDA model was significantly improved, with a prediction accuracy of 87.98%. The LSSVM model still achieved the best prediction accuracy of 88.52%. The results show that both the single features and their combination can improve the model performance.

Table 4. Prediction results of the model based on different spectral shape features.

Model	Pretreatment	Calibration Set			Prediction Set		
		Sample	Correct	Accuracy/%	Sample	Correct	Accuracy/%
LSSVM	SR	555	498	89.73	183	160	87.43
	SD	555	474	85.41	183	154	84.15
	NSID	555	492	88.65	183	157	85.79
	SR + SD + NSID	555	497	89.55	183	162	88.52
LDA	SR	555	494	89.01	183	154	84.15
	SD	555	487	87.75	183	153	83.61
	NSID	555	487	87.75	183	152	83.06
	SR + SD + NSID	555	506	91.17	183	161	87.98

The prediction result of the calibration model based on spectral shape features combined with LSSVM is shown in Table 5. The recall, precision, and F-Score were 86.89–90.16%, 84.38–92.98%, and 86.89–89.83%, respectively. The results show that the developed classifier can accurately predict the maturity level of apples. Compared with the prediction results of the model established directly using the raw spectra, the precision of the harvest mature increased from 77.14% to 84.38%. The reliability of the prediction results of harvest mature was greatly increased, indicating that the spectral shape features can be used to obtain more effective spectral information and improve the prediction performance of the model.

Table 5. Confusion matrix, recall, precision, and F-score based on the SR + SD + NSID-LSSVM model.

Model	Maturity Category	1	2	3	No.	Recall/%	Precision/%	F-Score/%	Accuracy/%
SR + SD + NSID-LSSVM	1	55	5	1	61	90.16	88.71	89.43	88.52%
	2	4	54	3	61	88.52	84.38	86.89	
	3	3	5	53	61	86.89	92.98	89.83	

3.6. Validation of the Device

After implementing the optimal LSSVM model into the device, the detection performance of the model was evaluated using the samples obtained in 2020. The validation result is shown in Figure 6. The validation accuracy was 84.72%. The recall, precision, and F-Score were 84.43–85.44%, 82.01–87.29%, and 83.21–85.83%, respectively. All the evaluation indicators were higher than 82%, indicating that the developed model can predict the maturity level of apples at relatively high precision. Further analysis of the validation results using the ROC curve shows that the area under curve (AUC) values of different maturity were all higher than 0.8972, indicating that the developed model still has good prediction performance under the effect of seasonal variation. The results show that the multispectral sensor combined with the data analysis strategy can be used as a cheap and portable alternative for a spectrometer in the case of apple maturity detection.

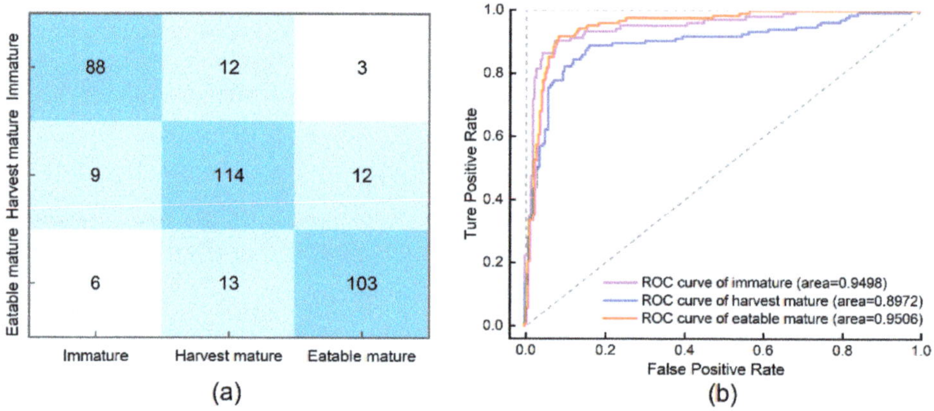

Figure 6. Validation results of the device: (**a**) confusion matrix, (**b**) ROC curve.

3.7. Discussion

In general, the multispectral sensors combined with spectral shape features can well distinguish the maturity levels of apples in this paper. The prediction accuracy of the optimal model, namely the SR + SD + NSID-LSSVM model, in the calibration set was 89.55%, the prediction accuracy of the prediction set was 88.52%, and the prediction accuracy of the external verification set was 84.72%. The quality of apples changed significantly during ripening, resulting in significant differences in spectral reflectance among apples with different maturity levels, which provided a theoretical basis for the prediction of apple ripening in this study.

However, non-destructive detection of apple maturity based on multispectral sensors was not easy. In this study, both traditional methods and spectral shape feature analysis can well distinguish immature and eatable mature samples. This result could be attributed to the fact that the contents of endogenous substances and spectral information of these two maturity levels of apples are of huge differences. However, through the analysis of the samples misclassified by the model, it is found that some of the samples are easily misclassified. The maturity of these samples is usually between two maturity levels, and there is little difference in quality and spectrum. A similar phenomenon was also observed by a previous study [10]. In fact, in the subsequent growth of these samples, the maturity will gradually change to the next level, especially for immature samples. Therefore, in order to solve this problem in the actual production, the mature time is still an important reference. Another reason for misclassification is that maturity is an overall attribute of apples, and the interaction model can only collect local information, so there will be misclassification of immaturity and edible maturity. For this kind of sample, the detection method needs to be improved, and the transmission spectrum is a reliable way to detect the internal properties of fruits [42]. In future research, transmission spectra can be collected for maturity detection to improve detection accuracy.

Multispectral sensors combined with spectral shape features have unique advantages. Because the absorption peaks of substances in VIS/NIR spectra are wide and overlapping, the spectral variables are collinear [43]. The reduction of spectral resolution can eliminate the collinear problem. The decrease in spectral resolution will lose part of the spectral information, which directly leads to the decline of the performance of multispectral sensors [19]. The spectral shape feature can be used to calculate the shape information contained in the spectral curve, which improves the detection accuracy of the multispectral sensor. In addition, the unique advantage of spectral shape features to eliminate the adverse effects of physical and biological variability on spectral information may also be the reason for the higher prediction accuracy [21,44].

In summary, a multispectral sensor combined with data analysis techniques can be used in the detection of fruit maturity and internal diseases instead of a spectrometer. However, the quantitative detection of fruit quality, such as SSC and firmness, needs more accurate and rich spectral information, so the application of multispectral sensors in fruit quality quantitative detection needs further research.

4. Conclusions

In this study, apple maturity can be detected rapidly and accurately at low cost by combining multispectral sensors with spectral shape features. There are significant differences in spectral information of apples with different maturity at the same wavelength due to significant differences in quality. Compared with the common pretreatment methods, spectral shape features were effective for processing multispectral information. The model based on the combination of SD, SR, NSID, and LSSVM had the highest accuracy, with 88.80% of prediction accuracy and 84.72% of validation accuracy. Our results proved that quality changes induce significant spectral changes in apples during ripening and the spectral changes can be detected by low-resolution multispectral sensors. As a tool for rapid and convenient detection of apple maturity, low-cost multispectral sensors could serve to determine the best harvest date and post-harvest processing strategy. As this study is aimed at 'Fuji' apples, the application of the developed model in apples of other cultivars should be further studied in the future and extend to practical fruit production.

Author Contributions: Conceptualization, M.Z., Y.P., H.L., B.Z., Z.Z. and J.Z.; Data curation, M.Z.; Funding acquisition, J.Z.; Methodology, M.Z., M.S., Y.P., H.L., B.Z. and Z.Z.; Supervision, X.R. and J.Z.; Validation, M.Z.; Writing—original draft, M.Z.; Writing—review & editing, M.S., X.R. and J.Z. All authors have read and agreed to the published version of the manuscript.

Funding: This study was financially supported by the National Natural Science Foundation of China (31701664), the Major Special Science and Technology Project of Shaanxi (2020zdzx03-05-01), the Key Research and Development Projects of Shaanxi (2017ZDXM-NY-017), and the Shaanxi Post-doctoral Science Foundation (2017BSHEDZZ141).

Institutional Review Board Statement: Not applicable.

Conflicts of Interest: The authors declare that they have no known competing financial interests or personal relationships that could have appeared to influence the work reported in this paper.

References

1. Tian, S.; Zhang, M.; Li, B.; Zhang, Z.; Zhao, J.; Zhang, Z.; Zhang, H.; Hu, J. Measurement orientation compensation and comparison of transmission spectroscopy for online detection of moldy apple core. *Infrared Phys. Technol.* **2020**, *111*, 103510. [CrossRef]
2. Pathange, L.P.; Mallikarjunan, P.; Marini, R.P.; O'Keefe, S.; Vaughan, D. Non-destructive evaluation of apple maturity using an electronic nose system. *J. Food Eng.* **2006**, *77*, 1018–1023. [CrossRef]
3. Van Beers, R.; Aernouts, B.; León Gutiérrez, L.; Erkinbaev, C.; Rutten, K.; Schenk, A.; Nicolaï, B.; Saeys, W. Optimal Illumination-Detection Distance and Detector Size for Predicting Braeburn Apple Maturity from Vis/NIR Laser Reflectance Measurements. *Food Bioprocess Technol.* **2015**, *8*, 2123–2136. [CrossRef]
4. Skic, A.; Szymańska-Chargot, M.; Kruk, B.; Chylińska, M.; Pieczywek, P.M.; Kurenda, A.; Zdunek, A.; Rutkowski, K.P. Determination of the Optimum Harvest Window for Apples Using the Non-Destructive Biospeckle Method. *Sensors* **2016**, *16*, 661. [CrossRef]
5. Guo, W.; Gu, J.; Liu, D.; Shang, L. Peach variety identification using near-infrared diffuse reflectance spectroscopy. *Comput. Electron. Agric.* **2016**, *123*, 297–303. [CrossRef]
6. Zhang, M.; Shen, M.; Li, H.; Zhang, B.; Zhang, Z.; Quan, P.; Ren, X.; Xing, L.; Zhao, J. Modification of the effect of maturity variation on nondestructive detection of apple quality based on the compensation model. *Spectrochim. Acta Part A Mol. Biomol. Spectrosc.* **2022**, *267*, 120598. [CrossRef]
7. Sohaib Ali Shah, S.; Zeb, A.; Qureshi, W.S.; Arslan, M.; Ullah Malik, A.; Alasmary, W.; Alanazi, E. Towards fruit maturity estimation using NIR spectroscopy. *Infrared Phys. Technol.* **2020**, *111*, 103479. [CrossRef]

8. Peirs, A.; Lammertyn, J.; Ooms, K.; Nicolaï, B.M. Prediction of the optimal picking date of different apple cultivars by means of VIS/NIR-spectroscopy. *Postharvest Biol. Technol.* **2001**, *21*, 189–199. [CrossRef]
9. Peirs, A.; Tirry, J.; Verlinden, B.; Darius, P.; Nicolaï, B.M. Effect of biological variability on the robustness of NIR models for soluble solids content of apples. *Postharvest Biol. Technol.* **2003**, *28*, 269–280. [CrossRef]
10. Zhang, M.; Zhang, B.; Li, H.; Shen, M.; Tian, S.; Zhang, H.; Ren, X.; Xing, L.; Zhao, J. Determination of bagged 'Fuji' apple maturity by visible and near-infrared spectroscopy combined with a machine learning algorithm. *Infrared Phys. Technol.* **2020**, *111*, 103529. [CrossRef]
11. Pourdarbani, R.; Sabzi, S.; Kalantari, D.; Karimzadeh, R.; Ilbeygi, E.; Arribas, J.I. Automatic non-destructive video estimation of maturation levels in Fuji apple (*Malus Malus pumila*) fruit in orchard based on colour (Vis) and spectral (NIR) data. *Biosyst. Eng.* **2020**, *195*, 136–151. [CrossRef]
12. Pourdarbani, R.; Sabzi, S.; Kalantari, D.; Paliwal, J.; Benmouna, B.; García-Mateos, G.; Molina-Martínez, J.M. Estimation of different ripening stages of Fuji apples using image processing and spectroscopy based on the majority voting method. *Comput. Electron. Agric.* **2020**, *176*, 105643. [CrossRef]
13. Choi, J.-H.; Chen, P.-A.; Lee, B.; Yim, S.-H.; Kim, M.-S.; Bae, Y.-S.; Lim, D.-C.; Seo, H.-J. Portable, non-destructive tester integrating VIS/NIR reflectance spectroscopy for the detection of sugar content in Asian pears. *Sci. Hortic.* **2017**, *220*, 147–153. [CrossRef]
14. Fan, S.; Wang, Q.; Tian, X.; Yang, G.; Xia, Y.; Li, J.; Huang, W. Non-destructive evaluation of soluble solids content of apples using a developed portable Vis/NIR device. *Biosyst. Eng.* **2020**, *193*, 138–148. [CrossRef]
15. Guo, W.; Li, W.; Yang, B.; Zhu, Z.; Liu, D.; Zhu, X. A novel noninvasive and cost-effective handheld detector on soluble solids content of fruits. *J. Food Eng.* **2019**, *257*, 1–9. [CrossRef]
16. Zhao, J.; Quan, P.; Zhang, M.; Tian, S.; Zhang, H.; Ren, X. Design of Apple Quality Integrated Non- destructive Testing Device Based on Multi- band LED Light Source. *Trans. Chin. Soc. Agric. Mach.* **2019**, *50*, 326–332. [CrossRef]
17. Abasi, S.; Minaei, S.; Jamshidi, B.; Fathi, D. Development of an Optical Smart Portable Instrument for Fruit Quality Detection. *IEEE Trans. Instrum. Meas.* **2021**, *70*, 7000109. [CrossRef]
18. Li, L.; Peng, Y.; Li, Y.; Yang, C.; Chao, K. Rapid and low-cost detection of moldy apple core based on an optical sensor system. *Postharvest Biol. Technol.* **2020**, *168*, 111276. [CrossRef]
19. Yang, B.; Zhu, Z.; Gao, M.; Yan, X.; Zhu, X.; Guo, W. A portable detector on main compositions of raw and homogenized milk. *Comput. Electron. Agric.* **2020**, *177*, 105668. [CrossRef]
20. Ma, J.; Sun, D.-W.; Pu, H. Spectral absorption index in hyperspectral image analysis for predicting moisture contents in pork longissimus dorsi muscles. *Food Chem.* **2016**, *197*, 848–854. [CrossRef]
21. Li, L.; Peng, Y.; Yang, C.; Li, Y. Optical sensing system for detection of the internal and external quality attributes of apples. *Postharvest Biol. Technol.* **2020**, *162*, 111101. [CrossRef]
22. Han, D.; Tu, R.; Lu, C.; Liu, X.; Wen, Z. Nondestructive detection of brown core in the Chinese pear 'Yali' by transmission visible–NIR spectroscopy. *Food Control* **2006**, *17*, 604–608. [CrossRef]
23. Moscetti, R.; Haff, R.P.; Aernouts, B.; Saeys, W.; Monarca, D.; Cecchini, M.; Massantini, R. Feasibility of Vis/NIR spectroscopy for detection of flaws in hazelnut kernels. *J. Food Eng.* **2013**, *118*, 1–7. [CrossRef]
24. Lleó, L.; Roger, J.M.; Herrero-Langreo, A.; Diezma-Iglesias, B.; Barreiro, P. Comparison of multispectral indexes extracted from hyperspectral images for the assessment of fruit ripening. *J. Food Eng.* **2011**, *104*, 612–620. [CrossRef]
25. Li, H.; Zhang, M.; Shen, M.; Zhang, Z.; Zhang, B.; Zhang, H.; Hu, J.; Ren, X.; Xing, L.; Zhao, J. Effect of ambient temperature on the model stability of handheld devices for predicting the apple soluble solids content. *Eur. J. Agron.* **2022**, *133*, 126430. [CrossRef]
26. Walsh, K.B.; Guthrie, J.A.; Burney, J.W. Application of commercially available, low-cost, miniaturised NIR spectrometers to the assessment of the sugar content of intact fruit. *Funct. Plant Biol.* **2000**, *27*, 1175–1186. [CrossRef]
27. Zhang, L.; Xu, H.; Gu, M. Use of signal to noise ratio and area change rate of spectra to evaluate the Visible/NIR spectral system for fruit internal quality detection. *J. Food Eng.* **2014**, *139*, 19–23. [CrossRef]
28. Blanpied, G.; Silsby, K.J. *Predicting Harvest Date Windows for Apples*; Cornell Cooperative Extension: Ithaca, NY, USA, 1992.
29. Gálvez-López, D.; Laurens, F.; Devaux, M.F.; Lahaye, M. Texture analysis in an apple progeny through instrumental, sensory and histological phenotyping. *Euphytica* **2012**, *185*, 171–183. [CrossRef]
30. Zhao, J.; Quan, P.; Liu, H.; Li, L.; Qi, S.; Zhang, M.; Zhang, B.; Li, H.; Zhao, Y.; Ma, B.; et al. Transcriptomic and Metabolic Analyses Provide New Insights into the Apple Fruit Quality Decline during Long-Term Cold Storage. *J. Agric. Food Chem.* **2020**, *68*, 4699–4716. [CrossRef]
31. Jie, D.; Zhou, W.; Wei, X. Nondestructive detection of maturity of watermelon by spectral characteristic using NIR diffuse transmittance technique. *Sci. Hortic.* **2019**, *257*, 108718. [CrossRef]
32. Li, B.; Lecourt, J.; Bishop, G. Advances in Non-Destructive Early Assessment of Fruit Ripeness towards Defining Optimal Time of Harvest and Yield Prediction-A Review. *Plants* **2018**, *7*, 3. [CrossRef] [PubMed]
33. Nagy, A.; Riczu, P.; Tamas, J. Spectral evaluation of apple fruit ripening and pigment content alteration. *Sci. Hortic.* **2016**, *201*, 256–264. [CrossRef]
34. Chauchard, F.; Cogdill, R.; Roussel, S.; Roger, J.M.; Bellon-Maurel, V. Application of LS-SVM to non-linear phenomena in NIR spectroscopy: Development of a robust and portable sensor for acidity prediction in grapes. *Chemom. Intell. Lab. Syst.* **2004**, *71*, 141–150. [CrossRef]

35. Li, X.; Wei, Y.; Xu, J.; Feng, X.; Wu, F.; Zhou, R.; Jin, J.; Xu, K.; Yu, X.; He, Y. SSC and pH for sweet assessment and maturity classification of harvested cherry fruit based on NIR hyperspectral imaging technology. *Postharvest Biol. Technol.* **2018**, *143*, 112–118. [CrossRef]
36. Ye, J. Least squares linear discriminant analysis. In Proceedings of the 24th international Conference on Machine Learning, Corvalis, OR, USA, 20–24 June 2007; pp. 1087–1093.
37. Otsu, N. A Threshold Selection Method from Gray-Level Histograms. *IEEE Trans. Syst. Man Cybern.* **1979**, *9*, 62–66. [CrossRef]
38. Xia, Y.; Huang, W.; Fan, S.; Li, J.; Chen, L. Effect of spectral measurement orientation on online prediction of soluble solids content of apple using Vis/NIR diffuse reflectance. *Infrared Phys. Technol.* **2019**, *97*, 467–477. [CrossRef]
39. Merzlyak, M.N.; Solovchenko, A.E.; Gitelson, A.A. Reflectance spectral features and non-destructive estimation of chlorophyll, carotenoid and anthocyanin content in apple fruit. *Postharvest Biol. Technol.* **2003**, *27*, 197–211. [CrossRef]
40. Wang, H.; Arakawa, O.; Motomura, Y. Influence of maturity and bagging on the relationship between anthocyanin accumulation and phenylalanine ammonia-lyase (PAL) activity in 'Jonathan' apples. *Postharvest Biol. Technol.* **2000**, *19*, 123–128. [CrossRef]
41. Zhao, Y.; Zhang, C.; Zhu, S.; Li, Y.; He, Y.; Liu, F. Shape induced reflectance correction for non-destructive determination and visualization of soluble solids content in winter jujubes using hyperspectral imaging in two different spectral ranges. *Postharvest Biol. Technol.* **2020**, *161*, 111080. [CrossRef]
42. Tian, S.; Wang, S.; Xu, H. Early detection of freezing damage in oranges by online Vis/NIR transmission coupled with diameter correction method and deep 1D-CNN. *Comput. Electron. Agric.* **2022**, *193*, 106638. [CrossRef]
43. Saeys, W.; Do Trong, N.N.; Van Beers, R.; Nicolai, B.M. Multivariate calibration of spectroscopic sensors for postharvest quality evaluation: A review. *Postharvest Biol. Technol.* **2019**, *158*, 110981. [CrossRef]
44. Tian, S.; Xu, H. Nondestructive Methods for the Quality Assessment of Fruits and Vegetables Considering Their Physical and Biological Variability. *Food Eng. Rev.* **2022**, 1–28. [CrossRef]

horticulturae

MDPI

Article

Comparative Transcriptome Analysis Reveals Sex-Biased Expression of Hormone-Related Genes at an Early Stage of Sex Differentiation in Red Bayberry (*Morella rubra*)

Huimin Jia [1,2,*], Lan Zhao [2], Yan Wang [2], Hongxia Wu [3], Haibo Zhao [2], Yifan Zhu [2], Yun Jiao [4], Guoyun Wang [5], Chaochao Zhou [5], Chunhui Huang [1], Huijuan Jia [2] and Zhongshan Gao [2,*]

1 College of Agronomy, Jiangxi Agricultural University, Nanchang 330045, China; lindahch@126.com
2 Fruit Science Institute, College of Agriculture and Biotechnology, Zhejiang University, Hangzhou 310058, China; 21716040@zju.edu.cn (L.Z.); 11816010@zju.edu.cn (Y.W.); zhaohaibo@zju.edu.cn (H.Z.); 21916154@zju.edu.cn (Y.Z.); jiahuijuan@zju.edu.cn (H.J.)
3 Key Laboratory of Tropical Fruit Biology, Ministry of Agriculture, South Subtropical Crops Research Institute, Chinese Academy of Tropical Agricultural Sciences, Zhanjiang 524091, China; whx1106@163.com
4 Institute of Forestry, Ningbo Academy of Agricultural Science, Ningbo 315400, China; jydyx@163.com
5 Yuyao Agricultural Technology Extension Service Station, Ningbo 315400, China; yywgy701218@163.com (G.W.); zhoucc1988@163.com (C.Z.)
* Correspondence: jiahuimin1988@163.com (H.J.); gaozhongshan@zju.edu.cn (Z.G.)

Abstract: The molecular mechanism of sex development and differentiation in the economically important dioecious fruit tree, red bayberry (*Morella rubra*), was revealed using next-generation transcriptome sequencing (NGS), and comparative analyses were used to identify differentially expressed genes (DEGs) in female and male flower buds. A total of 7029 of these DEGs were identified at two early development stages. KEGG pathway enrichment analysis revealed that plant hormone signal transduction was significantly overrepresented, and 91 genes related to hormones were identified. An analysis of 7029 DEGs revealed 161 hormone-related genes, with the 42 related to auxin and 26 related to ethylene being the most highly represented. A total of 62 genes were significantly up-regulated in females and 29 were in males, with 18 of them specifically expressed in females and 10 in males. A total of 415 transcription factors were identified, with 129 genes up-regulated in females and 53 in males. Moreover, 38 had female-specific expression and 18 had male-specific expression. Using weighted gene co-expression network analysis (WGCNA), two modules were found to be associated with sexual type. In the module coded light-green, there were five genes related to hormones, one to flower development and ten transcription factors with four genes specifically expressed in the males and four in females. The hub gene in the light-green module is *MR0TCONS_00017483.1* (*ACO*), which is involved in ethylene biosynthesis and had male-specific expression. Among the transcription factors, three of the four male-specific expressed genes involved in flavonoid biosynthesis, including the MYB gene *MR1TCONS_00020658.1* and two BHLH genes, *MR6G001563.1* and *MR8G020751.1*, played important roles in male floral differentiation. In the dark-cyan module, six hormone-related genes, five transcription factors and three flower development genes were identified with the hub gene *MR1G019545.1* (*ETR1*), which participates in the ethylene signaling pathway, and *MR4G023618.1*, which encodes the C3H zinc finger transcription factor. These results indicate that ethylene is the key hormone that interacts with other hormones and transcription factors to regulate sex differentiation in the red bayberry, which also provides new insights into the mechanism of sex determination and differentiation in the red bayberry.

Keywords: *Morella rubra*; sex determination and differentiation; hormones; ethylene; transcription factor

Citation: Jia, H.; Zhao, L.; Wang, Y.; Wu, H.; Zhao, H.; Zhu, Y.; Jiao, Y.; Wang, G.; Zhou, C.; Huang, C.; et al. Comparative Transcriptome Analysis Reveals Sex-Biased Expression of Hormone-Related Genes at an Early Stage of Sex Differentiation in Red Bayberry (*Morella rubra*). *Horticulturae* **2022**, *8*, 183. https://doi.org/10.3390/horticulturae8020183

Academic Editor: Dilip R. Panthee

Received: 5 January 2022
Accepted: 15 February 2022
Published: 21 February 2022

1. Introduction

The evolution of sex differentiation is an interesting topic because it is important to strategies for outbreeding, the prevention of inbreeding depression and a vital source of genetic variation. About 10% of flowering plants have separate unisexual flowers in different individuals (dioecy) or in the same individual (monoecy) [1]. For the mechanisms of unisexual flower development, two categories of unisexual flowers have been defined [2,3]. In type I, bisexual flowers become unisexual through the termination of the development of the androecium or gynoecium. The sex differentiation occurs at the particular developmental stage when one of the reproductive organs is arrested. Flowers of type II are unisexual from inception, and sex differentiation occurs before the formation of female or male organ primordia [3].

Sex determination has evolved multiple times, independently, suggesting various genetic mechanisms of unisexual flower formation are at play in different species [3,4]. The genetic mechanism of sex determination in dioecy is often thought to be controlled by a putative sex chromosome. This includes a dominant XY system (male heterogamety), such as the one found in diploid persimmon [5] and papaya [6], and, in a few cases, a ZW system (female heterogamety), as found in the red bayberry [7] and wild species of the allo-octoploid strawberry [8]. Sex-determining genes are located in the non-recombining region of the Y or W chromosome. For example, in garden asparagus [9,10] and kiwifruit [11,12], the two sex-determining genes are in the sex determination region (SDR), consistent with the classical 'two-mutation model' for the evolution of dioecy from hermaphroditism via gynodioecy.

Hormones and transcription factors are coded with genetic factors, and can also regulate sex differentiation [13,14]. Several sex-differentiating genes have been found to be involved in hormone biosynthesis or signaling pathways. In kiwifruit, a cytokinin response regulator *Shy Girl* (*SyGI*) has been identified as a sex-determining gene that acts as the suppressor of feminization [11]. In the cucumber, four of five genes controlling flower sex type encode the key enzymes involved in ethylene biosynthesis, including *CsACS1G* (*F*) [15], *CsACS2* (*M*) [16], *CsACS11* (*A*) [17] and one *CsACO2* [18]. In addition, the zinc finger transcription factor *WIP1* suppresses the female flower to control unisexual flower development in the cucumber and melon [18].

Morella rubra, commonly known as the red bayberry, is the only cultivated edible fruit species in the Myricaceae family, and it is widely distributed in south China [19,20]. Most species in Myricaceae are dioecious, with very few monoecious individuals [21,22], and with male, female and monoecious flowers on different plants. It is difficult to identify sex before flowering, as there is no obvious difference between males and females. The sex type of the red bayberry is controlled by the sex chromosomes: ZW for females, ZZ for males. A 59 kb non-recombining region in the W chromosomes harboring seven candidate genes has been identified [7]. The red bayberry flower is a type II unisexual flower, in which only female or male organ primordia are initiated. The male flowers are compound catkins, which are surrounded by bracts without receptacles or pedicels, and each of them has two stamens (Figure 1a). The female flowers are catkins (Figure 1b), and the ovary is unilocular with a Y-shaped stigma [20]. A time course experiment indicated that the critical period for flower bud development is from July to September. The red bayberry is also a good species for sex determination mechanistic studies because of its small genome size (320 Mb), ZW chromosome and viable WW genotype (super female) individuals [7,23]. Although seven female-specific gene and molecular markers linked to the sex type have been identified [7], the molecular mechanism and the associated and regulated genes that govern sex determination are not well understood in the red bayberry.

With the continuous progress observed in high-throughput sequencing, transcriptome analyses have been reported in many species to unravel the molecular process regulating the determination of unisexual flowers, such as in papaya [24], spinach [25,26] and *Jatropha curcas* L. [27]. Although a comparative analysis of gene expression between female and the monoecious mutant red bayberry flowers has been reported [28], the flower samples used were taken during late stages (December) of flower development, when sex differentiation is completed and the key genes regulating sex determination might be no longer expressed. In the present study, transcriptomes analyses of red bayberry floral buds from female and male trees were performed, before the initiation of reproductive organ primordia, aiming to capture useful information to understand the mechanism of sex differentiation and floral development in the red bayberry. The results of this study can also be used to guide cross-breeding between two cultivars by manually inducing male flowers in female plants.

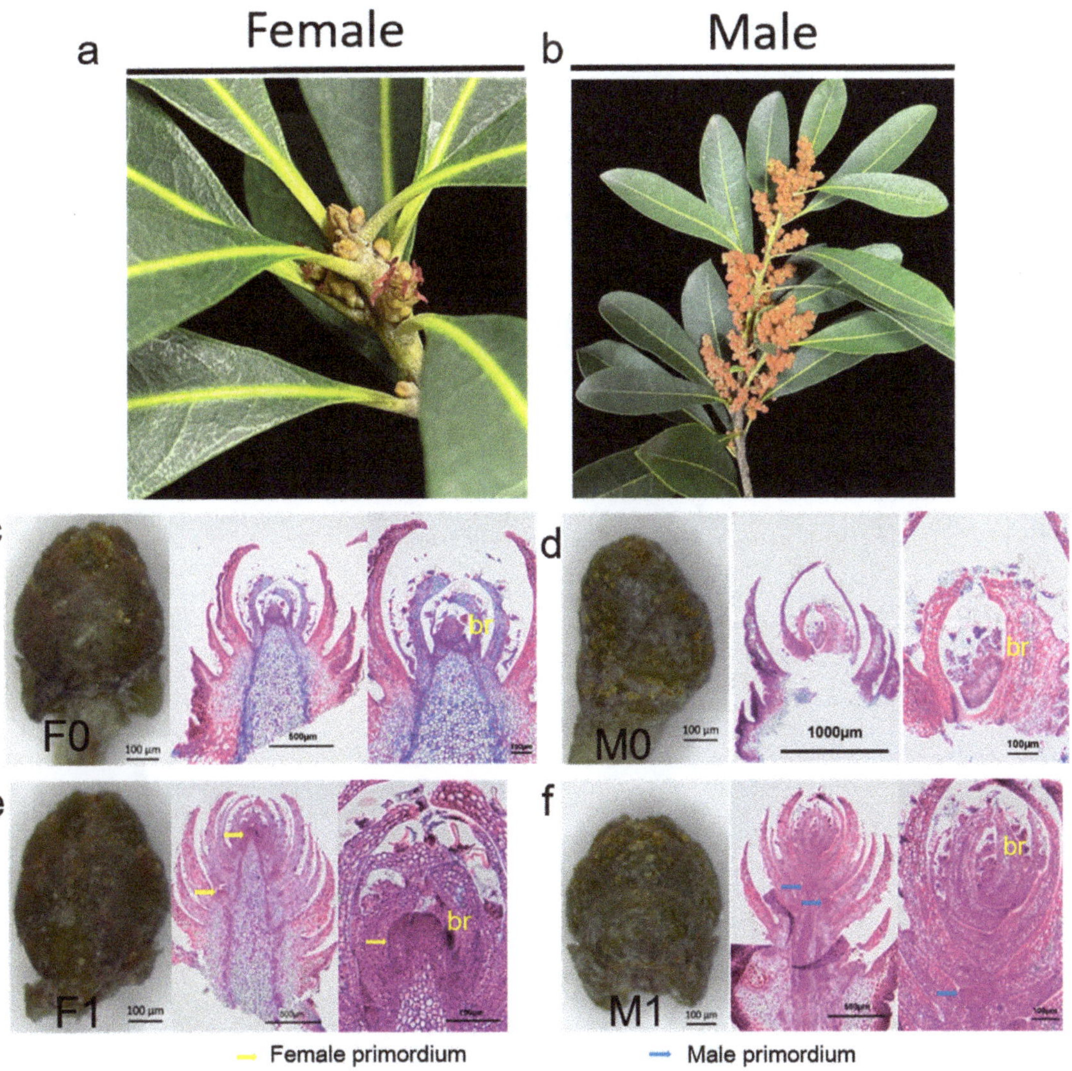

Figure 1. The male and female floral buds of the red bayberry. Female flower (**a**) and male flower (**b**) of the red bayberry. Male and female floral buds before sexual differentiation (**c**,**d**), and with female and male primordia (**e**,**f**). br, bracts. The figures (**e**,**f**) have been published in a previous study [7].

2. Materials and Methods

2.1. Flower Sample Collection and Morphological Observation

The male accession 'H2011-12' of a red bayberry, an approximately 30-year-old tree which was used for whole-genome sequencing [7], and the female individual 'H2011-11', grown at Zhejiang University Campus (Hangzhou, China), were used for all experiments.

Floral buds were collected on two dates in 2017: 13 July and 27 August. The male and female floral buds were sampled in three replicates. The samples were immediately frozen in liquid nitrogen and stored at −80 °C prior to RNA extraction. The male and female flowers were screened under a stereoscopic microscope. Paraffin sections were used as in a previous study [29] to confirm the developmental stages of male and female buds. We used floral buds before sexual differentiation (F0 for female and M0 for male floral buds) and the early emergence of female and male primordia (F1 for female and M1 for male floral buds) for transcriptomic analysis to uncover the molecular regulatory mechanisms of unisexual flower development and sexual differentiation in the red bayberry.

2.2. *RNA Extraction and Transcriptome Sequencing*

The total RNA of male and female flowers was extracted by using a Quick RNA isolation Kit (Waryong, Beijing, China) according to the manufacturer's manual. The quality of total RNA was checked on 1% denatured agarose gel and then assessed on an Agilent Bioanalyzer 2100 system (Agilent Technologies, Santa Clara, CA, USA). The pair-end cDNA sequencing libraries of four floral buds were prepared with three biological replicates by the Beijing Genome Institution (BGI, Beijing, China) and then sequenced at BGI using an Illumina HiSeq 2500 platform (Illumina) with PE125.

2.3. *RNA-Seq Data Analysis*

Adapter sequences, ambiguous nucleotides, and low-quality sequences in raw reads were removed with SOAPnuke (version1.4.0) [30] and Trimmomatic (v0.36) [31]. The paired-end clean reads from each library were mapped to the reference genome (https://www.ncbi.nlm.nih.gov/genome/?term=Morella+rubra, accessed on 15 February 2022) using HISAT2 (Hierarchical Indexing for Spliced Alignment of Transcripts) to calculate the mapping ration [32]. Clean reads were aligned to complete the reference using Bowtie2 [33], and RSEM (RNA-Seq by Expectation Maximization) was used to estimate the abundance of the genes [34]. Read counts were normalized by calculating the FPKM (Fragments Per Kilobase of exon model per Million mapped fragments) value of all the transcripts in the samples. Gene function was annotated based on NCBI nonredundant protein sequences (Nr), the protein family (Pfam), the Swiss-Prot protein database and the Kyoto Encyclopedia of Genes and Genomes (KEGG). Gene ontology (GO) annotation was used for sequences with a match in the Nr database by using Blast2GO v.3.0 [35]. Differential expression between the two groups was analyzed with the DESeq R package version 1.30.0 [36]. Genes with a minimal two-fold difference in expression ($|\log2 \text{ Ratio}| \geq 1$) and padj ≤ 0.01 were considered to be differentially expressed genes (DEGs). Heatmaps (scaled by row) were prepared with TBtools and used for GO enrichment analysis and KEGG pathway enrichment analysis [37]. For the weighted gene co-expression network analysis (WGCNA), an R package (version 1.0.7) [38] was used to construct a potential regulatory network of sex differentiation in the red bayberry.

2.4. *Validation of RNA-Seq Data*

Quantitative real-time PCR (qRT-PCR) was used to validate the RNA-seq data. The Total RNA of F0, F1, M0 and M1 were extracted by using the Quick RNA isolation Kit (Waryong, Beijing, China) according to the manufacturer's manual, and the concentration was checked on a BioDrop spectrophotometer (Biochrom, Cambridge, UK). First-strand cDNA synthesis was performed with 2 μg RNA with EasyScript® All-in-One First-Strand cDNA Synthesis SuperMix for qPCR (One-Step gDNA Removal) (TransGen Biotech, Beijing, China). The qRT-PCR primers were designed using an online tool, Primer 3 (https://primer3.ut.ee/, accessed on 15 February 2022). The PCR products of qRT–PCR were verified by sequencing. The qPCR was performed with the SYBR Premix Ex Taq™ (Takara, Ohtsu, Japan) and CFX Connect™ real-time PCR system (Bio-Rad, Hercules, CA, USA) with the reaction and program detailed as previously published by Ni et al. [39]. Actin was used for normalization, and the expression data were calculated with the $2^{-\Delta\Delta Ct}$ formula [40].

3. Results

3.1. Morphological Observation and Dissecting Red Bayberry Flower Buds

Red bayberry floral meristems were initiated in the bract axils of both male and female plants. At the early stage of floral development, the inflorescence rachis became elongated (Figure 1c,d), and this was followed by the flower initiation stage, with the female and male flower primordia forming between the two bracts (Figure 1e,f). At this stage, there was no obvious morphological difference between male and female flowers.

3.2. Sequencing of Red Bayberry Flower Buds

A total of 562.3 million raw sequencing reads were obtained (Table S1), and the clean reads were aligned to the red bayberry reference genome and used to quantify the expression levels of genes using HISAT and Bowtie2. About 83.5% of reads could be mapped to the reference genome (Table S1). A principal component analysis (PCA) based on FPKM values separated the samples into four distinct groups, with each sample making a separate group with its replicates (Figure S1). The total variation portion (46.2%) consists of two principal components (PC1: 27.6%, PC2: 18.6%), with PC1 separating females from males. The distributions of the number of genes at different expression levels among the four groups were similar (Figure 2a). The number of genes not expressed (FPKM < 1) accounted for the highest proportion, followed by those genes expressed at a low level ($1 \leq$ FPKM < 10) and mid-level ($10 \leq$ FPKM < 100). The highly expressed genes with FPKM $\geq$ 100 accounted for the smallest proportion. About 85% of genes were expressed in both female and male floral buds, and 1979 genes were specifically expressed in female and 2015 in male buds (Figure 2b).

3.3. Gene Ontology (GO) and KEGG Enrichment Analysis of Highly Expressed Genes Revealed Rapid Growth in Female and Male Buds

A total of 633, 677, 521 and 558 highly expressed genes were detected in F0, F1, M0 and M1 floral buds, respectively, and 61.3% of them were shared between female and male floral buds (Figure 2c). The GO enrichment analysis and KEGG pathway enrichment analysis of the highly expressed genes in female and male floral buds were performed. The results from the GO enrichment analysis of the highly expressed genes showed that genes involved in cell proliferation, such as the 'structural constituent of ribosome' and 'structural molecule activity' belonging to molecular function categories, 'ribosomal subunit' and 'ribosome' belonging to the cellular component and the 'peptide biosynthetic process' and 'peptide metabolic process' belonging to the biological process, were more abundant (Table S2). This was in agreement with the results of the KEGG enrichment analysis, with ribosome overrepresented in the four floral samples (Table S3). In these analyses for mid-level expressed genes in four floral samples (Tables S4 and S5), the KEGG enrichment results indicated that ribosome was overrepresented in the four floral samples (Table S5). These results suggest that floral buds at these two development stages undergo rapid cell division and expansion.

3.4. Identification of Differentially Expressed Genes

We compared female vs. male floral buds in pair-wise analyses of M0 vs. F0 and M1 vs. F1. We also compared floral buds within a sexual type, F1 vs. F0 and M1 vs. M0. In all, 8889 genes were significantly expressed, with $|\text{Log2FC}| \geq 1$ and p-value < 0.01 in four pairs (Figure S2), including 4970 DEGs in M0 vs. F0, 5421 DEGs in M1 vs. F1, 2659 DEGs in F1 vs. F0 and 2888 DEGs in M1 vs. M0 (Figure 2d). The number of DEGs between female and male was greater than those within one sex in an inner comparison, suggesting that a very complex physiological and biochemical process occurs during sexual differentiation and unisexual flower development.

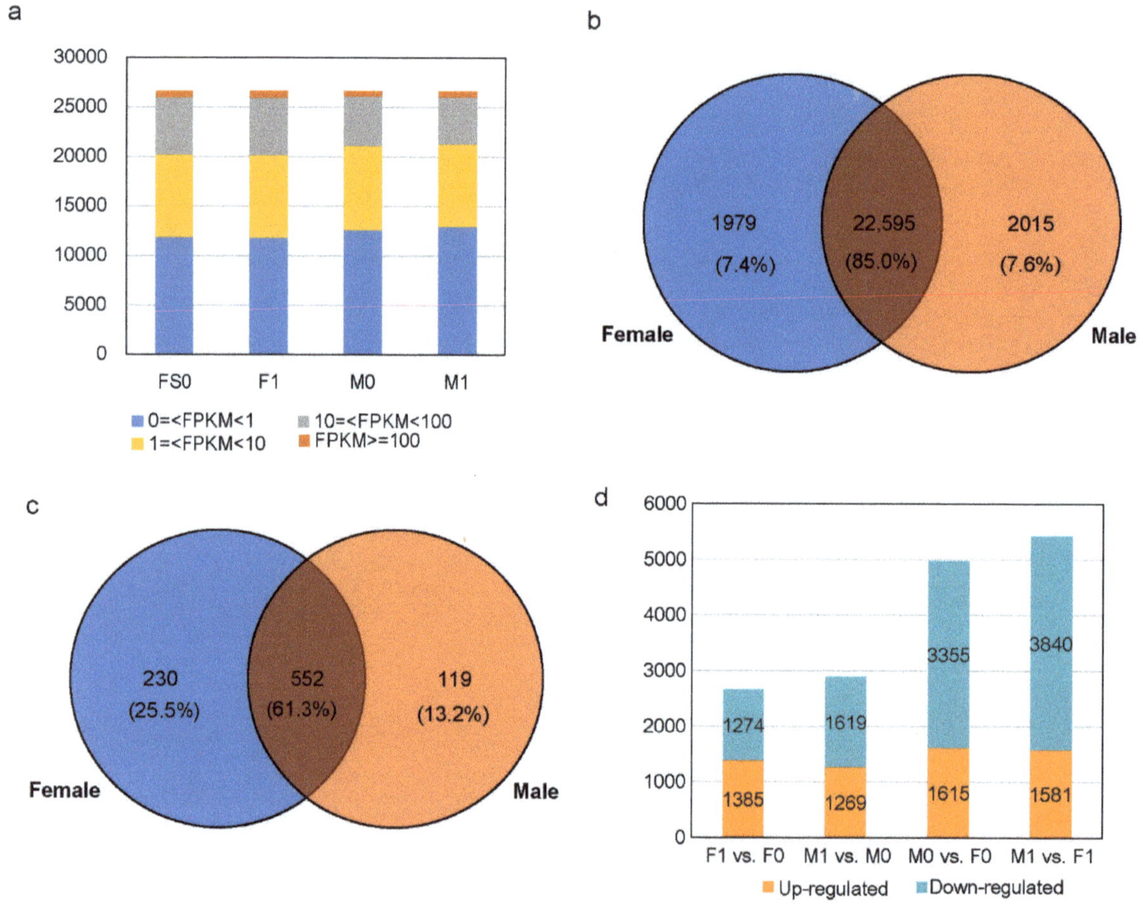

Figure 2. Overview of red bayberry M0, M1, F0 and F1 floral bud transcriptome. (**a**) Distribution of genes with high expression (FPKM $\geq$ 100), mid-level expression (100 > FPKM $\geq$ 10), low expression (10 > FPKM $\geq$ 1) and no expression (1 > FPKM $\geq$ 0). Venn diagram of genes expressed in female and male libraries (**b**) and genes with high expression in at least one library (**c**). (**d**) Distribution of up- and down-regulated DEGs in the pair-wise F1 vs. F0, M1 vs. M0, M0 vs. F0 and M1 vs. F1 analyses.

Because we aim to identify genes related to sexual differentiation in the red bayberry, we only focus on an analysis of DEGs between M0 vs. F0 and M1 vs. F1. We further used GO and KEGG enrichment to analyze DEGs between M0 vs. F0, M1 vs. F1, and all the 7029 DEGs of the two compared groups. GO enrichment results showed that catalytic activity was overrepresented in molecular functions (Table S6). KEGG pathway enrichment analysis showed that plant hormone signal transduction (ko04075) was significantly overrepresented in M0 vs. F0 and M1 vs. F1 and all 7029 DEGs (Figure S3), suggesting that plant hormones play a role in sex differentiation in the red bayberry. A total of 91 genes were identified in this category, 35 associated with auxin, 14 involved in CK, 10 with ABA and JA, and the remainder associated with gibberellin, BR, ethylene, and SA (Table S7).

3.5. DEGs Are Involved in Plant Hormone Biosynthesis and Signalling

To investigate hormone function during the process of sexual differentiation, hormone-related genes were identified in the red bayberry. A total of 161 hormone-related genes were identified in 7029 DEGs (Table S8), and the distribution for different hormones is shown in Figure 3a: 42 genes related to auxin were most represented, followed by 26 related to ethylene, 20 to CK, 18 to SA, 17 to ABA, 15 to GA and JA, and 8 to brassinosteroid. Among the 161 hormone-related genes, 62 DEGs were significantly up-regulated in females and 29 were in males (Table S8). Moreover, a total of 18 genes displayed female-specific expression and 10 genes were expressed only in males (Table S9).

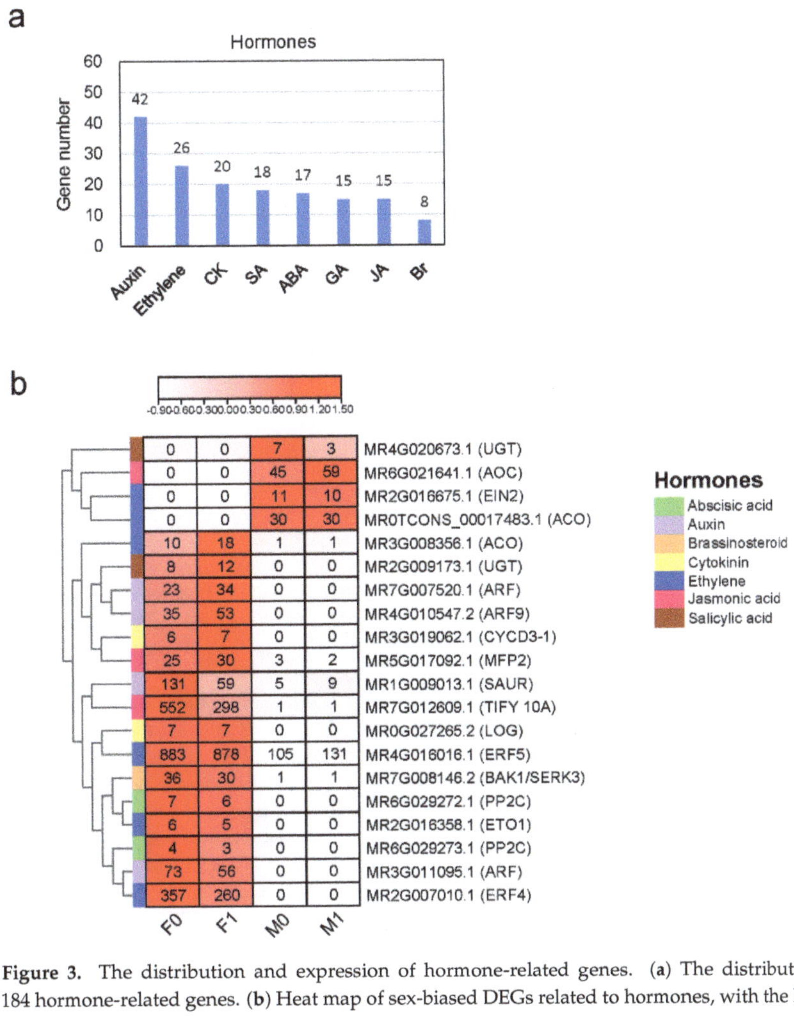

Figure 3. The distribution and expression of hormone-related genes. (**a**) The distribution of 184 hormone-related genes. (**b**) Heat map of sex-biased DEGs related to hormones, with the FPKM value of genes shown in the box.

The expression of 13 out of 42 auxin-related genes was significantly up-regulated in females. *MR1G009013.1* (*SAUR*) was expressed at a much higher level in females than in males (Figure 3b). Five of them were only expressed in females, including three *ARF* genes: *MR3G011095.1*, *MR4G010547.2* and *MR7G007520.1*; one IAA gene *MR3G001167.1* and one *YUC* gene *MR1G027448.1* (Table S9). The expression of six auxin-related genes was significantly up-regulated in males, including the *GH3.9* gene *MR8G002059.1*, which was only expressed in males. In the present study, 12 out of 26 ethylene-related genes were significantly up-regulated in females, including 4 female-specific expressed genes, *MR1G003948.1* (*ETO1*), *MR2G007010.1* (*ERF4*), *MR2G016358.1* (*ETO1*), and *MR3G008357.1*

(*ACO*) (Figure 3b and Table S9). The ethylene biosynthesis gene *ACO* (*MR3G008356.1*) and ethylene signal transduction gene *ERF5* (*MR4G016016.1*) were expressed at a much higher level in females than in males (Figure 3b). Four ethylene-related genes were significantly up-regulated in the male buds, including three male-specific expressed genes: *MR0TCONS_00017483.1* (*ACO*), *MR6G009324.1* (*ACO*) and *MR2G016675.1* (*EIN2*). The jasmonic acid biosynthesis gene *MR6G021641.1* (*AOC*) also had male-specific expression (Figure 3b). The cytokinin biosynthesis gene *MR0G027265.2* (*LOG*) and signal transduction gene *MR3G019062.1* (*CYCD3-1*) were only expressed in females, while two *UGT* genes involved in SA biosynthesis gene had the opposite expression pattern: *MR2G009173.1* was only expressed in females and *MR4G020673.1* was only expressed in males. The brassinosteroid signal transduction gene *BIM1* (*MR5G011875.1*) showed male-specific expression and BAK (*MR7G008146.2*) had much higher expression in females than in males (Figure 3b).

3.6. Transcription Factors Differentially Expressed in Male and Female Buds

Transcription factors (TFs) play important roles in flower development. A function annotation of the 7029 DEGs revealed that 415 transcription factors belong to 55 gene families (Table S10). The MYB family, AP2/EREBP family and bHLH family were over-represented (Figures 4a and S4). Among the transcription factors, the expression of 129 transcription factors genes were significantly up-regulated in females, and 53 were down-regulated. On further analysis of the expression of the sex-type-specific genes, 38 were found to be only expressed in females and 18 in males (Table S11).

Of the 58 MYB family genes, there was a significant female-biased expression of 12, and 6 had significant male-biased expression. The *MR4G023331.1* and *MR4G009025.1* gene expression was female-specific, while the *MR1TCONS_00020658.1* was only expressed in males; its homolog in *Arabidopsis AT3G13540.1* (*AtMYB5*) is involved in flavonoid biosynthesis. Among 36 genes in the bHLH gene family, 6 genes expressed were female-specific, including *MR1G020885.1*, *MR6G026964.1*, *MR7G020473.2*, *MR2G024394.1*, *MR4G009030.1* and *MR8TCONS_00067952.1*, and 3 genes were male-specific, including *MR5G011875.1*, *MR6G001563.1* and *MR8G020751.1* (Figure 4b). Four of AP2/EREBP genes displayed female-specific expression, including *MR0G006452.1*, *MR1G017646.1*, *MR2G007010.1* and *MR8G022041.1*. Moreover, *MR2G007010.1* was highly expressed in females with FPKM values of 357 and 260 for F0 and F1, respectively (Figure 4b).

MADS-box genes are involved in floral development and floral organ identity. In this study, we identified 10 MADS-box genes differentially expressed in the two groups. The *SOC1* gene *MR0G005926.1* was only expressed in males (Figure 4b); SOC1 activates the floral meristem identity gene *LEAFY* (*LFY*). Two C2C2-CO-like zinc finger genes, *MR6G018079.1* and *MR3G018958.1*, were specifically expressed in females, and *MR8G027384.1* was expressed only in males. We also found that two C2C2-Dof genes, *MR5G011870.1* and *MR3G009872.2*, and one C2C2-GATA gene, *MR6G010822.1*, were only expressed in females.

Among the 415 transcription factors, the expression of 24 of them was high or medium in females and were not expressed in males (Figure 4b). The hormone-related MYC, MYB, STRE, ABRE and TF transcription factor binding sites in the promoters were also enriched (Table S12).

3.7. Genes in the Non-Recombining Region of the Red Bayberry W Chromosome

The 59 kb FSR (female-specific region) had seven predicted genes, and the expression levels of seven female-specific genes are shown in Table 1. The expression of all the female-specific genes in F0 and F1 was very low, with the FPKM value < 2, except for *MrTFIID2*. The expression of these genes was consistent with those previously reported in red bayberry floral buds using qRT-PCR [7], and further confirmed that the gene expression profile of the RNA-seq data was reliable.

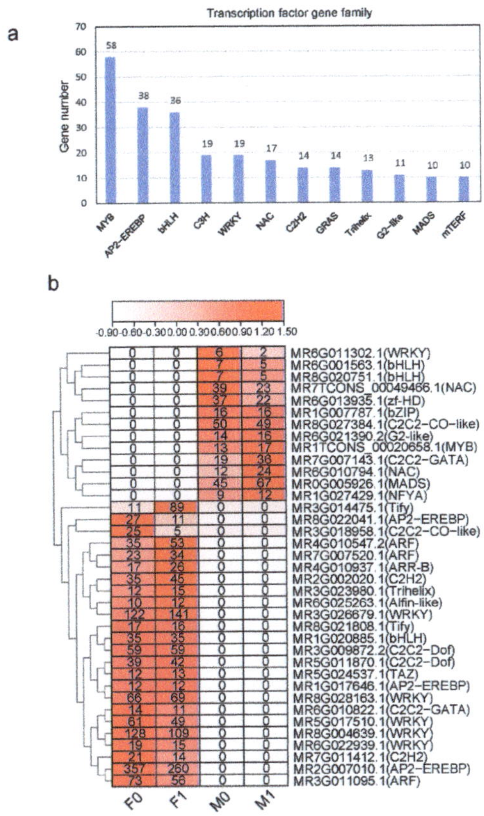

Figure 4. The distribution and expression of transcription factor genes in all 8889 DEGs. (**a**) The distribution of 547 transcription factor genes (top 12 families). (**b**) Heat map of sex-biased DEGs related to transcription. The FPKM value of genes is shown in the box.

Table 1. List of the expression level of seven female-specific genes in red bayberry floral buds.

Gene Name	Gene_Id	M0_FPKM	M1_FPKM	F0_FPKM	F1_FPKM
MrCPS2	*MR8G025874.1*	0	0	0.00	0.04
MrCKA2	*MR8G025875.1*	0	0	0.71	0.60
MrTFIID2	*MR8G025876.1*	0	0	4.90	4.33
MrASP2	*MR8G025877.1*	0	0	0.53	1.14
MrSAUR2	*MR8G025878.1*	0	0	0.44	2.06
MrLsd90-2	*MR8G025879.1*	0	0	0.43	0.59
MrFT2	*MR8G025880.1*	0	0	0.23	0.25

3.8. Co-Expression Networks of Female and Male Buds

We used weighted gene co-expression network analysis (WGCNA) [38] to construct a potential regulatory network of sex differentiation in the red bayberry. A total of 42 modules were established using WGCNA in four male and female flower buds at early development stages (Figure 5a). The genes in the same module had strong connectivity (Figure 5b). The expression pattern of each module was analyzed in all samples. The module–trait relationships showed that the light-green ($r = 0.97$, $P = 6 \times 10^{-5}$) and dark-cyan ($r = -0.75$, $P = 0.03$) modules were highly associated with the sex phenotype (Figure 5c), so these two modules may play important roles in regulating sex differentiation and flower development.

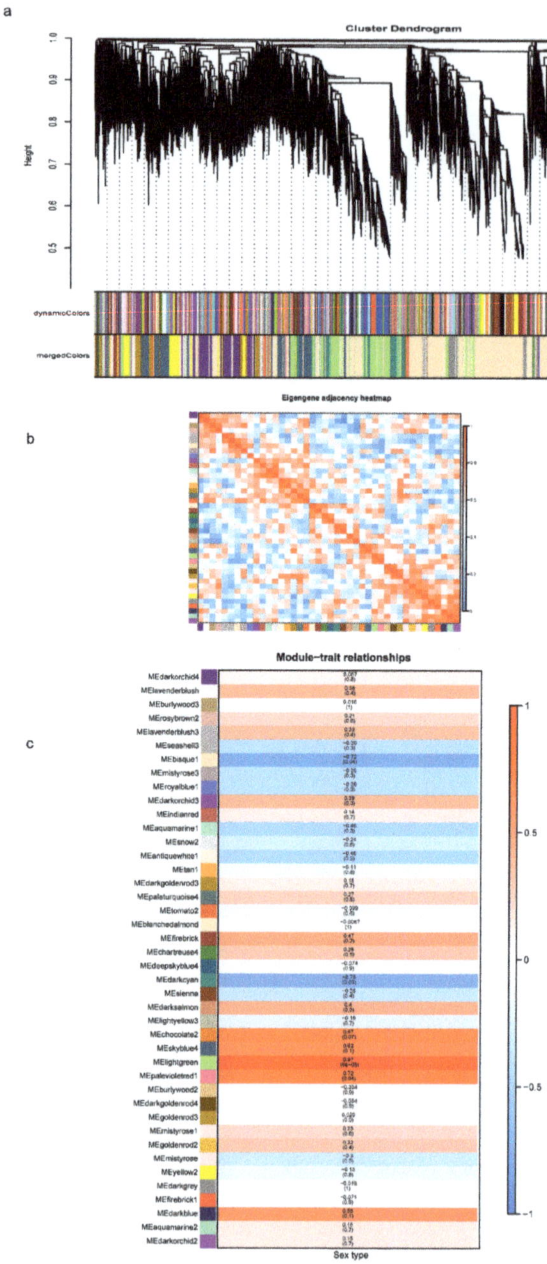

Figure 5. Weighted gene co-expression network analysis. (**a**) Hierarchical cluster tree showing co-expression modules identified using WGCNA; 'DynamicColors" represents modules divided according to clustering results, and "MergedColors" represents the merging of the module with a similar expression pattern according to the module similarity, with the analysis conducted according to the merged module. (**b**) Module gene correlation analysis; each row and column represent a gene; the darker the color of each point indicates a higher connectivity between the two genes. (**c**) Sample expression pattern analysis.

We screened the genes related to hormone and transcription factors in the two modules (Table S13). A total of 55 DEGs were selected from the light-green and 126 DEGs from the dark-cyan modules. As shown in Figure 6a and Table S14, the hub gene in the light-green module was *MR0TCONS_00017483.1*, which encodes 1-aminocyclopropane-1-carboxylate oxidase (ACO), which is a key gene in ethylene biosynthesis, and is only expressed in males (Figure 3b). In addition, the module included genes involved in ethylene biosynthesis (*MR2G016358.1* (*ETO1*), which is only expressed in females (Figure 3b)), ABA signaling (*MR8G022503.2* and *MR7G016748.1* (*PP2C*)) and CTK signaling (*MR7G024824.1* (*ARR5*), a male-specific expressed gene (Figure 6a)). Nine transcription factors were identified in the light-green module co-expression network, including three zinc finger genes (*MR5G011870.1* (Dof), *MR8G027384.1* (*COL4*) and *MR1G019332.1* (*C3H*)), one *MYB* gene (*MR1TCONS_00020658.1*), one *AP2/EREBP* (*MR2G007010.1*), one *PHD* gene (*MR5G024537.1*), one Tify (*MR8G021808.1*), one GRF (*MR4G021188.1*) and two *bHLH* genes (*MR6G001563.1* and *MR8G020751.1*). The floral meristem determinacy gene *MR6G008503.1* (*ULT1*) was also in the light-green module co-expression network (Figure 6a and Table 2).

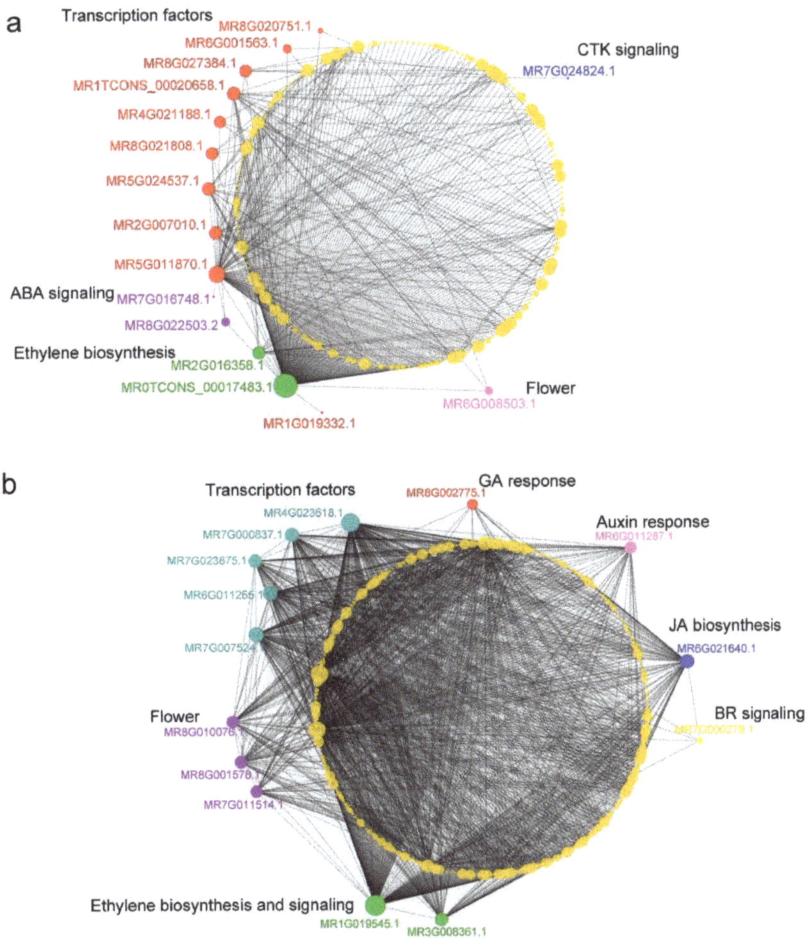

Figure 6. Visualization of co-expression network. (**a**) Co-expression network of genes in light-green modules. (**b**) Co-expression network of genes in dark-cyan modules.

Table 2. List of hormone-related genes, transcription factors and flowering-related genes in two modules.

Module	Gene ID	Gene Name	Specifically Expressed in	Involved Process
Light green	*MR0TCONS_00017483.1*	ACO	male	ET biosynthesis
	MR2G016358.1	ETO	female	
	MR8G022503.2	PP2C		ABA signaling
	MR7G016748.1	PP2C		
	MR7G024824.1	ARR5	male	CTK signalling
	MR5G011870.1	dof	female	Transcription factor
	MR1TCONS_00020658.1	MYB	male	
	MR2G007010.1	AP2/EREBP	female	
	MR5G024537.1	TAZ	female	
	MR8G021808.1	Tify	female	
	MR8G027384.1	C2C2-CO-like	male	
	MR4G021188.1	GRF		
	MR6G001563.1	bHLH106	male	
	MR8G020751.1	bHLH106	male	
	MR1G019332.1	C3H		
	MR6G008503.1	ULT	male	Flower development
Dark cyan	*MR1G019545.1*	ETR	-	ET biosynthesis and signaling
	MR3G008361.1	ACO		
	MR8G001578.1	SPL		Flower development
	MR7G011514.1	VRN1		
	MR8G010076.1	UGT		
	MR6G021640.1	AOC		Jasmonic acid biosynthesis
	MR6G011287.1	AUX22D		Auxin response
	MR8G002775.1	GASA1		GA response
	MR7G000279.1	ASK7		Gibberellin
	MR7G007524.1	AP2-EREBP		Transcription factor
	MR4G023618.1	C3H		
	MR7G000837.1	ZF-HD		
	MR7G023875.1	ZF-HD		
	MR6G011265.1	bzip		

The hub genes in the dark-cyan co-expression network include *MR1G019545.1*, which encodes ethylene response 1, with its homolog *AtETR1* (AT1G66340), which is involved in ethylene signaling pathways, and *MR4G023618.1*, which encodes the C3H zinc finger transcription factor. The dark-cyan module also included five hormone-related genes, four transcription factors and three flower development genes (Figure 6a and Table S15). The details are shown in Table 2: *MR3G008361.1* (*ACO*) is involved in ethylene biosynthesis; *MR6G021640.1* (*AOC*) is involved in JA biosynthesis; *MR7G000279.1* (*ASK7*) is involved in the brassinosteroid signal transduction pathway; and *MR6G011287.1* (AUX22D) and *MR8G002775.1* (*GASA1*) are involved in the auxin and gibberellin response, respectively. With regard to the transcription factors, *MR7G007524.1* belongs to the AP2/EREBF gene family and *MR6G011265.1* encodes the *bZIP* transcription factor. *MR7G000837.1* and *MR7G023875.1* belongs to the zinc finger gene family. There were three genes related to flower development: *MR8G010076.1*, which encodes homologs of *AtUGT87A2* (*AT2G30140*), which regulate flowering time via *FLOWERING LOCUS C*; *MR7G011514.1* the homolog of *Arabidopsis AtVRN1*; and the *MR8G001578.1* homolog in *Arabidopsis* is *AtSPL14* (*AT1G20980*). These results showed that ethylene biosynthesis and signaling pathways may play vital roles in sex differentiation and flower development, with the transcription factor playing vital roles in regulating flower buds.

3.9. Validation of Gene Expression Level in Male and Female Red Bayberry Floral Buds

To confirm the gene expression pattern of the male and female floral buds identified in the RNA-Seq data, ten DEGs were randomly selected for validation by using qRT-PCR and the primers were shown in Table S16. The results of the qPCR were consistent with the gene expression patterns from RNA-seq (Figure S5), which confirmed that the gene expression profile of the RNA-seq data was reliable.

4. Discussion

In the present study, we constructed female and male *Morella rubra* floral bud libraries representing two early development stages, and used transcriptomes to analyze the genes differentially expressed in female and male floral buds to identify sex differentiation- and flower development-related genes.

The identification of sex-biased genes related to flower development is helpful to reveal the sex differentiation mechanisms. Here, we found 8889 DEGs in pair-wise analyses of F1 vs. F0, M1 vs. M0, M0 vs. F0 and M1 vs. F1 that may be related to sex differentiation and flower development. KEGG analysis indicated these DEGs were functionally enriched in plant hormone signal transduction (ko04075) (Figure S3). Phytohormones regulate various developmental processes, including floral development and sex differentiation, and the effects of plant hormones on sex differentiation vary among plant species [41]. Ethylene has been demonstrated to play a key role in the sex determination of cucumber and melon [15,42,43]. In maize, GA suppresses stamen development, while brassinosteroid and jasmonate coordinately suppress tassel development [44,45]. The exogenous application of ethrel, an auxin inhibitor, can induce male-to-female sex reversal in papaya [24]. In the red bayberry, the exogenous application of a GA inhibitor can induce female-to-male sex reversal in the 'Dongkui' and other cultivars (http://patents.google.com/patent/CN107581062A/en, accessed on 15 February 2022).

Hormones can crosstalk to regulate floral organ development. For example, ethylene can control floral transition via the DELLA-dependent regulation of floral meristem identity genes LEAFY (LFY) and SUPPRESSOR OF OVEREXPRESSION OF CONSTANS 1 (SOC1) [46], and salicylic acid regulates flowering [47,48]. In this study we found 161 DEGs that were related to hormones, with 18 female-specific expressed genes and 10 male-specific expressed genes. Those genes, for instance, ethylene-related genes *MR2G016358.1* (*ETO1*) and *MR0TCONS_00017483.1* (*ACO*), are good candidates for studying the mechanism of red bayberry sex differentiation.

That transcription factors play vital roles in the regulation of sex differentiation and flower development has been reported in many plants. In the present study, we identified 415 transcription factors among the 7029 DEGs. Transcription factors interact with hormone-related genes to control sex identity. Moreover, the rest of three male-specific expressed genes, including one MYB gene (*MR1TCONS_00020658.1*) and two BHLH genes (*MR6G001563.1* and *MR8G020751.1*), have been reported to regulate flavonoid biosynthesis. Male floral differentiation has been found to be significantly related to flavonoid biosynthesis in *Jatropha curcas* L., and the inhibition of flavonoid biosynthesis causes male sterility in Petunia [27,49]. Our results suggest that transcription factors related to flavonoids may play roles in male floral differentiation.

For example, the sex *G* (*gynoecious*) gene *CmWIP1*, a C2H2-type zinc finger transcription factor, is a central integrator of the transcriptional networks, leading both to the inhibition of carpel development and the control of the expression of the stamen inhibitor *CmACS-7* in melon [50]. The female-specific expressed gene *MR2G002020.1* encodes the WIP2 protein identified in the present study. Based on the WGCNA results, a total of 15 transcription factors were identified, including four genes specifically expressed in females and four in males (Table 2). The homologue of the male-specific gene *MR8G027384.1* in *Arabidopsis* is *COL4* (*AT5G24930.1*), which can regulate flowering time [51]. The homolog of the zinc finger gene *MR3G018958.1*, also specifically expressed in the female (encoding a CONSTANS-like protein), in rice is *OsCOL16*, which represses flowering by up-regulating

Ghd7 expression [52]. We also identified the male-specific expression of certain genes. The homologs of the *MADS-box* gene *MR0G005926.1* in *Arabidopsis* is *SOC1*, which is a core gene in the floral induction pathway [53,54]. Those sex-specific expressed transcription factor genes may play a role in the induction of female and male primordia in the red bayberry. Clearly, this sex-biased expression of transcription factors may interact with other genes to regulate red bayberry sex differentiation, and we will verify their function in future research.

To narrow down the key genes regulating sex differentiation in the red bayberry, we constructed the regulation network of sex differentiation in the red bayberry using WGCNA (Figures 5 and 6). Interestingly, we found modules coded light green and dark cyan to be associated with sex type. Both hub genes in the two modules were ethylene-related genes, and the homolog of *MR0TCONS_00017483.1* in the light-green module network was *AtACO*, which is a key gene in ethylene biosynthesis. Moreover, *MR0TCONS_00017483.1* was expressed specifically in the male bud. In the dark-cyan module network, *MR1G019545.1* is homologous to *AtETR1*, which is involved in ethylene signaling pathways (Figure 6 and Table 2).

Sex differentiation genes have been found to be involved in ethylene biosynthesis and signaling pathways in Cucurbitaceae species. In the cucumber, sex differentiation is mainly controlled by three *ACS* (1-aminocyclopropane-1-carboxylate (ACC) synthase) genes, including the *F* (*CsACS1G*), *M* (*CsACS2*) and *A* (CsACS11) genes. Additionally, *ACO* is also essential for the development of female flowers [18]. A recent study has shown that the ethylene receptors *CpETR1A* and *CpETR2B*, highly homologous to *AtETR1* and *AtETR2*, cooperate in the control of sex determination in *Cucurbita pepo* [55]. In present study, the male-specific expressed gene *MR0TCONS_00017483.1* (*ACO*) and female-specific expressed gene *MR2G016358.1* were identified in light-green module. Therefore, we can speculate that ethylene may be the key hormone regulating red bayberry male and female flower differentiation. We found GA-, auxin-, ABA-, CTK-, SA- and Br-related genes in both modules and transcription factors. We also identified three genes involved in floral identity and development. The homolog of *MR8G010076.1* is *AtUGT87A2*, which can regulate flower time via *FLC* (*FLOWERING LOCUS C*) [56], while the *MR7G011514.1* homolog in *Arabidopsis*, *AtVRN1*, acts as a key gene in the vernalization pathway, mediated by the floral pathway integrator *FT* (*FLOWERING LOCUS T*) and targeted to *FLC* to regulate flower development [57]. The homolog of *MR8G001578.1* (*SPL*) in *Arabidopsis* is a negative regulator of inflorescence identity [58]. Those genes may play roles in red bayberry floral development.

Sex-determining genes are often located in the non-recombining region of sex chromosomes. In the red bayberry, our previous study showed seven putative genes in the 59 Kb female-specific region of the W chromosome [7]. We found that all the female-specific genes in F0 (before sexual differentiation) and F1 (flower initiation) displayed low expression with the FPKM value < 5 (Table 1). Hermaphrodite *35S-MrFT2*-transformed *Arabidopsis* plants exhibited a slight reduction in the numbers of stamens (data not shown). Based on the above results, we speculate that ethylene-related gene *MR0TCONS_00017483.1* (*ACO*) may be downstream of the sex-determining gene, and it may interact with the transcription factor *MR8G027384.1* (*COL4*) to control sex differentiation in the red bayberry. This hypothesis needs to be verified in future work and can be tested with external ethylene application in an attempt to alter sex. This could provide effective technical support for cross-breeding between varieties.

Supplementary Materials: The following supporting information can be downloaded at https://www.mdpi.com/article/10.3390/horticulturae8020183/s1, Figure S1. Principal component analysis (PCA) of red bayberry floral buds; Figure S2. Venn diagram of the 8889 DEGs in the pair-wise analyses of F1 vs. F0, M1 vs. M0, M0 vs. F0, and M1 vs. F1; Figure S3. KEGG pathway enrichment analysis of the DEGs in M0 vs. F0 (a), M1 vs. F1 (b) and all the 7029 DEGs in the pair-wise M0 vs. F0, and M1 vs. F1 comparisons (c); Figure S4. Distribution of the 415 transcription factors genes in 7029 DEGs; Figure S5. Validation of the expression patterns of ten DEGs via qRT-PCR. Capital letters above the black bars represent significant differences between the samples, and values are mean ± SE (n = 3); Table S1. Raw data of sequencing sample in the study; Table S2. Level 2 gene ontology annotation of highly expressed genes in four floral buds; Table S3. KEGG enrichment analysis of highly expressed genes in four floral buds; Table S4. Level 2 gene ontology annotation of middlingly expressed genes in four floral buds; Table S5. KEGG enrichment analysis of middlingly expressed genes in four floral buds; Table S6. Level 2 gene ontology annotation of the DEGs between M0 vs. F0, M1 vs. F1, and 7029 DEGs of the two compared groups in ontologies; Table S7. List of DEGs enriched in plant hormone synthesis and the signal transduction pathway; Table S8. List of hormone-related genes identified in the 7029 DEGs; Table S9. List of hormone-related genes specifically expressed in males and females; Table S10. List of transcription factors identified in the 7029 DEGs; Table S11. List of transcription factor genes specifically expressed in males and females; Table S12. The element enriched in promoter region of hormone-related genes; Table S13. List of genes in the light-green and dark-cyan networks; Table S14. Annotation of genes in light-green module; Table S15. Annotation of genes in dark-cyan module; Table S16. List of primers used for qRT-PCR.

Author Contributions: H.J. (Huimin Jia) and Z.G. designed this study and supervised the work. H.J. (Huimin Jia) performed the bioinformatic analyses and wrote the manuscript. L.Z., Y.W., H.W., H.Z. and H.J. (Huijuan Jia) performed the bioinformatic analyses. H.J. (Huijuan Jia), Y.Z., Y.J., G.W., C.Z. and C.H. collected the samples and generated the raw sequence data. All authors have read and agreed to the published version of the manuscript.

Funding: HM Jia was supported by the National Natural Science Foundation for Young Scientists of China (No. 31901985) and ZS Gao acknowledges support from the National Natural Science Foundation of China (No. 31972364).

Institutional Review Board Statement: Not applicable.

Informed Consent Statement: Not applicable.

Data Availability Statement: The sequencing data have been deposited in the NCBI Sequence Read Archive (http://www.ncbi.nlm.nih.gov/sra/, accessed on 15 February 2022), with the accession number SAMN18515266 and SAMN18515266 for female and male plants, respectively.

Conflicts of Interest: The authors declare no conflict of interest.

References

1. Renner, S.S. The relative and absolute frequencies of angiosperm sexual systems: Dioecy, monoecy, gynodioecy, and an updated online database. *Am. J. Bot.* **2014**, *101*, 1588–1596. [CrossRef]
2. Mitchell, C.H.; Diggle, P.K. The evolution of unisexual flowers: Morphological and functional convergence results from diverse developmental transitions. *Am. J. Bot.* **2005**, *92*, 1068–1076. [CrossRef] [PubMed]
3. Diggle, P.K.; Di Stilio, V.S.; Gschwend, A.R.; Golenberg, E.M.; Moore, R.C.; Russell, J.R.; Sinclair, J.P. Multiple developmental processes underlie sex differentiation in angiosperms. *Trends Genet. TIG* **2011**, *27*, 368–376. [CrossRef] [PubMed]
4. Chuck, G. Molecular mechanisms of sex determination in monoecious and dioecious plants. *Adv. Bot. Res.* **2010**, *54*, 53–83. [CrossRef]
5. Akagi, T.; Henry, I.M.; Tao, R.; Comai, L. A Y-chromosome-encoded small RNA acts as a sex determinant in persimmons. *Science* **2014**, *346*, 646–650. [CrossRef]
6. Wang, J.; Na, J.-K.; Yu, Q.; Gschwend, A.R.; Han, J.; Zeng, F.; Aryal, R.; VanBuren, R.; Murray, J.E.; Zhang, W.; et al. Sequencing papaya X and Y^h chromosomes reveals molecular basis of incipient sex chromosome evolution. *Proc. Natl. Acad. Sci. USA* **2012**, *109*, 13710–13715. [CrossRef]
7. Jia, H.M.; Jia, H.J.; Cai, Q.L.; Wang, Y.; Zhao, H.B.; Yang, W.F.; Wang, G.Y.; Li, Y.H.; Zhan, D.L.; Shen, Y.T.; et al. The red bayberry genome and genetic basis of sex determination. *Plant Biotechnol. J.* **2019**, *17*, 397–409. [CrossRef]
8. Tennessen, J.A.; Govindarajulu, R.; Liston, A.; Ashman, T.-L. Homomorphic ZW chromosomes in a wild strawberry show distinctive recombination heterogeneity but a small sex-determining region. *New Phytol.* **2016**, *211*, 1412–1423. [CrossRef]
9. Harkess, A.; Zhou, J.; Xu, C.; Bowers, J.E.; Van der Hulst, R.; Ayyampalayam, S.; Mercati, F.; Riccardi, P.; McKain, M.R.; Kakrana, A.; et al. The asparagus genome sheds light on the origin and evolution of a young Y chromosome. *Nat. Commun.* **2017**, *8*, 1279. [CrossRef]
10. Harkess, A.; Huang, K.; van der Hulst, R.; Tissen, B.; Caplan, J.L.; Koppula, A.; Batish, M.; Meyers, B.C.; Leebens-Mack, J. Sex Determination by Two Y-Linked Genes in Garden Asparagus. *Plant Cell* **2020**, *32*, 1790–1796. [CrossRef]
11. Akagi, T.; Henry, I.M.; Ohtani, H.; Morimoto, T.; Beppu, K.; Kataoka, I.; Tao, R. A Y-Encoded Suppressor of Feminization Arose via Lineage-Specific Duplication of a Cytokinin Response Regulator in Kiwifruit. *Plant Cell* **2018**, *30*, 780–795. [CrossRef] [PubMed]

12. Akagi, T.; Pilkington, S.M.; Varkonyi-Gasic, E.; Henry, I.M.; Sugano, S.S.; Sonoda, M.; Firl, A.; McNeilage, M.A.; Douglas, M.J.; Wang, T.; et al. Two Y-chromosome-encoded genes determine sex in kiwifruit. *Nat. Plants* **2019**, *5*, 801–809. [CrossRef]
13. Aryal, R.; Ming, R. Sex determination in flowering plants: Papaya as a model system. *Plant Sci.* **2014**, *217-218*, 56–62. [CrossRef] [PubMed]
14. Heikrujam, M.; Sharma, K.; Prasad, M.; Agrawal, V. Review on different mechanisms of sex determination and sex-linked molecular markers in dioecious crops: A current update. *Euphytica* **2015**, *201*, 161–194. [CrossRef]
15. Zhang, H.; Li, S.; Yang, L.; Cai, G.; Chen, H.; Gao, D.; Lin, T.; Cui, Q.; Wang, D.; Li, Z.; et al. Gain-of-function of the 1-aminocyclopropane-1-carboxylate synthase gene ACS1G induces female flower development in cucumber gynoecy. *Plant Cell* **2021**, *33*, 306–321. [CrossRef] [PubMed]
16. Li, Z.; Huang, S.; Liu, S.; Pan, J.; Zhang, Z.; Tao, Q.; Shi, Q.; Jia, Z.; Zhang, W.; Chen, H.; et al. Molecular Isolation of the M Gene Suggests That a Conserved-Residue Conversion Induces the Formation of Bisexual Flowers in Cucumber Plants. *Genetics* **2009**, *182*, 1381–1385. [CrossRef] [PubMed]
17. Boualem, A.; Troadec, C.; Camps, C.; Lemhemdi, A.; Morin, H.; Sari, M.-A.; Fraenkel-Zagouri, R.; Kovalski, I.; Dogimont, C.; Perl-Treves, R.; et al. A cucurbit androecy gene reveals how unisexual flowers develop and dioecy emerges. *Science* **2015**, *350*, 688–691. [CrossRef]
18. Chen, H.; Sun, J.; Li, S.; Cui, Q.; Zhang, H.; Xin, F.; Wang, H.; Lin, T.; Gao, D.; Wang, S.; et al. An ACC Oxidase Gene Essential for Cucumber Carpel Development. *Mol. Plant* **2016**, *9*, 1315–1327. [CrossRef]
19. Huguet, V.; Gouy, M.; Normand, P.; Zimpfer, J.F.; Fernandez, M.P. Molecular phylogeny of Myricaceae: A reexamination of host–symbiont specificity. *Mol. Phylogenet. Evol.* **2005**, *34*, 557–568. [CrossRef]
20. Chen, K.S.; Xu, C.J.; Zhang, B.; Ferguson, I.B. Red bayberry: Botany and horticulture. *Hortic. Rev.* **2004**, *30*, 83–114. [CrossRef]
21. Wilbur, R.L. The Myricaceae of the United States and Canada: Genera, subgenera, and series. *SIDA Contrib. Bot.* **1994**, *16*, 93–107.
22. Jia, H.M.; Jiao, Y.; Wang, G.Y.; Li, Y.H.; Jia, H.J.; Wu, H.X.; Chai, C.Y.; Dong, X.; Guo, Y.P.; Zhang, L.P.; et al. Genetic diversity of male and female Chinese bayberry (*Myrica rubra*) populations and identification of sex-associated markers. *BMC Genom.* **2015**, *16*, 394. [CrossRef] [PubMed]
23. Wang, Y.; Jia, H.-M.; Shen, Y.-T.; Zhao, H.-B.; Yang, Q.-S.; Zhu, C.-Q.; Sun, D.-L.; Wang, G.-Y.; Zhou, C.-C.; Jiao, Y.; et al. Construction of an anchoring SSR marker genetic linkage map and detection of a sex-linked region in two dioecious populations of red bayberry. *Hortic. Res.* **2020**, *7*, 53. [CrossRef] [PubMed]
24. Liu, J.; Chen, L.-Y.; Zhou, P.; Liao, Z.; Lin, H.; Yu, Q.; Ming, R. Sex biased expression of hormone related genes at early stage of sex differentiation in papaya flowers. *Hortic. Res.* **2021**, *8*, 147. [CrossRef] [PubMed]
25. Liu, Z.; Wang, H.; Xu, Z.; Zhang, H.; Li, G.; Wang, X.; Qian, W. Transcriptome profiling of differentially expressed genes of male and female inflorescences in spinach (*Spinacia olerace*a L.). *Genome* **2021**, *64*, 777–788. [CrossRef] [PubMed]
26. Li, N.; Meng, Z.; Tao, M.; Wang, Y.; Zhang, Y.; Li, S.; Gao, W.; Deng, C. Comparative transcriptome analysis of male and female flowers in *Spinacia oleracea* L. *BMC Genom.* **2020**, *21*, 850. [CrossRef]
27. Hui, W.; Yang, Y.; Wu, G.; Peng, C.; Chen, X.; Zayed, M.Z. Transcriptome profile analysis reveals the regulation mechanism of floral sex differentiation in *Jatropha curcas* L. *Sci. Rep.* **2017**, *7*, 16421. [CrossRef]
28. Zhang, Z.; Lin, Q.; Zhong, Q. Monoecious Mutant Reveals New Insights into Male and Female Inflorescence Development in the Chinese Bayberry (*Morella rubra*). *HortSci. Horts* **2017**, *52*, 343. [CrossRef]
29. Ao, C.Q. Developmental origins of the conjoined twin mature embryo sacs in Smilax davidiana, with special notes on the formation of their embryos and endosperms. *Am. J. Bot.* **2013**, *100*, 2509–2515. [CrossRef]
30. Chen, Y.; Chen, Y.; Shi, C.; Huang, Z.; Zhang, Y.; Li, S.; Li, Y.; Ye, J.; Yu, C.; Li, Z.; et al. SOAPnuke: A MapReduce acceleration-supported software for integrated quality control and preprocessing of high-throughput sequencing data. *GigaScience* **2017**, *7*, 1–6. [CrossRef]
31. Bolger, A.M.; Lohse, M.; Usadel, B. Trimmomatic: A flexible trimmer for Illumina sequence data. *Bioinformatics* **2014**, *30*, 2114–2120. [CrossRef] [PubMed]
32. Kim, D.; Langmead, B.; Salzberg, S.L. HISAT: A fast spliced aligner with low memory requirements. *Nat. Methods* **2015**, *12*, 357–360. [CrossRef] [PubMed]
33. Langmead, B.; Salzberg, S.L. Fast gapped-read alignment with Bowtie 2. *Nat. Methods* **2012**, *9*, 357–359. [CrossRef] [PubMed]
34. Li, B.; Dewey, C.N. RSEM: Accurate transcript quantification from RNA-Seq data with or without a reference genome. *BMC Bioinform.* **2011**, *12*, 323. [CrossRef]
35. Conesa, A.; Gotz, S.; Garcia-Gomez, J.M.; Terol, J.; Talon, M.; Robles, M. Blast2GO: A universal tool for annotation, visualization and analysis in functional genomics research. *Bioinformatics* **2005**, *21*, 3674–3676. [CrossRef]
36. Wang, L.; Feng, Z.; Wang, X.; Wang, X.; Zhang, X. DEGseq: An R package for identifying differentially expressed genes from RNA-seq data. *Bioinformatics* **2010**, *26*, 136–138. [CrossRef]
37. Chen, C.; Chen, H.; Zhang, Y.; Thomas, H.R.; Frank, M.H.; He, Y.; Xia, R. TBtools: An Integrative Toolkit Developed for Interactive Analyses of Big Biological Data. *Mol. Plant* **2020**, *13*, 1194–1202. [CrossRef]
38. Langfelder, P.; Horvath, S. WGCNA: An R package for weighted correlation network analysis. *BMC Bioinform.* **2008**, *9*, 559. [CrossRef]

39. Ni, J.; Bai, S.; Zhao, Y.; Qian, M.; Tao, R.; Yin, L.; Gao, L.; Teng, Y. Ethylene response factors Pp4ERF24 and Pp12ERF96 regulate blue light-induced anthocyanin biosynthesis in 'Red Zaosu' pear fruits by interacting with MYB114. *Plant Mol. Biol.* **2019**, *99*, 67–78. [CrossRef]
40. Livak, K.J.; Schmittgen, T.D. Analysis of relative gene expression data using real-time quantitative PCR and the 2(-Delta Delta C(T)) Method. *Methods* **2001**, *25*, 402–408. [CrossRef]
41. Khryanin, V.N. Role of Phytohormones in Sex Differentiation in Plants. *Russ. J. Plant Physiol.* **2002**, *49*, 545–551. [CrossRef]
42. Yamasaki, S.; Fujii, N.; Matsuura, S.; Mizusawa, H.; Takahashi, H. The M Locus and Ethylene-Controlled Sex Determination in Andromonoecious Cucumber Plants. *Plant Cell Physiol.* **2001**, *42*, 608–619. [CrossRef] [PubMed]
43. Tao, Q.; Niu, H.; Wang, Z.; Zhang, W.; Wang, H.; Wang, S.; Zhang, X.; Li, Z. Ethylene responsive factor ERF110 mediates ethylene-regulated transcription of a sex determination-related orthologous gene in two Cucumis species. *J. Exp. Bot.* **2018**, *69*, 2953–2965. [CrossRef] [PubMed]
44. Bensen, R.J.; Johal, G.S.; Crane, V.C.; Tossberg, J.T.; Schnable, P.S.; Meeley, R.B.; Briggs, S.P. Cloning and characterization of the maize An1 gene. *Plant Cell* **1995**, *7*, 75–84. [CrossRef] [PubMed]
45. Hartwig, T.; Chuck, G.S.; Fujioka, S.; Klempien, A.; Weizbauer, R.; Potluri, D.P.; Choe, S.; Johal, G.S.; Schulz, B. Brassinosteroid control of sex determination in maize. *Proc. Natl. Acad. Sci. USA* **2011**, *108*, 19814–19819. [CrossRef] [PubMed]
46. Achard, P.; Baghour, M.; Chapple, A.; Hedden, P.; Van Der Straeten, D.; Genschik, P.; Moritz, T.; Harberd, N.P. The plant stress hormone ethylene controls floral transition via DELLA-dependent regulation of floral meristem-identity genes. *Proc. Natl. Acad. Sci. USA* **2007**, *104*, 6484–6489. [CrossRef] [PubMed]
47. Martinez, C.; Pons, E.; Prats, G.; Leon, J. Salicylic acid regulates flowering time and links defence responses and reproductive development. *Plant J.* **2004**, *37*, 209–217. [CrossRef]
48. Jin, J.B.; Jin, Y.H.; Lee, J.; Miura, K.; Yoo, C.Y.; Kim, W.-Y.; Van Oosten, M.; Hyun, Y.; Somers, D.E.; Lee, I.; et al. The SUMO E3 ligase, AtS1Z1, regulates flowering by controlling a salicylic acid-mediated floral promotion pathway and through affects on FLC chromatin structure. *Plant J.* **2008**, *53*, 530–540. [CrossRef]
49. Van der Meer, I.M.; Stam, M.E.; van Tunen, A.J.; Mol, J.N.; Stuitje, A.R. Antisense inhibition of flavonoid biosynthesis in petunia anthers results in male sterility. *Plant Cell* **1992**, *4*, 253–262. [CrossRef]
50. Martin, A.; Troadec, C.; Boualem, A.; Rajab, M.; Fernandez, R.; Morin, H.; Pitrat, M.; Dogimont, C.; Bendahmane, A. A transposon-induced epigenetic change leads to sex determination in melon. *Nature* **2009**, *461*, 1135–1138. [CrossRef]
51. Steinbach, Y. The Arabidopsis thaliana CONSTANS-LIKE 4 (COL4)—A Modulator of Flowering Time. *Front. Plant Sci.* **2019**, *10*, 651. [CrossRef] [PubMed]
52. Wu, W.; Zheng, X.M.; Chen, D.; Zhang, Y.; Ma, W.; Zhang, H.; Sun, L.; Yang, Z.; Zhao, C.; Zhan, X.; et al. OsCOL16, encoding a CONSTANS-like protein, represses flowering by up-regulating Ghd7 expression in rice. *Plant Sci. Int. J. Exp. Plant Biol.* **2017**, *260*, 60–69. [CrossRef] [PubMed]
53. Lee, J.; Lee, I. Regulation and function of SOC1, a flowering pathway integrator. *J. Exp. Bot.* **2010**, *61*, 2247–2254. [CrossRef] [PubMed]
54. Moon, J.; Suh, S.S.; Lee, H.; Choi, K.R.; Hong, C.B.; Paek, N.C.; Kim, S.G.; Lee, I. The SOC1 MADS-box gene integrates vernalization and gibberellin signals for flowering in Arabidopsis. *Plant. J. Cell Mol. Biol.* **2003**, *35*, 613–623. [CrossRef] [PubMed]
55. García, A.; Aguado, E.; Martínez, C.; Loska, D.; Beltrán, S.; Valenzuela, J.L.; Garrido, D.; Jamilena, M. The ethylene receptors *CpETR1A* and *CpETR2B* cooperate in the control of sex determination in *Cucurbita pepo*. *J. Exp. Bot.* **2020**, *71*, 154–167. [CrossRef] [PubMed]
56. Wang, B.; Jin, S.-H.; Hu, H.-Q.; Sun, Y.-G.; Wang, Y.-W.; Han, P.; Hou, B.-K. UGT87A2, an Arabidopsis glycosyltransferase, regulates flowering time via FLOWERING LOCUS C. *New Phytol.* **2012**, *194*, 666–675. [CrossRef]
57. Levy, Y.Y.; Mesnage, S.; Mylne, J.S.; Gendall, A.R.; Dean, C. Multiple Roles of *Arabidopsis VRN1* in Vernalization and Flowering Time Control. *Science* **2002**, *297*, 243–246. [CrossRef]
58. Stone, J.M.; Liang, X.; Nekl, E.R.; Stiers, J.J. Arabidopsis AtSPL14, a plant-specific SBP-domain transcription factor, participates in plant development and sensitivity to fumonisin B1. *Plant J.* **2005**, *41*, 744–754. [CrossRef]

horticulturae

MDPI

Article

Identification of New Sources of Resistance to Anthracnose Caused by *Colletotrichum horii* among Persimmon Germplasms

Changfei Guan, Jie Hu, Yongkuan Li, Qinghui Che and Yong Yang *

State Key Laboratory of Crop Stress Biology for Arid Areas, College of Horticulture, Northwest A&F University, Yangling 712100, China; guanchangfei@nwafu.edu.cn (C.G.); 2019055166@nwafu.edu.cn (J.H.); liyongkuan@nwafu.edu.cn (Y.L.); cheqinghui@nwafu.edu.cn (Q.C.)
* Correspondence: yang_yong@nwsuaf.edu.cn

Abstract: Persimmon (*Diospyros kaki*) anthracnose, predominantly caused by *Colletotrichum horii*, is a destructive disease of persimmon. Thus, the evaluation of resistance resources is imperative for persimmon breeding and resistant variety deployment. In this study, the isolate from persimmon branches was identified as *C. horii* by using molecular and morphological characteristics. A total of 142 varieties were tested for anthracnose disease response by inoculation with the conidial suspension of *C. horii*. A significant variability was observed among the studied accessions. Only two accessions, which had a hypersensitive reaction with a slight lesion at the infection site at 14 days post inoculation, were highly resistant to anthracnose. A total of 7 and 24 accessions exhibited resistant and susceptible responses, respectively, to anthracnose. A total of 109 varieties with the highest proportion (76.76%) of accessions showed high susceptibility to *C. horii*. A total of 43 persimmon main cultivars were selected for the re-evaluation of the resistance level by counting the natural disease incidence of branches and fruits that showed similar resistant response. Moreover, nine of highly resistant or resistant accessions possessed desirable agronomic characters, including high fruit weight and strong growth potential. These resistant accessions could be used in the breeding of anthracnose-resistant persimmon varieties.

Keywords: persimmon anthracnose; germplasm resources; resistance; identification; *Colletotrichum horii*

Citation: Guan, C.; Hu, J.; Li, Y.; Che, Q.; Yang, Y. Identification of New Sources of Resistance to Anthracnose Caused by *Colletotrichum horii* among Persimmon Germplasms. *Horticulturae* **2022**, *8*, 180. https://doi.org/10.3390/horticulturae8020180

Academic Editor: Daniele Bassi

Received: 31 December 2021
Accepted: 18 February 2022
Published: 21 February 2022

1. Introduction

Persimmon (*Diospyros kaki*) is regarded to originate from China [1] and is predominantly cultivated in East Asia, including China, Japan, and Korea. China, which has the largest area and production harvest in the world, produces 3.427 million tons, accounting for 75% among all the countries (The Food and Agriculture Organization of the United Nations, FAO, 2021). The persimmon fruit is rich in vitamins, antioxidants, and trace elements that are vital for human health and has been used in various medicinal and chemical industries and for commercial fruit consumption [2,3]. Currently, the cultivation of the persimmon crop is rapidly expanding in Israel, Italy, and Spain, suggesting that persimmon is becoming one of the most popular fruits worldwide.

The collection and evaluation of persimmon germplasms support the safe preservation of genetic diversity and facilitate the selection and breeding of excellent resistant varieties of persimmon [4]. The National Field Genebank for Persimmon (NFGP) in China began to collect persimmon germplasms in 1962 and was first built in 1987 in Meixian County, Shannxi, China [5]. Currently, the NFGP saves more than 1000 persimmon resources from different regions in China and a relatively high number of varieties with a high genetic diversity from other countries [6,7].

Persimmon anthracnose is a severe fungal disease, causing branches and leaves to wither, fruit to rot, and possibly the loss of the entire plant, gravely jeopardizing the development of the persimmon industry [8,9]. Previously, the pathogen that causes anthracnose was reported as *Gloeosporium kaki* by Shotaro Hori and Seiya Ito [10,11]. Maffei [12] described the leaf spot pathogen of persimmon from a specimen collected in Italy as *Colletotrichum kaki*. Von [13] later reported that the fungi previously described are the same and synonymous with *C. gloeosporioides*. However, in 2010, after examining the molecular and morphological characteristics of specimens from China, Japan, and New Zealand, Weir and Johnston renamed the persimmon pathogen as *C. horii* [14]. Additionally, the persimmon anthracnose can be caused by *C. siamense, C. karstii, C. fructicola, C. nymphaeae,* and *C. melonis* [15–18]. Persimmon anthracnose is a dangerous disease in China and in other nations across the world [15–18]. Carraro reported that *C. fructicola, C. nymphaeae,* and *C. melonis* cause persimmon anthracnose in Brazil [16]. Moreover, *C. siamense* and *C. nymphaeae* cause persimmon anthracnose in Korea [15,18]. With increased rain, the anthracnose in persimmon orchards becomes serious because many main varieties are susceptible to anthracnose. In addition to chemical control and cultivation management, breeding anthracnose-resistant varieties based on resistance evaluation is the most effective method to control anthracnose. Thus, screening resistance resources is imperative for breeding resistant varieties.

Abundant *Diospyros* germplasm resources are in the NFGP of China, and the evaluation and identification of persimmon germplasm resistance to anthracnose can provide new sources for resistance breeding programs. In the present study, we aim to (a) isolate and identify the pathogen of persimmon anthracnose, (b) evaluate the resistance level of 142 persimmon varieties in the NFGP, and (c) identify some anthracnose-resistant varieties for the future application of persimmon breeding.

2. Materials and Methods

2.1. Sample Collection and Pathogen Fungus Isolation

A total of 142 persimmon germplasm resources without infections and insect pests were preserved in the NFGP, Yangling, Shaanxi, China (34°16′56.24″ N, 108°04′27.95″ E; 420 m altitude).

For candidate pathogen isolation, symptomatic twigs of 'Fuping Jianshi' persimmon were randomly collected from NFGP at Yangling, China (34°17′52.55″ N, 108°04′05.58″ E) and brought to the laboratory for the isolation of putative causal agents. Samples were surface disinfected for 90 s with 70% ethanol and 1% sodium hypochlorite, rinsed twice with sterile distilled water. Sterilized small branch pieces (5 mm × 5 mm) were placed on PDA plates and then incubated at 25 °C with a photoperiod of 12 h until fungal growth was observed. Pathogen isolates were purified with single-spore culturing prior to use in subsequent experiments. The isolated pathogen fungus was named 'FJ 3'.

For the microscopic morphology observation of the incidence of tissue samples after infection, the collected branches and fruit samples with different incidence degrees in different diseased parts were cut into small pieces of tissue (5–8 mm) at the junction of disease and health. After cutting, samples were quickly put into FAA fixative. Paraffin sections were prepared in accordance with the reported method [19].

2.2. Molecular Characterization of the Fungus

DNA was extracted from the isolated pathogen, and the genomic DNA sequences of the ITS region of rDNA, beta-tubulin (*TUB2*) gene, glyceraldehyde-3-phosphate dehydrogenase (*GAPDH*), chitin synthase (*CHS-1*), and a partial sequence of the actin (*ACT*) were amplified with primers of ITS1F/ITS4, T1/Bt2b, GDF1/GDRl, ACT-512F/ACT-783R, and CHS-79F/CHS-345R, respectively [20] (Table S1). The amplified reaction system contained a total volume of 25 µL, including 2.5 µL of 2 × Reaction Mix, 0.2 µL Golden DNA Polymerase, 1 µL template DNA, 1 µL of 10 µmol/L primers, and 9.3 µL ddH_2O. PCR conditions were as follows: pre-denaturation at 95 °C for 3 min, denaturation at 95 °C for 30 s, annealing at 56 °C for 30 s, and extension at 72 °C for 1 min for 42 cycles. The last extension was at 72 °C for 5 min. PCR products were detected by 1% agarose gel electrophoresis. Subsequently, products were cloned into the T-Vector pMD19 (TaKaRa, Dalian, China), and three positive clones for each fragment were sequenced (Aoke Technology Co., Ltd, Beijing, China). A phylogenetic tree was obtained using the MEGA5 software [21].

2.3. Artificial Inoculation and Natural Disease Incidence

After 18 days of culturing for purified pathogen isolation, conidia suspensions were prepared by flooding the culture plates and removing the fungal mycelium. The number of conidia was determined using a hemocytometer, and the conidia suspension was adjusted to a 1×10^6/mL concentration as a backup.

For artificial inoculation, an inoculum concentration of 1.0×10^6 conidia/mL for the 'FJ 3' fungus was used to infect the uninjured young branches ($n = 3$) by 5 μL suspensions with a short interval of 1.5 cm (Figure 1). The growth of diseased spots was recorded using a Vernier caliper for 15 consecutive days. The experiment was conducted twice in 2019 and 2020.

Figure 1. Anthracnose symptoms and plant tissue sectioning caused by *C. horri* on persimmon. Disease lesions on (**A**) fruit, (**B**) branch, and (**C**) petiole. The red ellipse indicates the infection site of anthracnose, "a" indicates the location of the disease, and "×" means magnification.

For natural disease incidence in field, diseased fruits (≥20) and branches (≥20) with uniform growth potential were recorded in four different directions of the persimmon tree. The surface or sepals of the fruit exhibiting typical black spots were counted, and the incidence was presented as percent anthracnose infection. Branches showing typical anthracnose lesions were evaluated as infected.

2.4. Disease Resistance Criterion

For the artificial inoculation, the standard of disease classification is shown in Table 1. Disease index (DI) = Σ (number of branches infected in the rank × corresponding disease rank) × 100 / total number of infected branches × highest rank. On the basis of the range of DI, persimmon accessions were classified into highly resistant (0 < DI ≤ 10.0), resistant (10.0 < DI ≤ 30.0), susceptible (30.0 < DI ≤ 50.0), and highly susceptible (50.0 < DI ≤ 100.0, Table 2). If the disease severity varied between two years, a high disease grade of the year should be used as a rating scale of the accession resistance.

Table 1. Branch lesion standard for persimmon anthracnose.

Rank	Description
0	No lesion on branches
1	0% < Ratio of total lesion diameter to total branch length ≤ 20%
2	20% < Ratio of total lesion diameter to total branch length ≤ 40%
3	40% < Ratio of total lesion diameter to total branch length ≤ 60%
4	60% < Ratio of total lesion diameter to total branch length ≤ 80%
5	80% < Ratio of total lesion diameter to total branch length ≤ 100%

Table 2. Levels and types of resistance to persimmon anthracnose.

Resistance Levels	Disease Index (DI)	Resistance
1	0 < DI ≤ 10.0	Highly Resistant (HR)
3	10.0 < DI ≤ 30.0	Resistant (R)
5	30.0 < DI ≤ 50.0	Susceptible (S)
7	50.0 < DI ≤ 100.0	Highly Susceptible (HS)

Under natural field conditions, the severity of fruits and branches was categorized using a 1–4 scale, where highly resistant ≤ 2%, 2% ≤ resistant < 5%, 5% ≤ susceptible < 20%, and highly susceptible ≥ 20%. Fruit and branch severity values were presented as percent infected samples. At least 20 samples were measured for each persimmon variety.

3. Results

3.1. Symptoms of Persimmon Anthracnose Disease

The anthracnose fungus may attack the fruits, twigs, and leaf petioles in susceptible persimmon. When the fruit is infected with anthracnose, black spots appear on the surface of the fruit first and diseased spots slowly expand and dent downwards, finally causing the fruit to soften (Figure 1A). The lesion location had a modest number of stone cells, the peel was softened, and the infection gradually spread throughout the flesh.

Anthracnose symptoms on young twigs first appeared in dark spots, and minute spots developed into dark brown specks, showing a clear dividing line between diseased and symptomless tissues. The anthracnose fungus still attacked the xylem of branches and led to collapse. With the spread of disease spots, adjacent lesions connected until the entire twig was infected (Figure 1B).

The fungus infected petioles and caused similar dark brown spots (Figure 1C). With the development of the disease, leaves gradually fell off.

3.2. Morphological and Molecular Characterization of *C. horri*

For fungal isolation, symptomatic twigs of persimmon were randomly collected and brought to the laboratory for the isolation of putative causal agents. Samples were surface disinfected for 90 s with 70% ethanol and 1% sodium hypochlorite, then rinsed twice with sterile distilled water. Sterilized small branch pieces (5 mm × 5 mm) were placed on PDA plates. Representative purified isolates of the 'FJ 3' fungus colonies resembling *Colletotrichum* spp. were selected for further phylogenetic and morphological analyses. The 'FJ 3' fungal strain grew rapidly on PDA, thereby filling the entire Petri dish (90 mm) after

11 days at 25 °C and 12 h photoperiod. Colonies of 'FJ 3' were initially white and gradually became grayish on PDA with dark concentric zonation and regular margins in reverse (Figure 2). Few conidia (no conidial mass) were observed across the colony after 15 days of incubation at 25 °C on PDA. Conidia were hyaline, single-celled, straight, cylindrical apices and measured up to 10.5 to 17.1 μm × 4.0 to 5.56 μm (n = 100, average = 13.6 μm × 4.6 μm, Figure 2). The 'FJ 3' isolate showed similar morphological traits with the previously reported *C. horri* [22].

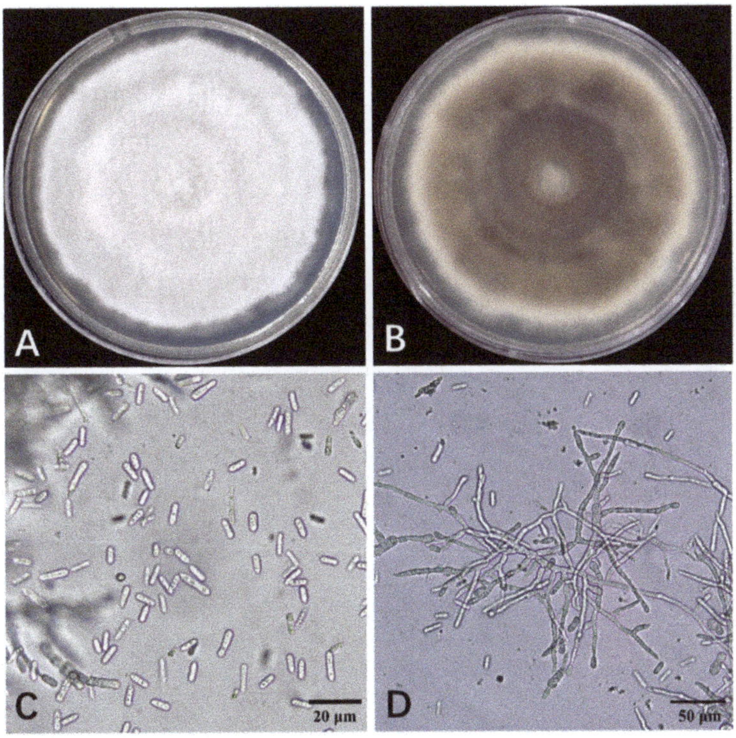

Figure 2. Cultural and morphological characteristics of the *C. horii* 'FJ 3' isolation growing on PDA: (**A**) the upper surface, (**B**) the lower surface, (**C**) conidia (bar = 20 μm), and (**D**) mycelial growth (bar = 50 μm) of *C. horii*.

Five isolates obtained from the 'Fuping Jianshi' persimmon were identified as the same fungal species on the basis of multigene phylogenetic and morphological analyses (Figure 3). The persimmon anthracnose 'FJ 3' fungus was selected for further phylogenetic analysis. Then, genomic DNA sequences were amplified with ITS1F/ITS4, T1/Bt2b, GDF1/GDRl, ACT-512F/ACT-783R, and CHS-79F/CHS-345R (Weir et al., 2012). These amplicons were submitted to GenBank with accession numbers of OL347726 for ITS, OL364188 for GAPDH, OL364190 for CHS-1, OL364191 for TUB2, and OL364189 for ACT. The BLAST search in GenBank revealed that sequences showed high similarity to those of *C. horii*. Amplification sequences from the 'FJ 3' fungus were identical to JX010450 for CHS-1, 98.34% to GQ329690 for ITS, 98.34% to JX010450 for GAPDH, 99.29% to JX009438 for ACT, and 99.74% to JX010450 for TUB2. The maximum likelihood tree was obtained from a concatenated dataset of *ITS, ACT, CHS, GAPDH,* and *TUB2* gene sequences of the *C. gloeosporioides* species complex with 1000 bootstrapping, showing that the present 'FJ 3' isolate clustered with the reference isolate *C. horii* (NBRC 7478) with high bootstrap support. Thus, the current 'FJ 3' isolate was identified as *C. horii* on the basis of cultural characteristics and sequence similarity data (Figure 3).

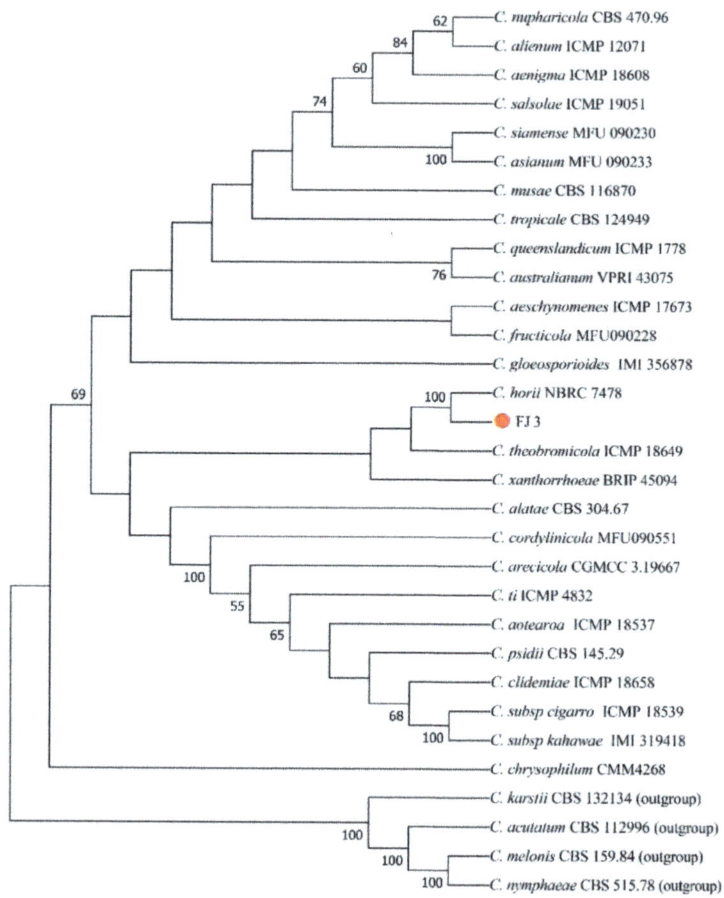

Figure 3. Phylogenetic analysis using the neighbor-joining method through comparative studies of nucleotide sequences of the internal transcribed spacer (ITS) region of rDNA, glyceraldehyde-3-phosphate dehydrogenase (*GAPDH*), a partial sequence of the actin (*ACT*), chitin synthase (*CHS-1*), and beta-tubulin (*TUB2*) genes from the present isolates with those of other *Colletotrichum* spp. in the *C. gloeosporioides* species complex retrieved from GenBank. The 'FJ 3' isolate is emphasized in the red circle. Numbers above the branches represent bootstrap values.

3.3. Evaluation of Anthracnose Resistance among Persimmon Germplasms by Artificial Infection

Persimmon accessions were grown in NFGP at Yangling, China (34°17′52.55″ N, 108°04′05.58″ E). In 2019 and 2020, this area received total rainfall values of 656 and 683 mm, respectively, and average temperatures of 13.0 °C and 13.2 °C, respectively. Rainfall varied between months of the same year, whereas the temperature of the area fluctuated little, especially between different years. The rainfall from July to September accounts for nearly half of the annual rainfall, and this period represents the hottest three months (highest average temperature (26.1–26.3 °C) is observed in July). The temperature and relative humidity from July to September are conducive to the occurrence of the persimmon anthracnose disease.

The resistance level was defined in four categories, including highly resistant, resistant, susceptible, and highly susceptible, on the basis of the DI of inoculated branches in 2019 and 2020 (Figures 4 and 5). A significant variability was observed among the studied accessions. Only two accessions, which had a hypersensitive reaction with a slight lesion at the injection site at 15 days postinoculation, were highly resistant to anthracnose. Seven cultivars exhibited a resistant response to anthracnose and accounted for 4.93% of the 142 accessions (Figure 4). The susceptible response was observed in 24 accessions, which accounted for 16.90% of the total. A total of 109 varieties with the highest proportion (76.76%) of the accessions showed a highly susceptible response to *C. horii* 'FJ 3' (Figure 4 and Table 3).

Figure 4. Representative varieties of 'Qianxian Muwashi' and 'Hebei Lianhuashi' exhibiting high susceptibility and susceptibility, respectively, to *C. horii* after artificial infection in (**A**) 2019 and (**B**) 2020.

Figure 5. Representative varieties of 'Chengsushi' and 'Yuanqu Bayuehong' exhibiting resistant and highly resistant responses, respectively, to *C. horii* after artificial infection in (**A**) 2019 and (**B**) 2020.

3.4. Evaluation of the Anthracnose Resistance for Selected Persimmon Accessions by Natural Disease Incidence

A total of 43 persimmon main cultivars were selected for the re-evaluation of anthracnose resistance in the nature field in 2018 and 2019 (Table 4). Four accessions were highly resistant and had no symptomatic anthracnose in branches and fruits, and nine accessions were resistant with average percent disease incidence ≤ 5. In 34 remaining accessions, 5 and 29 were susceptible and highly susceptible, respectively, with percent disease incidence values of 5–20% and 20–100%, respectively. The data of repeated evaluation experiments showed that persimmons performed consistent resistance by artificial infection and natural disease incidence, except for two accessions, i.e., 'Zaozhuang Ehuangshi' and 'Ningbo Tongpenshi'.

Table 3. Anthracnose resistance level based on the average lesions of persimmon germplasm inoculated with *C. horri* isolate.

	Group	2019				2020				
Number	Variety	Average Lesions (mm)	Disease Index	Resistance Level	Resistance	Average Lesions (mm)	Disease Index	Resistance Level	Resistance	Identification Result
1	Xiangfen Qiyuehong	14.32 ± 0.6	100	7	HS	14.36 ± 0.23	100	7	HS	HS
2	Jishan Hanbanjin	13.25 ± 0.98	96	7	HS	11.94 ± 1.19	88	7	HS	HS
3	Jishan Banshi	8.54 ± 1.69	68	7	HS	4.67 ± 0.4	40	5	S	HS
4	Yongji Qingshi	11.67 ± 0.77	80	7	HS	6.52 ± 1.38	52	7	HS	HS
5	Baishi	11.97 ± 1.99	84	7	HS	8.42 ± 1.05	68	7	HS	HS
6	Meixian Qinghuamanaitou	11.85 ± 3.14	88	7	HS	13.99 ± 0.53	100	7	HS	HS
7	Qingxuanshi	14.88 ± 0.24	100	7	HS	7.87 ± 1.91	56	7	HS	HS
8	Yaoxian Wuhuashi	6.84 ± 0.66	56	7	HS	6.71 ± 1.79	56	7	HS	HS
9	Wenxi Pingdingshi	14.03 ± 1.52	96	7	HS	14.65 ± 0.71	100	7	HS	HS
10	Lintong Jiandingshi	13.47 ± 1.62	92	7	HS	11.46 ± 1.14	84	7	HS	HS
11	Lintong Fangshi	5.99 ± 1.4	44	5	S	12.55 ± 0.69	96	7	HS	HS
12	Lintong Huojing	8.89 ± 2.63	72	7	HS	9.5 ± 2.58	72	7	HS	HS
13	Chengou Huishi	6.66 ± 4.94	56	7	HS	5.9 ± 4.54	48	5	S	HS
14	Weiboshi	6.45 ± 1.98	56	7	HS	5.52 ± 3.12	44	5	S	HS
15	Boai Bayuehuang	11.18 ± 1.08	84	7	HS	12.01 ± 2.17	88	7	HS	HS
16	Shiyangshi	14.45 ± 0.46	100	7	HS	10.28 ± 3.01	84	7	HS	HS
17	Wanghoushi	12.63 ± 2.37	92	7	HS	13.64 ± 1	100	7	HS	HS
18	Meipishi	10.33 ± 2.46	76	7	HS	8.54 ± 1.69	68	7	HS	HS
19	Ernizi	11.86 ± 0.34	80	7	HS	12.92 ± 1.63	92	7	HS	HS
20	Hongxuanshi	12.39 ± 2.12	88	7	HS	13.09 ± 2.23	92	7	HS	HS
21	Meixian Niuxinshi	13.71 ± 0.88	100	7	HS	12.09 ± 1.67	96	7	HS	HS
22	Yichuanling	11.37 ± 1.94	80	7	HS	10.19 ± 1.45	76	7	HS	HS
23	Xunyang Huoshi	7.37 ± 1.13	56	7	HS	7.28 ± 1.08	56	7	HS	HS
24	Xunyang Guanguanshi	10.24 ± 3.31	80	7	HS	8.53 ± 2.46	64	7	HS	HS
25	Nanzhang Jianshi	14.19 ± 0.24	100	7	HS	11.09 ± 2.89	84	7	HS	HS
26	Luoyang Guilianqing	14.87 ± 0.19	100	7	HS	6.77 ± 3.78	52	7	HS	HS
27	Liuba Huoshi	9.63 ± 2.65	76	7	HS	8.19 ± 3.07	60	7	HS	HS
28	Binxian Shuishi	7.13 ± 3.27	52	7	HS	12.97 ± 1	96	7	HS	HS
29	Huaitaishi	6.99 ± 0.69	60	7	HS	8.02 ± 2.11	64	7	HS	HS
30	Heixinshi	14.35 ± 0.37	100	7	HS	13.94 ± 0.75	100	7	HS	HS
31	Denglongshi	10.85 ± 3.02	80	7	HS	3.9 ± 4.37	32	5	S	HS
32	Yidu Tuoshi	11.44 ± 2.7	84	7	HS	7.26 ± 3.37	56	7	HS	HS

Table 3. *Cont.*

Number	Group	2019				2020				
	Variety	Average Lesions (mm)	Disease Index	Resistance Level	Resistance	Average Lesions (mm)	Disease Index	Resistance Level	Resistance	Identification Result
33	Zhouqu Huoshi	7.95 ± 1.89	64	7	HS	4.69 ± 1.81	40	5	S	HS
34	Changan Fudingjian	13.95 ± 0.8	100	7	HS	14.32 ± 0.82	100	7	HS	HS
35	Shuhuangshi	7.59 ± 2.22	64	7	HS	10.97 ± 0.58	80	7	HS	HS
36	Jincheng Gaishi	9.6 ± 3.42	72	7	HS	9.47 ± 3.52	72	7	HS	HS
37	Tongguan Lianhuashi	10.15 ± 3.06	76	7	HS	11.1 ± 1.56	84	7	HS	HS
38	Licheng Mianshi	12.36 ± 2.28	72	7	HS	14.43 ± 0.18	100	7	HS	HS
39	Matian FangSs	11.95 ± 0.95	92	7	HS	14.07 ± 0.61	100	7	HS	HS
40	Heishi	10.49 ± 0.62	80	7	HS	11.23 ± 3.34	88	7	HS	HS
41	Xiaoyi Niuxinshi	13.16 ± 2.22	96	7	HS	6.01 ± 0.76	48	5	S	HS
42	Xiaoerzao	13.53 ± 1.15	96	7	HS	11.89 ± 1	88	7	HS	HS
43	Cangshan Niuxinshi	6.38 ± 2.97	48	5	S	7.51 ± 1.27	64	7	HS	HS
44	Mianrangshi	14.81 ± 0.1	100	7	HS	14.09 ± 1.09	100	7	HS	HS
45	Xingyang Shuishi	9.98 ± 2.11	76	7	HS	12.05 ± 3.56	88	7	HS	HS
46	Suxian Niuxinshi	9.13 ± 1.41	72	7	HS	11.77 ± 2.76	88	7	HS	HS
47	Qujing Shuishi	11.42 ± 1.27	88	7	HS	6.9 ± 1.38	60	7	HS	HS
48	Qujing Xiaoshuishi	13.1 ± 2.35	92	7	HS	12.55 ± 2.04	92	7	HS	HS
49	Biantashi	12.91 ± 1.52	92	7	HS	12.01 ± 2.84	88	7	HS	HS
50	Fuzhengbing	13.49 ± 1.45	96	7	HS	12.97 ± 1.88	96	7	HS	HS
51	Qiuzhengbing	11.04 ± 1.42	84	7	HS	10.04 ± 2.08	80	7	HS	HS
52	Chikelang	11.62 ± 1.12	88	7	HS	9.12 ± 1.59	68	7	HS	HS
53	Hongmiandan	12.91 ± 0.87	100	7	HS	15 ± 0	100	7	HS	HS
54	Huxian Huangmiandan	15 ± 0	100	7	HS	14.87 ± 0.27	100	7	HS	HS
55	Xichou Huoshi	15 ± 0	100	7	HS	15 ± 0	100	7	HS	HS
56	Jiro	8.27 ± 1.52	64	7	HS	9.93 ± 0.22	80	7	HS	HS
57	Luoyang Shuiniuxin	14.92 ± 0.14	100	7	HS	15 ± 0	100	7	HS	HS
58	Lantian Shuishi	13.17 ± 0.88	96	7	HS	8.82 ± 5.08	68	7	HS	HS
59	Nanjing Gaozhuangshi	7.49 ± 1.87	60	7	HS	3.69 ± 3.65	32	5	S	HS
60	Lantian Dafangshi	10.57 ± 1.28	84	7	HS	9.7 ± 2.44	72	7	HS	HS
61	Binxian Jiandingshi	14.28 ± 0.68	100	7	HS	14.78 ± 0.47	100	7	HS	HS
62	Huxian Dashi	13.71 ± 1.3	96	7	HS	13.97 ± 0.6	100	7	HS	HS
63	Sifangshi	14.76 ± 0.13	100	7	HS	14.35 ± 0.35	100	7	HS	HS

Table 3. *Cont.*

Number	Group	2019				2020				
	Variety	Average Lesions (mm)	Disease Index	Resistance Level	Resistance	Average Lesions (mm)	Disease Index	Resistance Level	Resistance	Identification Result
64	Niutoushi	12.42 ± 0.8	92	7	HS	8.73 ± 0.33	64	7	HS	HS
65	Guyangshi	8.84 ± 0.72	68	7	HS	5.76 ± 0.4	44	5	S	HS
66	Yangshuo Niuxinshi	8.14 ± 1.54	64	7	HS	3.7 ± 2.52	36	5	S	HS
67	Shieryueshi	12.15 ± 1.34	92	7	HS	7.88 ± 2.7	64	7	HS	HS
68	Baoshan Dashuishi	7.79 ± 1.25	64	7	HS	4.64 ± 4.2	36	5	S	HS
69	Dafeng Niuxinshi	11 ± 1.28	88	7	HS	8.59 ± 1.61	68	7	HS	HS
70	Longhui Ruanzao	10.24 ± 2.8	76	7	HS	10.61 ± 3.81	80	7	HS	HS
71	Liuheshi	13.63 ± 0.97	100	7	HS	11.18 ± 1.74	84	7	HS	HS
72	Jurong Bianshi	14.41 ± 0.57	100	7	HS	13.13 ± 1.18	96	7	HS	HS
73	Suqian Bianshi	6.63 ± 1.03	52	7	HS	6.02 ± 1.23	52	7	HS	HS
74	Pixian Biangangshi	11.77 ± 3.41	84	7	HS	11.08 ± 2.23	88	7	HS	HS
75	Silenggaishi	11.55 ± 2.3	88	7	HS	10.15 ± 0.98	76	7	HS	HS
76	Huixian Dashi	11.87 ± 1.7	88	7	HS	13.06 ± 5.89	64	7	HS	HS
77	Huixian Shuishi	11.94 ± 1.58	92	7	HS	5.6 ± 2.83	48	5	S	HS
78	Pijianglou	8.35 ± 3.61	64	7	HS	6.9 ± 1.94	56	7	HS	HS
79	Ichidagaki	5.91 ± 0.67	52	7	HS	5.24 ± 1.98	44	5	S	HS
80	Shougatsu	7.19 ± 1.76	60	7	HS	4.16 ± 0.95	36	5	S	HS
81	Sakugosho	12.02 ± 1.76	88	7	HS	13.09 ± 1.23	96	7	HS	HS
82	Changan Shuishi	10.89 ± 3.28	80	7	HS	12.24 ± 2.15	92	7	HS	HS
83	Changan Shaoshi	14.49 ± 0.28	100	7	HS	14.19 ± 0.65	100	7	HS	HS
84	Huxian Tieguoshi	6.94 ± 2.76	56	7	HS	5.45 ± 0.74	44	5	S	HS
85	XiaoShuishi	14.29 ± 0.87	100	7	HS	11.66 ± 1.69	92	7	HS	HS
86	Xingyang Huoguan	14.39 ± 0.65	100	7	HS	13.76 ± 1.09	100	7	HS	HS
87	Qingchutou	12.02 ± 2.03	88	7	HS	8.84 ± 4.81	76	7	HS	HS
88	Yanshitiansheng-01	12.27 ± 1.22	92	7	HS	12.64 ± 1.72	92	7	HS	HS
89	Shahe Huoshi	14.61 ± 0.34	100	7	HS	13.85 ± 0.69	100	7	HS	HS
90	Jingjing Gaishi	9.01 ± 2.77	68	7	HS	10.63 ± 2.43	80	7	HS	HS
91	Pingshan Niujinshi	13.65 ± 1.52	96	7	HS	13.3 ± 2.19	96	7	HS	HS
92	Huaxian Baixuanshi	14.09 ± 1.13	100	7	HS	13.65 ± 0.98	100	7	HS	HS
93	Heibei Lianhuashi1	12.99 ± 2.35	92	7	HS	12.29 ± 2.92	92	7	HS	HS
94	Shiyueshi	14.54 ± 0.27	100	7	HS	13.73 ± 1.06	100	7	HS	HS
95	Lintong Diaoshi	8.9 ± 4.66	64	7	HS	6.34 ± 1.46	52	7	HS	HS

Table 3. *Cont.*

Number	Group	2019				2020				
	Variety	Average Lesions (mm)	Disease Index	Resistance Level	Resistance	Average Lesions (mm)	Disease Index	Resistance Level	Resistance	Identification Result
96	Laoshigou	11 ± 1.28	88	7	HS	8.59 ± 1.61	68	7	HS	HS
97	Baokangshi	10.24 ± 2.8	76	7	HS	10.61 ± 3.81	80	7	HS	HS
98	Daezi	13.63 ± 0.97	100	7	HS	11.18 ± 1.74	84	7	HS	HS
99	Changsha Shuishi	15 ± 0	100	7	HS	8.96 ± 4.79	71	7	HS	HS
100	Fuping Jianshi	15 ± 0	100	7	HS	15 ± 0	100	7	HS	HS
101	Chaoyang Yuanxiaoshi	8.46 ± 1.34	66	7	HS	9.61 ± 0.92	80	7	HS	HS
102	Conghuashi	14.31 ± 0.52	100	7	HS	15 ± 0	100	7	HS	HS
103	Dabaoshi	13.09 ± 0.76	92	7	HS	8.79 ± 5.18	68	7	HS	HS
104	Dabiegaishi	7.86 ± 1.78	60	7	HS	3.71 ± 3.53	32	5	S	HS
105	Dali Daqiyuehuang	10.66 ± 1.32	85	7	HS	9.65 ± 2.32	73	7	HS	HS
106	Dali Qiyuehuang	14.51 ± 0.76	100	7	HS	14.85 ± 0.55	100	7	HS	HS
107	Damianhu	13.73 ± 1.36	97	7	HS	13.97 ± 0.6	100	7	HS	HS
108	Dangshan Ehuangshi	14.76 ± 0.13	100	7	HS	14.35 ± 0.45	100	7	HS	HS
109	Dangshan Niuxinshi	13.12 ± 1.6	91	7	HS	8.52 ± 0.95	73	7	HS	HS
110	Yongji Mushi	3.57 ± 1.81	32	5	S	3.46 ± 1.35	28	3	R	S
111	Hangzhou Niuxinshi	5.52 ± 0.65	44	5	S	4.79 ± 0.53	40	5	S	S
112	Chetoushi	4.03 ± 0.99	36	5	S	3.69 ± 1.04	36	5	S	S
113	Shagu2	3.89 ± 0.54	32	5	S	3.51 ± 1.03	28	3	R	S
114	Fuyang Gongshi	4.37 ± 0.93	40	5	S	4.59 ± 0.46	40	5	S	S
115	Qianxian Muwashi	5.94 ± 2.49	48	5	S	4.86 ± 1.66	40	5	S	S
116	Putian Goushi	2.37 ± 0.32	20	3	R	5.09 ± 0.55	40	5	S	S
117	Xiaobaxianshi	5.36 ± 0.39	40	5	S	5.81 ± 2.21	44	5	S	S
118	Guangzhou Niuxinshi	5.41 ± 0.78	44	5	S	3.22 ± 0.82	28	3	R	S
119	Zhengyang Baheshi	5.53 ± 0.93	48	5	S	4.98 ± 0.52	40	5	S	S
120	Gongcheng Shuishi	2.76 ± 1.16	28	3	R	3.5 ± 0.78	36	5	S	S
121	Sibanmianshi	3.21 ± 1.59	28	3	R	5.89 ± 3.39	48	5	S	S
122	Zhoushan Changshi	3.79 ± 0.34	40	5	S	4.1 ± 1.88	40	5	S	S
123	Tangcunshi	2.44 ± 0.47	24	3	R	3.28 ± 0.6	36	5	S	S
124	Caojishi	4.38 ± 1.83	40	5	S	4.32 ± 2.51	32	5	S	S
125	Songyang Bianshi	4.85 ± 0.42	40	5	S	4.18 ± 1.01	36	5	S	S
126	Mancheng Niuxinshi	6.1 ± 0.92	48	5	S	3.34 ± 1.42	32	5	S	S
127	Taian Jingmianshi	3.8 ± 3.45	36	5	S	1.53 ± 3.32	16	3	R	S

Table 3. *Cont.*

Group		2019				2020				
Number	Variety	Average Lesions (mm)	Disease Index	Resistance Level	Resistance	Average Lesions (mm)	Disease Index	Resistance Level	Resistance	Identification Result
128	Baoshan Hongshi	3.76 ± 1.28	38	5	S	3.53 ± 0.98	37	5	S	S
129	Baoshan Shuishi	3.67 ± 1.71	31	5	S	5.92 ± 3.49	49	5	S	S
130	Ganmaokui	3.97 ± 0.56	42	5	S	4.09 ± 1.91	40	5	S	S
131	Changan Fudingshi	2.49 ± 0.66	23	3	R	3.38 ± 0.66	37	5	S	S
132	Huxian Jiandingshi	4.54 ± 1.68	39	5	S	4.36 ± 2.29	32	5	S	S
133	Youhou	3.11 ± 1.06	29	3	R	4.24 ± 1.01	37	5	S	S
134	Ningbo Tongpenshi	1.05 ± 0.33	12	3	R	1.31 ± 1.22	12	3	R	R
135	Zhaoan Yuanxiaoshi	1.87 ± 0.32	16	3	R	2.34 ± 0.33	24	3	R	R
136	Chengsushi	2.74 ± 1.09	24	3	R	2 ± 0.36	20	3	R	R
137	Baoshan Dafangshi	1.6 ± 3.89	16	3	R	2.41 ± 3.31	20	3	R	R
138	Hyakume	1.41 ± 0.51	20	3	R	2.33 ± 1.33	24	3	R	R
139	Yanjing Yingshi	1.36 ± 1.74	16	3	R	3.31 ± 9.5	24	3	R	R
140	Zaozhuang Ehuangshi	1.87 ± 0.71	16	3	MR	0 ± 0	0	1	HR	R
141	Kangbing Jianshi	0 ± 0	0	1	HR	0 ± 0	0	1	HR	HR
142	Yuanqu Bayuehong	0.92 ± 0.25	8	3	HR	0 ± 0	0	1	HR	HR

Table 4. Anthracnose resistance level for 43 selected persimmon germplasms based on the severity of fruits and branches in the nature field.

Group		2018				2019				
Number	Variety	Infected Fruits (%)	Infected Braches (%)	Average (%)	Resistance	Infected Fruits (%)	Infected Braches (%)	Average (%)	Resistance	Identification Result
1	Jishan Hanbanjin	50.00%	0.00%	20.00%	HS	11.54%	35.00%	25.62%	HS	HS
2	Yongji Qingshi	x	24.29%	24.29%	HS	x	31.90%	31.90%	HS	HS
3	Lintong Jiandingshi	12.09%	26.00%	20.44%	HS	40.00%	16.46%	25.87%	HS	HS
4	Lintong Fangshi	30.00%	14.29%	20.57%	HS	41.00%	23.08%	30.25%	HS	HS
5	Lintong Huojing	50.00%	2.83%	21.70%	HS	50.00%	1.03%	20.62%	HS	HS
6	Boai Bayuehuang	15.38%	26.09%	21.80%	HS	18.42%	55.00%	40.37%	HS	HS
7	Meixian Niuxinshi	30.20%	7.00%	16.28%	S	50.00%	1.43%	20.86%	HS	HS
8	Luoyang Guilianqing	57.89%	1.85%	24.27%	HS	x	21.05%	21.05%	HS	HS
9	Denglongshi	34.38%	12.50%	21.25%	HS	35.29%	21.81%	27.20%	HS	HS
10	Yidu Tuoshi	29.82%	18.57%	23.07%	HS	100.00%	11.43%	46.86%	HS	HS

Table 4. *Cont.*

Number	Group	2018				2019				Identification Result
	Variety	Infected Fruits (%)	Infected Braches (%)	Average (%)	Resistance	Infected Fruits (%)	Infected Braches (%)	Average (%)	Resistance	
11	Zhouqu Huoshi	33.33%	11.76%	20.39%	HS	62.50%	4.44%	27.70%	HS	HS
12	Changan Fudingjian	x	16.00%	16.00%	S	6.90%	30.00%	20.76%	HS	HS
13	Shuhuangshi	37.50%	11.58%	21.95%	HS	x	52.50%	31.50%	HS	HS
14	Jincheng Gaishi	40.00%	10.00%	22.00%	HS	x	45.00%	45.00%	HS	HS
15	Tongguan Lianhuashi	0.00%	40.00%	24.00%	HS	x	32.50%	32.50%	HS	HS
16	Licheng Mianshi	12.50%	30.00%	23.00%	HS	4.44%	40.00%	25.78%	HS	HS
17	Matian FangSs	56.00%	29.05%	39.83%	HS	100.00%	0.00%	40.00%	HS	HS
18	Xingyang Shuishi	33.33%	17.14%	23.62%	HS	70.00%	21.00%	40.60%	HS	HS
19	Fuzhengbing	100.00%	8.33%	45.00%	HS	75.00%	17.50%	40.50%	HS	HS
20	Hongmiandan	40.54%	10.00%	22.21%	HS	100.00%	14.29%	48.57%	HS	HS
21	Lantian Shuishi	50.00%	2.56%	21.54%	HS	58.10%	61.00%	59.84%	HS	HS
22	Sifangshi	18.50%	17.14%	17.68%	S	79.32%	19.50%	43.43%	HS	HS
23	Gongcheng Shuishi	21.02%	12.50%	15.91%	S	100.00%	14.29%	48.57%	HS	HS
24	Baoshan Dashuishi	60.00%	1.67%	25.00%	HS	90.00%	18.33%	47.00%	HS	HS
25	Jurong Bianshi	51.02%	8.25%	25.36%	HS	x	50.00%	50.00%	HS	HS
26	Suqian Bianshi	100.00%	14.29%	48.57%	HS	100.00%	21.65%	52.99%	HS	HS
27	Huixian Dashi	100.00%	16.30%	49.78%	HS	61.54%	50.00%	54.62%	HS	HS
28	Huaxian Baixuanshi	x	18.60%	18.60%	HS	x	55.00%	55.00%	HS	HS
29	Fuping Jianshi	x	50.00%	50.00%	HS	x	60.00%	60.00%	HS	HS
30	Youhou	7.55%	3.03%	4.84%	R	0.00%	20.00%	12.00%	S	S
31	Putian Goushi	8.33%	5.80%	6.81%	S	7.69%	15.00%	12.08%	S	S
32	Sibanmianshi	5.88%	7.50%	6.85%	S	8.46%	14.74%	12.23%	S	S
33	Zhoushan Changshi	14.29%	4.79%	8.59%	S	28.57%	5.00%	14.43%	S	S
34	Tangcunshi	0.00%	8.33%	5.00%	R	21.74%	10.00%	14.50%	S	S
35	Ningbo Tongpenshi	x	0.00%	0.00%	HR	5.00%	0.00%	2.00%	HR	HR
36	Zhaoan Yuanxiaoshi	x	0.00%	1.39%	R	5.06%	0.00%	2.03%	R	R
37	Chengsushi	x	2.33%	0.00%	R	1.22%	2.63%	2.07%	R	R
38	Baoshan Dafangshi	x	0.55%	1.23%	R	0.00%	3.54%	2.12%	R	R
39	Hyakume	x	0.00%	1.67%	R	x	2.13%	2.13%	R	R
40	Yanjing Yingshi	0.00%	0.00%	0.00%	HR	1.37%	2.63%	2.13%	R	R
41	Zaozhuang Ehuangshi	0.00%	0.00%	0.00%	HR	0.00%	1.67%	1.00%	HR	HR
42	Kangbing Jianshi	x	0.00%	0.00%	HR	0.00%	0.00%	0.00%	HR	HR
43	Yuanqu Bayuehong	x	0.00%	0.00%	HR	0.00%	0.00%	0.00%	HR	HR

3.5. Agronomic Characteristics of Blast Resistant Finger Millet Genotypes

The agronomic traits of nine anthracnose-resistant germplasms were analyzed, and results showed that highly resistant genotypes, i.e., 'Kangbing Jianshi' and 'Yuanqu Bayuehuang', had strong growth potential, orange-yellow peel color, glutinous texture, and none had a cross-concave form on the fruit apex (Table S2). 'Hyakume', which could naturally remove astringency on the tree during the ripening stage, was classified under the pollination-variant nonastringent type, whereas others were classified under the pollination-constant astringent type. Moreover, 'Hyakume' had the highest average fruit weight of 192 g and maximum fruit weight of 386 g. 'Zhaoan Yuanxiaoshi' and 'Baoshan Dafangshi' had the most days of maintaining fruit crispness, with 27 days. All varieties could be used as female parent materials for resistance breeding. In addition, the 'Hyakume' persimmon, with occasional male flowers, could be used as a male parent for the breeding program (Table S2).

4. Discussion

The *C. gloeosporioides* complex is the most common and polyphagous species complex of the genus. Weir et al. [20] reported that *C. gloeosporioides* is a complex consisting of 22 species. To date, 57 species have been described [23]. The genus has been voted as one of the top 10 fungal plant pathogens in molecular plant pathology [24]. On the basis of multigene phylogenetic analysis and morphological characteristics, the pathogen causing persimmon anthracnose has been renamed as *C. horii*, which also belongs to the *C. gloeosporioides* species complex [14]. To the best of our knowledge, most individual species in the *C. gloeosporioides* species complex, such as *C. horii*, *C. fructicola*, and *C. siamense*, may cause persimmon anthracnose [15,16,25] (Figure 1). Moreover, some different plants can be infected by the same anthrax species [14,26], *C. acutatum* from pepper can cause typical persimmon anthracnose symptoms, showing black spot disease in Korea [27,28]. Thus, these species (*C. gloeosporioides* is a species complex) are extremely difficult to distinguish on the basis of morphological characteristics only. Thus, pathogenic strains from persimmon anthracnose should be identified on the basis of the multigene phylogenetic analysis. In the present study, the 'FJ 3' isolate from persimmon was identified as *C. horri* by the combined analysis of molecular and morphological characteristics. Moreover, the 'FJ 3' isolate was tested for pathogenicity on persimmon twigs in the laboratory and nature field to confirm the isolate's ability to cause the anthracnose disease. Therefore, the identification of the anthracnose pathogen in the national persimmon germplasm resource nursery is helpful to further carry out and deepen the research on the pathogenesis and defense mechanism of this anthracnose and provide a basis for the prevention and control of persimmon anthracnose.

Persimmon anthracnose, a serious disease occurring in many countries where persimmon is cultivated, is predominantly caused by *C. gloeosporioides* [29]. Persimmon anthracnose symptoms first appear in the spring as spots which then develop into dark lesions. Under high temperature and humidity conditions, adjacent lesions may integrate together, increasing in size until the entire twig is infected [30]. In the present study, the anthracnose pathogen can result in dark lesions in the fruit, calyx, branches, and leaf of persimmon, showing similar results with previous reports [9,15,22,30]. In addition, the *C. gloeosporioides* species complex consists of important plant pathogens that can cause anthracnose in economic crops, such as chili [31], mango [32], strawberry [33], apples [34], grapes [35], bananas [36,37], and tomato [38]. As a destructive disease of persimmon trees, *C. horii* can cause leaf defoliation, fruit rot, and even the death of the whole plant [39]. Unfortunately, most persimmon varieties are not resistant to anthracnose. For example, 'Gongcheng Shuishi' and 'Fuping Jianshi', which have the largest cultivation area in China, are susceptible to persimmon anthracnose. In some persimmon-growing areas in China, anthracnose can reduce the yield of persimmon by more than 50% [40].

Anthracnose-resistant persimmon varieties are important for the breeding of new resistant cultivars and the development of the persimmon industry. However, the evaluation and identification of persimmon germplasms with anthracnose resistance remain unknown. Persimmon is believed to originate in China and has abundant genetic diversity, with at least 1000 varieties [2,7]. Therefore, the identification of excellent persimmon resources resistant to anthracnose is necessary for resistance breeding. On the basis of the artificial infection, two accessions (i.e., 'Kangbing Jianshi' and 'Quyuan Bayuehuang'), which had a hypersensitive reaction with a slight lesion at the injection site at 14 days post inoculation, were highly resistant to anthracnose, and six cultivars (i.e., 'Yeshi', 'Pucheng Fangshi', 'Bayuehuang', 'Lushi 5', and ' Junqianzi) exhibited a resistant response to anthracnose. Furthermore, 43 main cultivars of persimmon were selected for the re-evaluation of anthracnose resistance in the nature field in 2018 and 2019. The resistance level of germplasm resources was consistent with the previous evaluation results of artificial infection. Moreover, through the evaluation of agronomic characters, the above eight resistance resources showed excellent characters and could be used in resistance breeding in the future. For example, 'Kangbing Jianshi' is suitable for making dried persimmon due to the heart-shaped fruit, and 'Quyuan Bayuehuang' is early maturing with high yield and strong growth potential.

5. Conclusions

Persimmon anthracnose, a destructive disease for persimmon, is predominantly caused by *C. horii* pathogen, resulting in shoot wilting, fruit rotting, and even death of the whole plant. In this study, the isolate 'FJ 3' from persimmon trees in the NFGP of China was further identified as *C. horii* by using morphological characteristics and multigene phylogenetic analysis. A total of 142 persimmon varieties were tested for anthracnose disease response by inoculation with the conidial suspension of *C. horii* in the NFGP. Resistance level was defined in four categories, including highly resistant, resistant, susceptible, and highly susceptible, on the basis of the DI of the inoculated branch. A significant variability was observed among the studied accessions. Only two varieties, which had a hypersensitive reaction with a slight lesion at the injection site at 14 days post inoculation, were highly resistant to anthracnose. Seven cultivars exhibited a resistant response to anthracnose and accounted for 4.93% of the 142 accessions. A susceptible response was observed for 24 accessions, which accounted for 16.90% of the total. A total of 109 varieties with the highest proportion (76.76%) of the accessions showed a highly susceptible response to *C. horii*. A total of 43 persimmon main cultivars were selected for the re-evaluation of anthracnose resistance by natural disease incidence in 2018 and 2019 and showed similar resistance levels. Moreover, eight highly resistant or resistant accessions possessed desirable agronomic characters. The use of these resistance sources in persimmon breeding should aid in expanding the genetic diversity and in the development of new resistant varieties.

Supplementary Materials: The following supporting information can be downloaded at: https://www.mdpi.com/article/10.3390/horticulturae8020180/s1. These amplicons have been submitted to GenBank with accession numbers of OL347726 for ITS, OL364188 for *GAPDH*, OL364190 for *CHS-1*, OL364191 for *TUB2*, and OL364189 for *ACT*. Table S1: Primers for pathogen identification; Table S2: The agronomic characteristics of anthracnose resistant persimmon genotypes.

Author Contributions: Conceptualization, C.G. and Y.Y.; methodology, Q.C. and Y.L.; software, Q.C. and J.H.; validation, C.G., Y.Y., and J.H.; writing—original draft preparation, C.G. and J.H.; writing—review and editing, C.G.; visualization, Y.Y.; supervision, Y.Y.; funding acquisition, Y.Y. and C.G. All authors have read and agreed to the published version of the manuscript.

Funding: This research was supported by National Key Research and Development Program of China (2019YFD1000600) and National Horticulture Germplasm Resources Center (NHGRC2020-NH06).

Institutional Review Board Statement: Not applicable.

Informed Consent Statement: Not applicable.

Data Availability Statement: Data are contained within the article.

Conflicts of Interest: The authors declare no conflict of interest.

References

1. Guan, C.; Liu, S.; Wang, M.; Ji, H.; Ruan, X.; Wang, R.; Yang, Y. Comparative transcriptomic analysis reveals genetic divergence and domestication genes in *Diospyros*. *BMC Plant Biol.* **2019**, *19*, 227. [CrossRef] [PubMed]
2. Wang, R.; Luo, Z. Persimmon in China: Domestication and traditional utilizations of genetic resources. *Adv. Hortic. Sci.* **2008**, *22*, 239–243.
3. Woolf, A.B.; Ben-Arie, R. Persimmon (*Diospyros kaki* L.). In *Postharvest Biology and Technology of Tropical and Subtropical Fruits*; Yahia, E.M., Ed.; Woodhead Publishing: Cambridge, UK, 2011; Volume 4, pp. 166–193.
4. Greene, S.; Morris, J. The case for multiple-use plant germplasm collections and a strategy for implementation. *Crop Sci.* **2001**, *41*, 886–892. [CrossRef]
5. Wang, R. Chinese persimmon germplasm resources. *Acta Hortic.* **1997**, *436*, 43–50.
6. Guan, C.; Zhang, P.; Hu, C.; Chachar, S.; Riaz, A.; Wang, R.; Yang, Y. Genetic diversity, germplasm identification and population structure of *Diospyros kaki* Thunb. from different geographic regions in China using SSR markers. *Sci. Hortic.* **2019**, *251*, 233–240. [CrossRef]
7. Guan, C.; Zhang, Y.; Zhang, P.; Chachar, S.; Wang, R.; Du, X.; Yang, Y. Germplasm conservation, molecular identity and morphological characterization of persimmon (*Diospyros kaki* Thunb.) in the NFGP of China. *Sci. Hortic.* **2020**, *272*, 109490. [CrossRef]
8. Lee, J.-H.; Han, K.-S.; Lee, S.-C.; Shim, C.-K.; Bae, D.-W.; Kim, D.-K.; Kim, H.-K. Early detection of epiphytic anthracnose inoculum on phyllosphere of *Diospyros kaki* var. domestica. *Plant Pathol. J.* **2004**, *20*, 247–251. [CrossRef]
9. Zhang, J.-Z. Anthracnose of persimmon caused by *Colletotrichum gloeosporioides* in China. *Asian Australas J. Plant Sci. Biotechnol.* **2008**, *2*, 50–54.
10. Hori, S. Kaki no Shinbyogai Tansobyo. *Engei No Tomo* **1910**, *6*, 58–61.
11. Ito, S. Gloeosporiose of the Japanese Persimmon. *J. Plant Res.* **1911**, *25*, 197–202.
12. Maffei, L. Una malattia delle foglie del 'Kaki' dovuta al *Colletotrichum Kaki* n. sp. *Riv. Patol. Veg.* **1921**, *11*, 116–118.
13. Von, J. A revision of the fungi classified as *Gloeosporium*. *Bibliof. Mycol.* **1970**, *24*, 203.
14. Weir, B.S.; Johnston, P.R. Characterisation and neotypification of Gloeosporium kaki Hori as *Colletotrichum horii* nom. nov. *Mycotaxon* **2010**, *111*, 209–219. [CrossRef]
15. Chang, T.; Hassan, O.; Jeon, J.; Shin, J.; Oh, N.; Lim, T. First report of anthracnose of persimmon (*Diospyros kaki* L. f.) caused by *Colletotrichum siamense* in Korea. *Plant Dis.* **2018**, *102*, 443. [CrossRef]
16. Carraro, T.; Lichtemberg, P.; Michailides, T.; Pereira, W.; Figueiredo, J.; May-De Mio, L. First report of *Colletotrichum fructicola*, *C. nymphaeae*, and *C. melonis* causing persimmon anthracnose in Brazil. *Plant Dis.* **2019**, *103*, 2692. [CrossRef]
17. Wang, J.; Ai, C.; Yu, X.; An, M.; Sun, S.; Gao, R. First report of *Colletotrichum karstii* causing anthracnose on persimmon leaves in China. *Plant Dis.* **2016**, *100*, 532. [CrossRef]
18. Hassan, O.; Lee, D.; Chang, T. First report of anthracnose of persimmon caused by *Colletotrichum nymphaeae* in Korea. *Plant Dis.* **2019**, *103*, 1772. [CrossRef]
19. Fischer, A.H.; Jacobson, K.A.; Rose, J.; Zeller, R. Cutting sections of paraffin-embedded tissues. *CSH Protoc.* **2008**, *3*, 4987. [CrossRef]
20. Weir, B.; Johnston, P.; Damm, U. The *Colletotrichum gloeosporioides* species complex. *Stud. Mycol.* **2012**, *73*, 115–180. [CrossRef]
21. Tamura, K.; Peterson, D.; Peterson, N.; Stecher, G.; Nei, M.; Kumar, S. MEGA5: Molecular evolutionary genetics analysis using maximum likelihood, evolutionary distance, and maximum parsimony methods. *Mol. Biol. Evol.* **2011**, *28*, 2731–2739. [CrossRef]
22. Hassan, O.; Jeon, J.Y.; Chang, T.; Shin, J.S.; Oh, N.K.; Lee, Y.S. Molecular and morphological characterization of Colletotrichum species in the *Colletotrichum gloeosporioides* complex associated with persimmon anthracnose in South Korea. *Plant Dis.* **2018**, *102*, 1015–1024. [CrossRef] [PubMed]
23. Talhinhas, P.; Baroncelli, R. Colletotrichum species and complexes: Geographic distribution, host range and conservation status. *Fungal Divers.* **2021**, *110*, 109–198. [CrossRef]
24. Dean, R.; Van Kan, J.A.; Pretorius, Z.A.; Hammond-Kosack, K.E.; Di Pietro, A.; Spanu, P.D.; Rudd, J.J.; Dickman, M.; Kahmann, R.; Ellis, J. The Top 10 fungal pathogens in molecular plant pathology. *Mol. Plant Pathol.* **2012**, *13*, 414–430. [CrossRef] [PubMed]
25. Jeon, J.; Hassan, O.; Chang, T.; Lee, D.; Shin, J.; Oh, N. Anthracnose of persimmon (*Diospyros kaki*) caused by *Colletotrichum horii* in Sangju, Korea. *Plant Dis.* **2017**, *101*, 1035. [CrossRef]
26. Siddiqui, Y.; Ali, A. Colletotrichum gloeosporioides (Anthracnose). In *Postharvest Decay. Control Strategies*; Bautista-Baños, S., Ed.; Academic Press: Cambridge, MA, USA; Elsevier Inc.: London, UK, 2014; pp. 337–371, ISBN 978-0-12-411552-1.
27. Kim, H.R.; Lim, T.H.; Kim, J.-H.; Kim, Y.H.; Kim, H.T. Potential of cross-infection of *Colletotrichum* species causing anthracnose in persimmon and pepper. *Plant Pathol. J.* **2009**, *25*, 13–20. [CrossRef]
28. Kwon, J.H.; Kim, J.W. First report of fruit black spot of *Diospyros kaki* caused by *Colletotrichum acutatum* in Korea. *Plant Pathol. J.* **2011**, *27*, 100. [CrossRef]
29. Tong, Z.J.X. Various stages and amount of *Cottetotrichum gloeosporioides* on overwintering twigs of persimmon. *J. Plant Protect.* **2003**, *30*, 437–438.

30. Xie, L.; Zhang, J.Z.; Cai, L.; Hyde, K.D. Biology of *Colletotrichum horii*, the causal agent of persimmon anthracnose. *Mycology* **2010**, *1*, 242–253. [CrossRef]
31. Than, P.P.; Prihastuti, H.; Phoulivong, S.; Taylor, P.W.; Hyde, K.D. Chilli anthracnose disease caused by *Colletotrichum* species. *J. Zhejiang Univ. Sci.* **2008**, *9*, 764–778. [CrossRef]
32. Vitale, A.; Alfenas, A.C.; Siqueira, D.L.D.; Magistà, D.; Perrone, G.; Polizzi, G. Cultivar resistance against *Colletotrichum asianum* in the world collection of mango germplasm in southeastern Brazil. *Plants* **2020**, *9*, 182. [CrossRef]
33. Xiao, C.; MacKenzie, S.; Legard, D. Genetic and pathogenic analyses of *Colletotrichum gloeosporioides* isolates from strawberry and noncultivated hosts. *Phytopathology* **2004**, *94*, 446–453. [CrossRef]
34. González, E.; Sutton, T.B. Population diversity within isolates of *Colletotrichum* spp. causing Glomerella leaf spot and bitter rot of apples in three orchards in North Carolina. *Plant Dis.* **2004**, *88*, 1335–1340. [CrossRef] [PubMed]
35. Kono, A.; Nakaune, R.; Yamada, M.; Nakano, M.; Mitani, N.; Ueno, T. Effect of culture conditions on conidia formation by Elsinoë ampelina, the causal organism of grapevine anthracnose. *Plant Dis.* **2009**, *93*, 481–484. [CrossRef] [PubMed]
36. Zhimo, V.Y.; Dilip, D.; Sten, J.; Ravat, V.K.; Bhutia, D.D.; Panja, B.; Saha, J. Antagonistic yeasts for biocontrol of the banana postharvest anthracnose pathogen *Colletotrichum musae*. *J. Phytopathol.* **2017**, *165*, 35–43. [CrossRef]
37. Riera, N.; Ramirez-Villacis, D.; Barriga-Medina, N.; Alvarez-Santana, J.; Herrera, K.; Ruales, C.; Leon-Reyes, A. First report of banana anthracnose caused by *Colletotrichum gloeosporioides* in Ecuador. *Plant Dis.* **2019**, *103*, 763. [CrossRef]
38. Pardo-De la Hoz, C.; Calderón, C.; Rincón, A.; Cárdenas, M.; Danies, G.; López-Kleine, L.; Restrepo, S.; Jiménez, P. Species from the *Colletotrichum acutatum, Colletotrichum boninense* and *Colletotrichum gloeosporioides* species complexes associated with tree tomato and mango crops in Colombia. *Plant Pathol.* **2016**, *65*, 227–237. [CrossRef]
39. Zang, J.; Xu, T. Cytological characteristics of the infection in different species, varieties and organs of persimmon by *Colletotrichum gloeosporioides*. *Mycosystema* **2005**, *24*, 116–122.
40. Deng, Q.E.; Ding, X.Y.; Li, J.A.; Cui, L.K.; Xu, J.Q. Morphological characteristics and genetic diversity of *Colletotrichum horii* infecting persimmon tree in China. *Eur. J. Plant Pathol.* **2020**, *156*, 437–449. [CrossRef]

horticulturae

Article

Comparative Transcriptomic Analysis Provides Insight into the Key Regulatory Pathways and Differentially Expressed Genes in Blueberry Flower Bud Endo- and Ecodormancy Release

Yongqiang Li [1,2,*], Rui Ma [1], Ruixue Li [1], Qi Zhao [1], Zhenzhen Zhang [3], Yu Zong [1,2], Linbo Yao [1,2], Wenrong Chen [1,2], Li Yang [1,2], Fanglei Liao [1,2], Youyin Zhu [4] and Weidong Guo [1,2,*]

[1] College of Chemistry and Life Sciences, Zhejiang Normal University, Jinhua 321004, China; marui7202@163.com (R.M.); lrx2497809053@163.com (R.L.); zhaoqi17353517996@163.com (Q.Z.); yzong@zjnu.cn (Y.Z.); linboyao@zjnu.edu.cn (L.Y.); cwr@zjnu.cn (W.C.); yangli@zjnu.cn (L.Y.); fangleiliao@zjnu.cn (F.L.)
[2] Zhejiang Provincial Key Laboratory of Biotechnology on Specialty Economic Plants, Zhejiang Normal University, Jinhua 321004, China
[3] Dinghai Agricultural Technology Extension Center, Zhoushan 316000, China; zhangzhenzhen1983@sina.com
[4] School of Agriculture, Jinhua Polytechnic, Jinhua 321007, China; zhuyouyin@jhc.edu.cn
* Correspondence: lyq@zjnu.cn (Y.L.); gwd@zjnu.cn (W.G.)

Abstract: Endodormancy is the stage that perennial plants must go through to prepare for the next seasonal cycle, and it is also an adaptation that allows plants to survive harsh winters. Blueberries (*Vaccinium* spp.) are known to have high nutritional and commercial value. To better understand the molecular mechanisms of bud dormancy release, the transcriptomes of flower buds from the southern highbush blueberry variety "O'Neal" were analyzed at seven time points of the endo- and ecodormancy release processes. Pairwise comparisons were conducted between adjacent time points; five kinds of phytohormone were identified via these processes. A total of 12,350 differentially expressed genes (DEGs) were obtained from six comparisons. Gene Ontology analysis indicated that these DEGs were significantly involved in metabolic processes and catalytic activity. KEGG pathway analysis showed that these DEGs were predominantly mapped to metabolic pathways and the biosynthesis of secondary metabolites in endodormancy release, but these DEGs were significantly enriched in RNA transport, plant hormone signal transduction, and circadian rhythm pathways in the process of ecodormancy release. The contents of abscisic acid (ABA), salicylic acid (SA), and 1-aminocyclopropane-1-carboxylate (ACC) decreased in endo- and ecodormancy release, and the jasmonic acid (JA) level first decreased in endodormancy release and then increased in ecodormancy release. Weighted correlation network analysis (WGCNA) of transcriptomic data associated with hormone contents generated 25 modules, 9 of which were significantly related to the change in hormone content. The results of this study have important reference value for elucidating the molecular mechanism of flower bud dormancy release.

Keywords: blueberry; transcriptome; endodormancy; ecodormancy; hormone; transcription factor; WGCNA

Citation: Li, Y.; Ma, R.; Li, R.; Zhao, Q.; Zhang, Z.; Zong, Y.; Yao, L.; Chen, W.; Yang, L.; Liao, F.; et al. Comparative Transcriptomic Analysis Provides Insight into the Key Regulatory Pathways and Differentially Expressed Genes in Blueberry Flower Bud Endo- and Ecodormancy Release. *Horticulturae* **2022**, *8*, 176. https://doi.org/10.3390/horticulturae8020176

Academic Editor: Luigi De Bellis

Received: 19 December 2021
Accepted: 18 February 2022
Published: 20 February 2022

1. Introduction

Bud dormancy can be divided into three types: paradormancy, endodormancy, and ecodormancy [1]. During paradormancy, also named "summer dormancy" [2], the break of the axillary bud is repressed by the surrounding organs, similarly to apical dominance. In late fall, plants constantly respond to changes in photoperiod and temperature, and buds enter the endodormancy stage. Endodormancy is the evolutionary adaptation of perennial woody plants to survive the harsh winter. A certain amount of chilling accumulation during the winter is crucial for the transition from endodormancy to ecodormancy. Upon the arrival of warm temperatures in spring, bud break open and flowering begins, so dormancy throughout the bud life cycle ensures that buds can break at the right time. Correct flowering timing is a critical determinant of the adaptation of plants to different environments [3].

Endodormancy is a special characteristic of perennial woody plants from temperate zones, and chilling accumulation is vital for bud endodormancy release. Global climate warming has led to insufficient chilling in certain years and an inability to completely break endodormancy. It can also lead to early bud break during transient warm spells in the middle of winter. Subsequent freezing spells can then damage opening flower buds, which are more vulnerable to freezing damage than closed endodormant or ecodormant buds. These kinds of events can result in large economic losses in fruit tree yield; thus, bud dormancy research has been a major focus for decades. The study of hormones has also been important in the understanding of dormancy, especially abscisic acid (ABA) catabolism and gibberellic acid (GA) biosynthesis in bud or seed dormancy release [4,5]. In recent years, with the development and application of RNA sequencing (RNA-seq) technology, transcriptome studies have been widely applied in investigation of bud dormancy release, including in pear [6,7], cherry [8,9], peach [10], grape [11,12], Japanese apricot [4], and litchi [13]. Identifying differentially expressed genes (DEGs) and their functions is an important part of dormancy research. Significant progress in functional genomics research on transcription factors has been made recently through the application of RNA interference, transgenic plants, and protein–protein interaction approaches. The well-known *dormancy-associated MADS box* genes (*DAM* genes) can be regulated by *CRT-binding factor* (*CBF*) [14,15], *ABRE-binding factors 3* (*ABF3*) [16], and *TEOSINTE BRANCHED1/CYCLOIDEA/PROLIFERATING CELL FACTOR transcription factor 20* (*TCP20*) [17], and can promote the expression of *9-CIS-EPOXYCAROTENOID DIOXYGENASE* (*NCED*) and *BRANCHED 1* (*BRC1*) [18] to maintain bud dormancy. Nevertheless, more details on gene function need to be confirmed, and more key pathways should be studied in relation to dormancy induction, maintenance, and release. Due to the complex genomes of woody plants and the low transformation efficiency for many perennial woody plants, research on bud dormancy remains quite limited. Therefore, research on bud dormancy induction and release is still a young field.

Blueberries are known to have high nutritional and commercial value. The cultivated area of highbush blueberries has expanded dramatically in recent years [19]. Some northern highbush blueberry (NHB) cultivars, such as "Chandler" and "Bluecrop", have fruits that are large and delicious; however, the climate of Zhejiang Province (southeastern China) fails to meet their chilling requirements (CR) during winter, resulting in bud break delay and low bud break rate. Although most southern highbush blueberry (SHB) cultivars can be grown well in southern China, some outstanding cultivars, such as "Emerald" and "Springhigh", exhibit early flowering in winter. Fluctuating warm and cold spells in winter can cause flowers to suffer frost damage, which not only consumes nutrients in the bush but also causes pathogen attacks on the remaining flowers, leading to disease outbreaks and decreased fruit quality and yield. This is a major problem affecting various kinds of fruit trees and bushes [20,21].

To date, although the mechanisms and the functions of some key genes in dormancy release have been revealed, the temporal variations in key metabolic pathways from endodormancy to flowering have not been studied in detail. This research focused on the process of blueberry flower bud endo- and ecodormancy release through transcriptome sequencing and explored the similarities and differences in metabolic pathways, as well as the DEGs between the two types of dormancy release. These results were used to identify the key genes for which expression matched the timeline of flower bud maturity from endodormancy to bud break. By performing weighted correlation network analysis (WGCNA), we explored modules that are highly correlated with hormones and searched for new gene regulation patterns. The research findings can be used in the future to generate new cultivars with different chilling requirements through genetic breeding. Moreover, these results can provide candidate genes for screening early-generation materials through crossbreeding. These results will also be of great help in the regulation of endodormancy to cope with climate change and improve fruit production.

2. Materials and Methods

2.1. Plant Materials

Shoots and flower buds of "O'Neal", one of the most popular southern highbush blueberry (SHB), were collected from the blueberry orchard (N 29°1′39.05″, E 119°44′18.17″) of the Zhejiang Provincial Key Laboratory of Biotechnology on Specialty Economic Plants in China. Healthy shoots with a length of 25 cm and 8–10 flower buds were pruned from blueberry tree. 100 shoots were sampled respectively on the following dates: 19 November, and 1, 8, 15, 22, 29 December in 2018, and 7 January in 2019. Shoots collected from multiple plants were pooled as a single sample, and samples collected from different locations in the field were used to create three biological replicates [22]. Flower buds were frozen immediately in liquid nitrogen after being detached from the shoots and were then stored at −80°C until RNA isolation.

2.2. Identification of the Chilling Requirement and Dormancy Status of O'Neal

The annual temperature and humidity in the orchard were recorded using a temperature recorder (DSR-THUA, ZOGLAB China), and the chilling accumulation was calculated using the Utah model [23]. To determine the optimal period of endodormancy release for detection of associated gene expression, we measured the percentage of bud break using the following method [8]. Pruned shoots were placed in a 500 mL glass beaker with 5 cm length at the bottom immersed in water. Those shoots were kept under the following conditions: light/darkness 16/8 h, light intensity: 320 $\mu mol \cdot m^{-2} \cdot s^{-1}$, and relative air humidity 75%. The water in the glass beaker was changed every 2 days, and the ends of the shoots were recut. Dormancy status were evaluated according to the percentage of bud break after 25 days. The end of endodormancy was considered to have been reached when the percentage of bud break was 50% [24]. Three biological replicates were performed for all of the above treatments.

2.3. Artificial Warming Treatment

For artificial warming accumulation, on 29 December 2018, 1 year old shoots (flower buds at the endodormancy stage) of O'Neal trees were collected, incubated at 15 °C for 3 d, and then transferred to 25 °C. The shoots were kept following the conditions described in Section 2.2. After 6, 12, 18, and 24 h, 30 flower buds were collected and frozen in liquid nitrogen. Tissues were stored at −80 °C for later use. Other shoots were exposed to forcing conditions for bud break measurements. Three biological replicates were performed for all the above treatments.

2.4. Preparation of the RNA-seq Library

Total RNA was extracted with an isolation kit (Foregene Co., Ltd., Chengdu, China), and mRNA was enriched with magnetic beads containing oligo(dT). The mRNA was fragmented into short segments of 150~200 nt by adding fragmentation buffer. Short segments of mRNA were used as templates to synthesize cDNA using random six-base hexamers. A two-stranded synthesis reaction system was then prepared to synthesize double-stranded cDNA. The purified cDNA was first repaired with an adhesive end, followed by adding an A base to the 3′ end and connecting the cDNA to the connector. Finally, fragment size selection and PCR amplification were performed.

The RNA from 21 samples (19 November, 8 December, 29 December, 6 h, 12 h, 18 h, and 24 h; 7 bud development stages × 3 replicates) was sequenced on an Illumina HiSeqTM 4000 (Illumina, San Diego, CA, USA) after the libraries were qualified using an Agilent 2100 Bioanalyzer (Agilent Technologies, Inc., Santa Clara, CA, USA) and ABI Step One Plus Real-Time PCR System (Applied Biosystems, Waltham, MA, USA).

2.5. Data Quality Control and Functional Annotation

The original FASTQ reads (raw reads) were filtered for nonconforming sequence reads, such as those containing adapters, those with poly-N greater than 5%, and low-quality reads (the number of bases with a mass value $Q \leq 10$ accounting for more than 20% of the whole read). After the data quality assessment was performed, clean reads were compared to the *V. corymbosum* (NHB "Draper") reference genome sequence (https://www.vaccinium.org (accessed on 2 September 2019)) using HISAT2 (version 2.0.4) to evaluate the overall quality of the sequencing library. The clean reads were deposited in the National Center for Biotechnology Information (NCBI) Sequence Read Archive (accession number PRJNA762194).

2.6. Screening and Analysis of Differentially Expressed Genes

The number of reads aligned to each gene was counted using HTSeq V0.6.1 [25]. RSEM software was used for quantitative analysis, and the level of gene expression was measured using the FPKM (fragments per kb per million fragments) value. According to the expression results for all genes in each sample, DESeq2 software was used to screen DEGs between samples, with cutoffs of $|\log2FC|$ 1 or higher and FDR < 0.05 representing significant DEGs. Noiseq [26] and DESeq2 [27] were used to analyze the differences between repeated samples.

Gene Ontology (GO) analysis was performed with Blast2GO software (V2.5.0) [28,29]. Kyoto Encyclopedia of Genes and Genomes (KEGG) pathway annotation was performed with Blastall [30]. After the GO annotation of each unigene was obtained, the GO function of all unigenes was classified by WEGO to investigate the gene function of distributed species [31]. GO enrichment analysis was performed via the GO seq R package. When the GO term padj (corrected *p*-value) was <0.05, it was considered to be significantly enriched. KOBAS software was used for enrichment analysis of the DEGs in KEGG pathways.

2.7. Hormone Profiling

The phytohormone (ABA: abscisic acid, SA: salicylic acid, JA: jasmonic acid, ACC: 1-aminocyclopropane-1-carboxylic acid, and CTK: cytokinin) extraction was done following a previously described method [32]. A 100 mg aliquot of ground tissues was mixed with 1 mL of 80% methanol with internal standards (45 pg of $^2H_5$-tZ, $^2H_5$-tZR, $^2H_6$-iP, $^2H_6$-iPR, and 100 pg of [$^2H_6$] ABA), and subsequently extracted twice using a laboratory rotator for 2 h at 4 °C. After centrifugation (10 min, 15,000× *g*, and 4 °C), the supernatant was collected and dried with nitrogen gas. The pellet was then resolved in 300 μL of 30% methanol and filtered through a filter membrane (0.22 μm).

The phytohormones were separated and analyzed using an Exion LC™ (AB SCIEX, Framingham, MA, USA) equipped with an Acquity UPLC BEH C18 column (2.1 × 100 mm, particle size of 1.7 μm), following a method previously described by [33]. The column was maintained at 40 °C and the mobile phases for cytokinins were composed of water (A) and MeOH (B) using a multistep linear gradient elution: 5% B at 0–2.5 min, 5–20% B at 2.5–3 min, 20–50% B at 3–12.5 min, 50–100% B at 12.5–13 min, 100% B at 13–15 min, 100–5% B at 15–15.2 min, and 5% B at 15.2–18 min. The mobile phases were composed of water (A) with 0.1% formic acid and MeOH (B) with 0.1% formic acid using a multistep linear gradient elution: 20% B at 0 to 1 min, 20 to 100% B at 1 to 7 min, 100% B at 7 to 9 min, 100 to 20% B at 9 to 9.3 min and 20% B at 9.3 to 12 min. The flow rate was 0.3 mL min^{-1}.

The phytohormones were analyzed using the triple quadrupole mass spectrometer QTRAP 5500 system (AB SCIEX, Framingham, MA, USA) following a method previously described [34]. The optimized conditions were as follows: curtain gas, 40 psi; ion spray voltage, 5000 V (CTK, ACC) for positive ion mode; 4500 V (ABA, SA, and JA) for negative ion mod [35]; turbo heater temperature, 600 °C; nebulizing gas (Gas 1), 60 psi; heated gas (Gas 2), 60 psi. Analyst software (version1.6.3, AB SCIEX, Framingham, MA, USA) was used. The data analysis was processed using MultiQuant software (version3.0.2, AB SCIEX, Framingham, MA, USA). Hormones were accurately quantified through internal standards.

2.8. WGCNA

Weighted gene correlation network analysis (WGCNA) was performed using the "WGCNA" R package as previously described [36]. Changes in the contents of five hormones (ABA, SA, JA, ACC, and CTK) at six time points (19 November, 8 December, 29 December, 6 h, 18 h, and 24 h) of dormancy release were used as phenotypic data, and the gene expression level based on RNA-seq was used as the database. The soft threshold, β, was calculated using the Pick Soft Threshold function in the WGCNA package to make the coexpression network conform to the scale-free topology.

2.9. Quantitative RT–PCR (qRT–PCR) Validation

Total RNA extraction and 1st-strand cDNA synthesis were performed according to a previous report [37]. Synthesized first-strand cDNAs were diluted 3-fold for qPCR validation. Specific primers (Supplementary Table S1) of 20 randomly selected DEGs were designed using Primer-BLAST (https://www.ncbi.nlm.nih.gov/tools/primer-blast/ (accessed on 6 October 2019)), and *VcGAPDH* was used to normalize the amount of cDNA among the samples [38]. qPCRs were performed on an ABI Step One Plus^{TM} RT–PCR system (Applied Biosystems, USA). The PCR system, procedures, and data analysis were performed according to [39]. Data analyses were performed using the relative quantitative method ($2^{-\Delta\Delta Ct}$).

2.10. Statistical Analysis

Statistical analyses were performed using SPSS Statistics 21 software. The data are expressed as the mean value and standard deviation (mean $\pm$ SD, $n \geq 3$). Statistical significance was evaluated via an independent sample t-test (confidence interval = 95%) using Statistics 21 software. Figures were plotted using Prism and Tbtools software.

3. Results

3.1. Identification of the Dormancy Stage of Blueberry Flower Buds

Shoots were cut from field plants after different periods of chilling and warm temperature treatment, and bud development was monitored (Figure 1c) to determine the best treatments for RNA-seq. As of 19 November 2018, the chilling accumulation for field plants of the O'Neal variety was 0 CU. The flower buds were at the endodormant stage, and the sample was marked as Nov. 19. On 1 December 2018, chilling accumulation had reached 72.9 CU, and the final bud break percentage was 41.85%. On 8 December 2018, chilling accumulation reached 102.6 CU, the sample was labeled Dec. 8, and the final bud break rate was 65.64% (Figure 1a), exceeding 50% for the first time. Therefore, the time to break endodormancy was reached between 1 December and 8 December 2018. On 29 December 2018, the bud break percentage of the samples reached 62.31%, and the bud break rate did not improve with longer chilling accumulation. After that point, the flower buds of the blueberries had strong sprouting potential, and the bud stage was considered to be ecodormancy; the sample was labeled Dec. 29. When the ecodormant buds were transferred to 25 °C treatment for 2 d, 25.62% of the flower buds sprouted (Figure 1b); considering that the expression of genes occurs earlier than morphological changes, it was decided that samples would be taken every 6 h (6 h, 12 h, 18 h, and 24 h) after warm temperature treatment to screen for key DEGs involved in the process of ecodormancy release. Flower buds collected at the above seven time points (19 November, 8 December, 29 December, 6 h, 12 h, 18 h, and 24 h) from endodormancy to flowering were used as experimental materials for subsequent transcriptome sequencing.

Figure 1. Bud break percentage of blueberries after different periods of chilling and warming accumulation. (**a**) Comparison of the flower bud break percentage of O'Neal in 2018–2019. (**b**) Bud break percentage of blueberry flower buds on 29 December under warming accumulation. (**c**) Flower bud break process. Stages III and above are criteria for bud break statistics.

3.2. RNA Sequencing Data Analysis

The range of raw reads from a single library produced by transcriptome sequencing was 40,117,154-41,356,720 (Table 1). Raw reads were filtered to remove low-quality reads. The range of high-quality clean reads (Phred value > Q30) was 40,111,158~41,351,040, accounting for 99.99% of the raw data; subsequent analyses were based on clean reads. The GC content was approximately 47%, and the percentage of bases with a quality value of up to Q30 was more than 93%. Following quality assessment, HisAT2 software was used to compare the clean reads to blueberry genome sequences, and the percentage of mapped clean reads was 72.04%~79.71%. The above results indicated that we obtained high-quality transcriptomes of blueberry flower bud dormancy release that could be used for subsequent analyses.

Table 1. Statistical results of data quality.

Sample Name	Raw Data	Clean Data	Clean Data Ratio (%)	Clean Data Q20 (%)	Clean Data Q30 (%)	GC Content (%)	Mapped Reads	Percentage of Mapped Reads	Mapped Genome (%)	Mapped Gene (%)
Nov. 19-1	41,082,384	41,078,456	99.99	97.53	93.91	46.84	29,594,336	72.04%	72.04%	59.60%
Nov. 19-2	40,638,858	40,635,402	99.99	97.48	93.82	47.12	32,048,468	78.87%	78.87%	68.31%
Nov. 19-3	41,003,184	40,999,156	99.99	97.47	93.81	47.3	32,364,244	78.94%	78.94%	68.54%
Dec. 8-1	40,359,734	40,354,476	99.99	97.36	93.63	47.8	29,287,920	72.58%	72.58%	61.09%
Dec. 8-2	40,379,528	40,373,466	99.98	97.27	93.47	47.71	30,525,004	75.61%	75.61%	63.26%
Dec. 8-3	40,117,154	40,111,158	99.99	97.57	94.06	47.59	30,338,448	75.64%	75.64%	62.35%
Dec. 29-1	41,356,720	41,351,040	99.99	97.67	94.28	47.28	32,701,506	79.08%	79.08%	65.14%
Dec. 29-2	41,127,706	41,122,630	99.99	97.62	94.13	47.28	32,066,614	77.98%	77.98%	63.66%
Dec. 29-3	41,169,840	41,163,932	99.99	97.61	94.11	47.64	32,503,736	78.96%	78.96%	66.26%
6 h-1	40,431,514	40,428,068	99.99	97.83	94.5	46.61	29,676,104	73.40%	73.40%	59.28%
6 h-2	40,489,884	40,486,186	99.99	97.23	93.16	46.69	30,991,372	76.55%	76.62%	65.72%
6 h-3	40,486,658	40,482,478	99.99	97.19	93.08	47.3	31,017,546	76.62%	76.55%	64.33%
12 h-1	41,321,014	41,316,892	99.99	97.31	93.31	47.08	32,476,742	78.60%	78.60%	68.04%
12 h-2	40,172,084	40,168,200	99.99	97.3	93.31	47.24	30,641,344	76.28%	76.28%	65.73%
12 h-3	41,155,516	41,149,572	99.99	97.2	93.16	47.77	32,798,908	79.71%	79.71%	68.83%
18 h-1	40,407,682	40,404,328	99.99	97.41	93.53	46.63	29,553,020	73.14%	73.14%	61.94%
18 h-2	40,400,518	40,397,044	99.99	97.31	93.3	46.66	29,263,828	72.44%	72.44%	61.06%

3.3. Screening of Differentially Expressed Genes during Dormancy Release

The goal of the differential expression analysis was to identify the genes that were significantly differentially expressed among the samples. According to the FPKM values of all the genes in each sample (Supplementary Table S2), DESeq2 software was used to screen DEGs between the samples. The adjacent expression libraries were pair-matched: Dec. 8 vs. Nov. 19; Dec. 29 vs. Dec. 8; 6 h vs. Dec. 29; 12 h vs. 6 h; 18 h vs. 12 h; 24 h vs. 18 h. According to a cutoff of $|\log 2FC| \geq 1$ with an FDR < 0.05 threshold, significant DEGs in the process of flower bud dormancy release were filtered.

As shown in Figure 2, during the process of endodormancy release, there were 8733 DEGs in Dec. 8 vs. Nov. 19, among which 4284 genes were upregulated and 4449 genes were downregulated. There were 538 DEGs in Dec. 29 vs. Dec. 8, of which 206 were upregulated and 332 were downregulated, indicating that the gene expression changes were most obvious upon endodormancy release. However, the gene expression changes were not obvious after entering the ecodormancy stage. In the process of ecodormancy release, 3259 genes were upregulated and 2172 genes were downregulated in 6 h vs. Dec. 29. There were 750 DEGs in 12 h vs. 6 h, while the number of DEGs did not exceed 700 in subsequent comparisons, indicating that the release from ecodormancy does not involve as many DEGs.

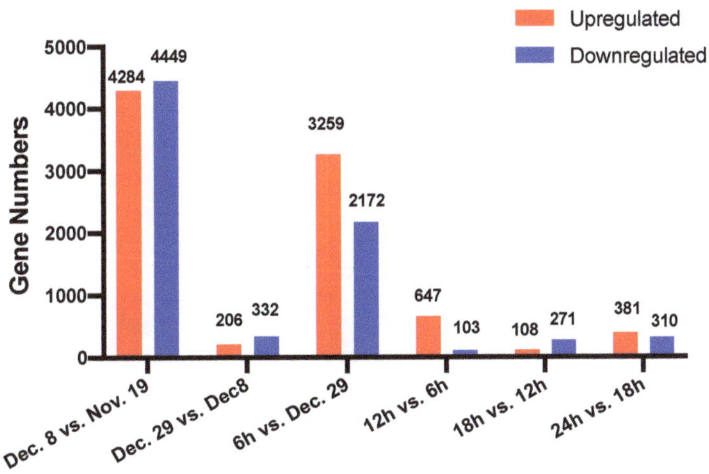

Figure 2. Statistical analyses of the quantity of differentially expressed genes (DEGs) during different dormancy stages. DEGs were identified by pairwise comparisons of the seven libraries.

3.4. GO Enrichment Analysis of Differentially Expressed Genes

To analyze the metabolic pathways and biological functions of the DEGs, GO enrichment analysis was performed (Figure 3 and Supplementary Table S3). The results are displayed for biological processes, cellular components, and molecular functions. In the process of endodormancy release, the most significantly enriched GO term was metabolic processes in the biological process category (3582 DEGs in Dec. 8 vs. Nov. 19). Among the cellular components, the top four enriched GO terms were cell, cell part, membrane, and membrane part. For molecular functions, DEGs were significantly enriched in binding and catalytic activity, and the number of DEGs in the catalytic activity term was greater than that of the binding term. Throughout the GO classification, metabolic processes and catalytic activity were significantly enriched, suggesting that the process of endodormancy release requires the activation of many energy metabolic pathways and related enzymes. During the process of ecodormancy release, the results of GO enrichment were similar to those for endodormancy release (Figure 3c–e).

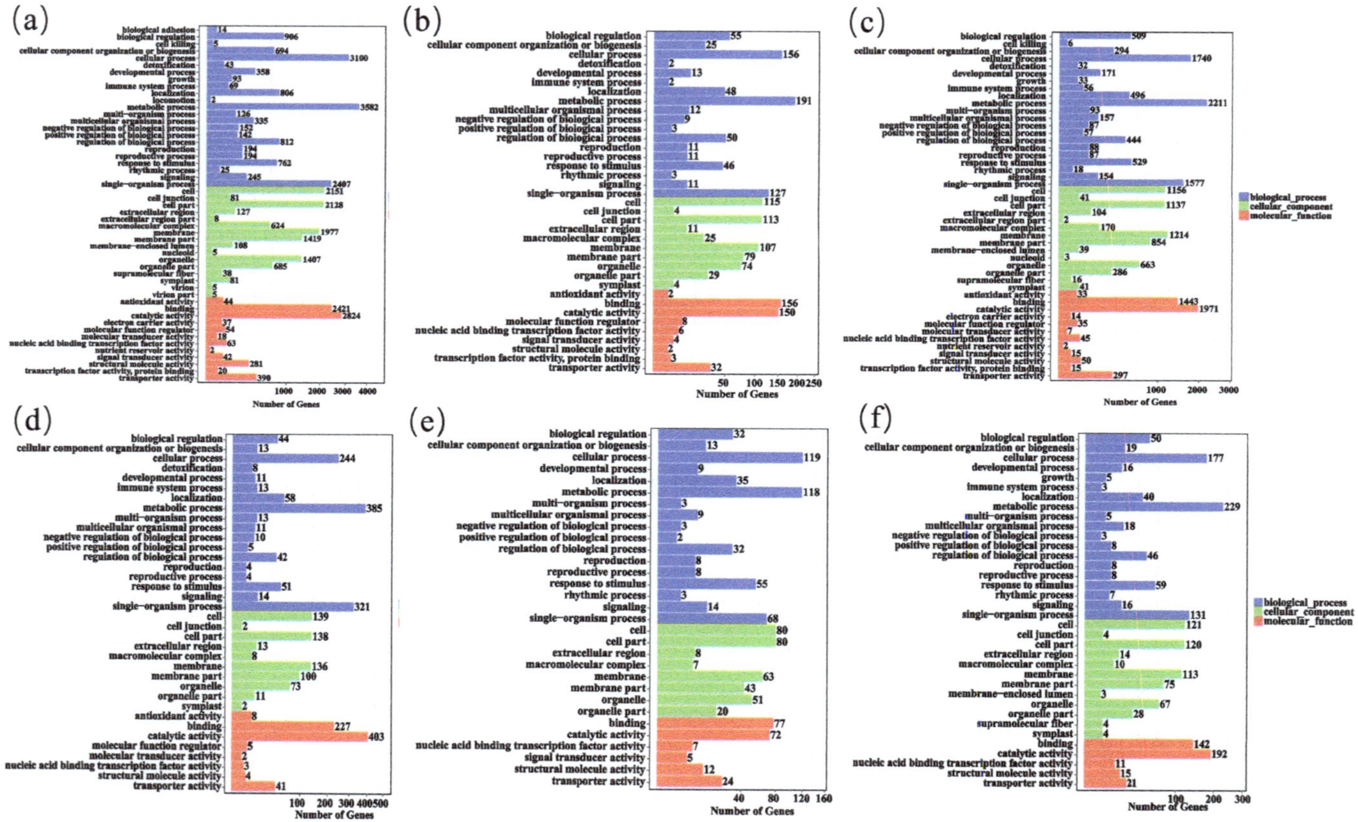

Figure 3. GO enrichment diagrams. (**a**–**f**) represents the secondary GO term annotation for DEGs in Dec. 8 vs. Nov. 19, Dec. 29 vs. Dec. 8, 6 h vs. Dec. 29, 12 h vs. 6 h, 18 h vs. 12 h, and 24 h vs. 18 h respectively.

3.5. KEGG Enrichment Analysis of the DEGs

During the endodormancy release process, the enrichment results showed that 9372 and 636 DEGs were mapped to 133 and 89 KEGG pathways in Dec. 8 vs. Nov. 19 and Dec. 29 vs. Dec. 8, respectively (Supplementary Table S4). Figure 4 shows the top 20 pathways in the KEGG classification. The smaller the *p*-value, the more reliable the enrichment significance of DEGs in the pathway. The rich factor refers to the ratio between the number of DEGs annotated to a certain pathway and the number of all genes annotated to this pathway. The higher the enrichment factor, the more significant the enrichment level of DEGs in this pathway. Dec. 8 vs. Nov. 19 was a critical phase of endodormancy release, and 5222 (55.7%) DEGs were significantly associated with 30 pathways ($p < 0.05$). The most mapped pathways were metabolic pathways and the biosynthesis of secondary metabolites; others included ribosomes, photosynthesis, amino acid biosynthesis, carbon metabolism, plant circadian rhythms, terpenoid biosynthesis, starch and sucrose metabolism, pentose phosphate pathway, galactose metabolism, glycolysis and glycosylation, vitamin B6 metabolism, flavonoid biosynthesis, and peroxisomes. The comparison of the four sequential libraries (Dec. 29 vs. Dec. 8, 6 h vs. Dec. 29, and 12 h vs. 6 h) showed that the metabolic pathway of DEG enrichment was the same as that for Dec. 8 vs. Nov. 19. However, the proportion of DEGs enriched in the metabolic pathways and biosynthesis of secondary metabolites based on the total number of DEGs increased significantly.

The comparison of the two libraries for warming accumulation for 18 h and 12 h (18 h vs. 12 h) showed that there were significant differences from the previous four comparisons. The top four metabolic pathways with DEG enrichment were RNA transport, protein processing in the endoplasmic reticulum, plant hormone signal transduction, and circadian rhythm. In the comparison between 24 h and 18 h of warming accumulation (24 h vs. 18 h), the metabolic pathways with DEG enrichment were similar to the previous comparison. Among the top four metabolic pathways with the most enrichment, the MAPK signaling pathway-plant replaced protein processing in the endoplasmic reticulum. The results indicated that there were significant metabolic differences in flower buds after 18 h of warming accumulation, which may represent the initiation of flowering.

3.6. Analysis of Differentially Expressed Genes in Circadian Rhythm and Key Metabolic Pathways during Dormancy Release

The circadian rhythm is thought to be important in the regulation of dormancy onset and release in some species, and we found DEGs that were significantly enriched in the circadian rhythm pathway during the process of dormancy release. Consequently, we focused on genes involved in the circadian rhythm to validate their roles in blueberry flower bud dormancy release. A total of 85 DEGs related to circadian rhythm were found in the comparison of Dec. 8 vs. Nov. 19 (Supplementary Table S5). We selected 11 representative genes and analyzed their expression patterns; the results indicated that the expression levels of a large number of circadian-rhythm-related genes differed from endodormancy to endodormancy release, e.g., *FT*: *Vaccdscaff28-snap-gene-259.27*; *ELF3*: *VaccDscaff36-snap-gene-15.50*; *HY5*: *VaccDscaff28-snap-gene-90.32*; and *CO*: *VaccDscaff1-augustus-gene-404.27*. As shown in Figure 5a, the expression patterns of circadian rhythm genes were grouped into two categories, with one group downregulated and the other upregulated during endodormancy release, for which *CO* and *FT* are examples, respectively. The expression level of *CO* was high in the endodormancy stage, and that of *FT* was low. The expression of the *CO* gene was downregulated during the endodormancy release, while the change of expression of *FT* was opposite. The results showed that their expression patterns are negatively correlated and *FT* may be inhibited by *CO*. During the transition of ecodormancy release, the expression levels of *COP1*, *TCP21*, and *HY5* were downregulated from 18 h to 24 h, while other genes' expression levels were upregulated with warm treatment.

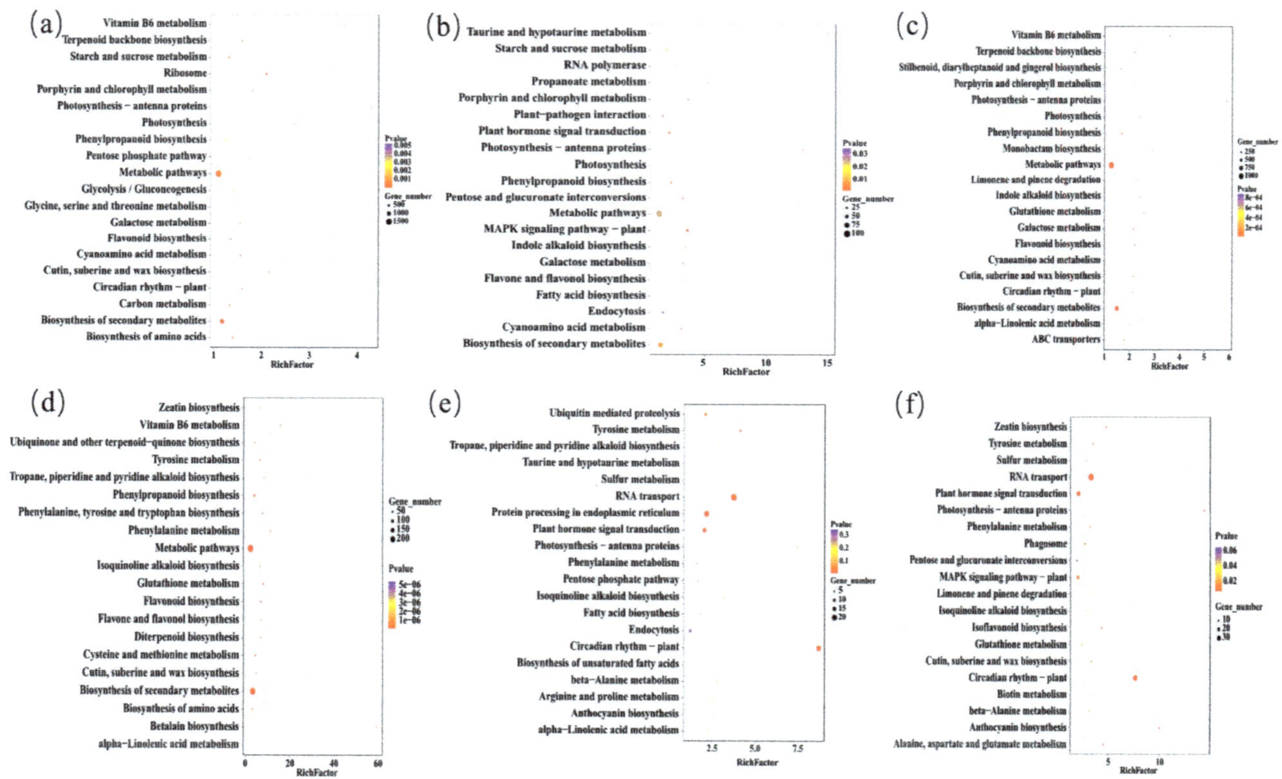

Figure 4. KEGG pathway enrichment results of DEGs. The ordinate represents pathway entry, and the abscissa represents the rich factor. Dots represent significant DEG classes, and the size of a dot represents the number of significant DEGs in that class. The larger the dot, the greater the number of significant DEGs. Different colored dots represent different p values. (**a**–**f**) represents the pathways for DEGs in Dec. 8 vs. Nov. 19, Dec. 29 vs. Dec. 8, 6 h vs. Dec. 29, 12 h vs. 6 h, 18 h vs. 12 h, and 24 h vs. 18 h respectively.

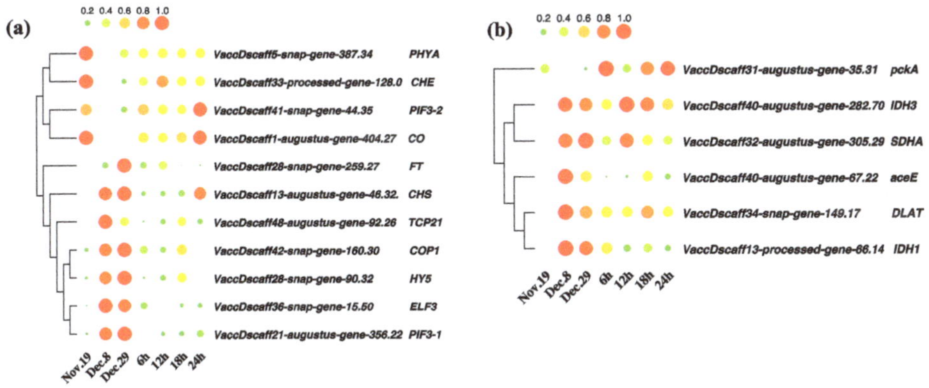

Figure 5. Analysis of changes in the expression levels of key genes in different metabolic pathways. Each column represents different sampling points, and each row represents a gene. The mean expression values were calculated using zero to one. Genes and expression patterns were hierarchically clustered based on the average Pearson's metric. Red indicates high expression, and green indicates low expression. (**a**) Key DEGs in circadian rhythm. (**b**) Key DEGs in the TCA cycle.

It takes substantial energy for flower buds to proceed from endodormancy to bloom, and we also found DEGs that were significantly enriched in the metabolic pathway during the process of dormancy release. The TCA cycle, also known as the citric acid cycle, is the final metabolic pathway of three classes of biological macromolecules (carbohydrates, lipids, and amino acids), as well as the metabolic hub and connection for sugars, lipids, and amino acids. Accordingly, we focused on genes involved in the TCA cycle to validate their roles in blueberry flower bud dormancy release. We selected six representative genes for analysis (Supplementary Table S6). As shown in Figure 5b, from Nov. 19 to Dec. 8, nearly all genes related to the TCA cycle pathway appeared to be upregulated, indicating that the metabolism of energy was active during the process of endodormancy release. There was only one DEG (*citrate synthase gene: VaccDscaff11-augustus-gene-378.31*) related TCA pathway in the comparison of Dec. 29 vs. Dec. 8, indicating that the expression levels of TCA-pathway-related genes in the ecodormant stage were similar to those during endodormancy release. The expression levels of *pckA*, *IDH3*, *SDHA*, and *DLAT* were upregulated during the process of ecodormancy release, which lasted from Dec. 29 until warm treatment for 24 h.

3.7. Changes in Hormone Levels during the Dormancy Release of Flower Buds

Hormone levels were measured during flower bud dormancy release, as shown in Figure 6. ABA levels were higher during the endodormant stage but decreased rapidly with the accumulation of low temperatures. On 8 December, the level of ABA in flower buds decreased significantly and was only one-quarter of that in the previous stage. With the continuous accumulation of low temperatures, the ABA level on 29 December (ecodormant stage) decreased slightly compared with that on 8 December, but there was no significant difference. The change in the JA level was similar to that in the ABA level, which decreased significantly upon endodormancy release, but JA levels increased during ecodormancy release. Although the gaseous hormone ethylene has a simple structure, it plays an important role in plant growth, development, and stress responses, and has been the focus of recent research. ACC is a precursor to ethylene synthesis and is used to reflect the change in ethylene content. Within the flower bud endodormancy release process, the ACC level decreased; however, unlike the lower levels of ABA and JA, the level of ACC was consistently lower with low temperature accumulation, and all three stages had significant differences, demonstrating that when flower bud endodormancy was released to ecodormancy, ethylene was still involved in metabolism and gene regulation. The level of SA did not change significantly during the process of endodormancy release but decreased from ecodormancy to flowering. CTK levels did not change significantly during the process of endodormancy but increased at 18 h in the process of ecodormancy release.

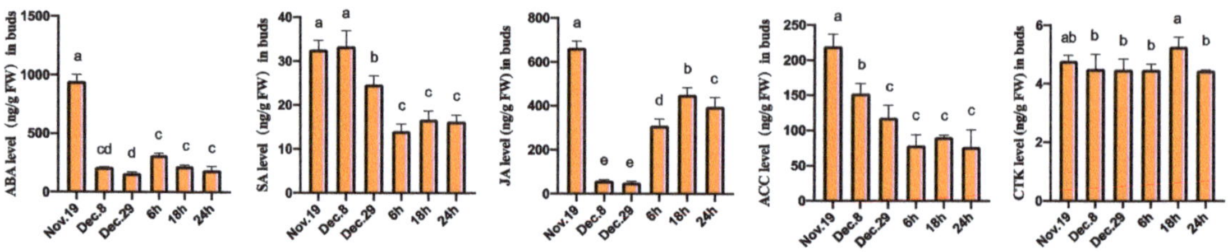

Figure 6. Hormone and ACC contents (ng/g FW) during flower bud endo- and ecodormancy release. Hormone types: ABA, abscisic acid; SA, salicylic acid; JA, jasmonic acid; CTK, cytokinin. ACC, 1-aminocyclopropane-1-carboxylic acid, an ethylene precursor. Values are the means of three biological replicates of flower buds from different plants. Error bars represent ± SD. Data were analyzed using one-way ANOVA followed by Duncan's test. a, b, c, d and e indicate significant differences at $p < 0.05$.

3.8. Key Differentially Expressed Genes in Hormone Anabolism and Signal Transduction Pathways

Due to significant changes in hormone contents (Figure 6), some DEGs in hormone anabolism and signal transduction pathways were significantly enriched during the process of dormancy release. We investigated genes involved in the hormone anabolism and signal transduction to validate their expression patterns in blueberry flower bud dormancy release. The carotenoid biosynthesis pathway plays an important role in plant growth and development and is directly involved in the biosynthesis of abscisic acid. A total of 31 DEGs of this pathway were found in the comparison of Dec. 8 vs. Nov. 19, among which 18 genes were upregulated and 13 genes were downregulated (Supplementary Table S7). We selected seven representative genes of them and showed their expressions in the dormancy release. For example, ABA biosynthetic genes such as *15-Cis-phytoene synthase* (*VaccDscaff41-augustus-gene-247.13*), *15-cis-phytoene desaturase* (*Vaccdscaff45-processed-gene-82.3*), and *NCED* (*Vaccdscaff10-processed gene-43.7*), as shown in Figure 7a: the expression levels of both genes were significantly downregulated during endodormancy release. In contrast, the expression level of the ABA-degrading enzyme gene *CYP707A* (*Vaccdscaff14-snap-gene-19.34*) was significantly increased during the same process. Transcriptome data showed that there were only three DEGs of the pathway in Dec. 29 vs. Dec. 8; among them, the beta-carotene 3-hydroxylase gene (*Vaccdscaff33-processed-gene-283.1*) was downregulated while the zeaxanthin cyclase gene (*VaccDscaff25-snap-gene-19.42*) was upregulated. In the process of ecodormancy release, some of the ABA biosynthetic genes were upregulated, but the specific functions of these genes in the process remain to be studied.

Figure 7b shows the expressions of selected DEGs involved in ethylene, JA, brassinolide (BR) biosynthesis. ACS is a key enzyme for ACC synthesis, and its gene *Vaccdscaff2-augustus-gene-429.20* was downregulated in Dec. 8 vs. Nov. 19. ACC oxidase (ACO) (*Vaccdscaff15-snap-gene-205.28*), which is a key enzyme in ethylene biosynthesis, is involved in the conversion of ACC to ethylene, and the expression level of *ACO* was also significantly decreased in Dec. 8 vs. Nov. 19. The genes involved in the precursor and final synthesis of ethylene were downregulated in the process of chilling accumulation, which was consistent with the results of ACC content levels described in Section 3.7. *CYP90* is a BR synthesis gene. In Dec. 8 vs. Nov. 19, the CYP90 family genes *Vaccdscaff33-snap-gene-210.34*, *Vaccdscaff46-augustus-gene-90.16*, and *Vaccdscaff4-augustus-gene-417.20* were all downregulated (Figure 7b). *Jasmonate O-methyltransferase* is a key gene for JA synthesis; with the chilling accumulation, the expression of *Vaccdscaff32-processed-gene-197.0* was significantly downregulated. After entering ecodormancy, the expression levels of the above genes increased with warming accumulation, which was consistent with the changes in the measured hormone contents.

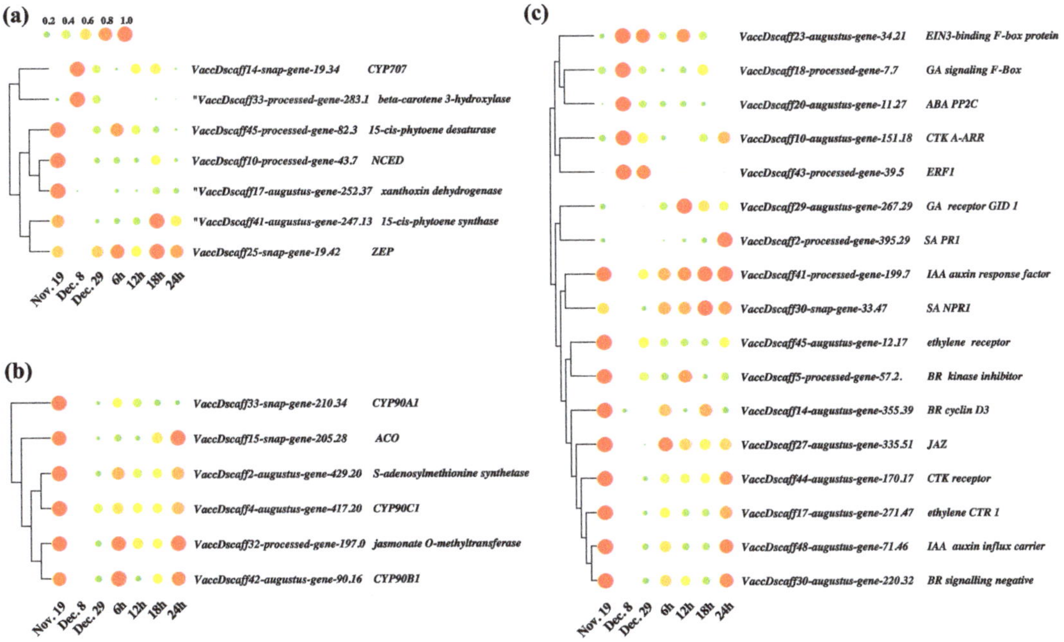

Figure 7. Analysis of changes in the expression levels of key genes in different hormone metabolic pathways. Each column represents a different sampling point, and each row represents a gene. The mean expression values were calculated using 0 to 1. Genes and expression patterns were hierarchically clustered based on the average Pearson's metric. Red indicates high expression and blue indicates low expression. (**a**) Key differentially expressed genes in the ABA biosynthesis pathway. (**b**) Key differentially expressed genes in the ethylene, JA, and BR biosynthesis pathways. (**c**) Key differentially expressed genes in the plant hormone signal transduction pathway.

There were 184 DEGs in the plant hormone signal transduction pathway (Supplementary Table S8), covering the signal transduction of auxin, gibberellin, ABA, ethylene, brassinolide, JA, and SA. In ABA signal transduction, PP2C protein phosphatase is a serine/threonine residue protein phosphatase that negatively regulates the ABA signal transduction pathway. In Dec. 8 vs. Nov. 19, the expression of *P2CC*: *Vaccdscaff20-augustus-gene-11.27* was significantly upregulated, indicating not only that the ABA level was decreased, but also that the signal transduction pathway was inhibited during the endodormancy release of flower buds. In ethylene signal transduction, the Dec. 8 vs. Nov. 19 comparison showed that *ETR*(*Vaccdscaff45-augustus-gene-12.17*), *CTR1*(*VaccDscaff17-augustus-gene-271.47*), and *MPK6*(*Vaccdscaff36-augustus-gene-293.29*), which are upstream of ethylene signal transduction, were all downregulated; however, the downstream genes *EBF1/2*(*Vaccdscaff23-augustus-gene-34.21*), *EIN3-bingding F-box protein* (*VaccDscaff23-augustus-gene-34.21*), and *ERF1* (*Vaccdscaff43-processed gene-39.5*) were all upregulated.

3.9. Coexpression Analysis Identified Hormone-Related Genes in Dormancy Release

Through coexpression clustering, we were able to better understand the effect of gene expression modules on the change in hormone contents during the process of dormancy release. The expression matrix was 98,994 rows and six columns. Because there were too many genes, and the expression levels of most genes were very low and their effects on the phenotype were minimal, we extracted genes with an FPKM average of more than 10 for subsequent analysis. The final total was 15,201 genes. When the β value was 12, the gene relationship network was closest to the scale-free distribution (Supplementary Figure S1).

After multiple parameter adjustments, 25 effective coexpression modules were obtained by clustering (Supplementary Figure S2). As shown in Figure 8a, the number of genes in the modules ranged from 50 (MEorange) to 4333 (MEturquoise). According to the correlation value (Figure 8b), we determined that MEdarkred (82 genes) was the most positively correlated with ABA, and MEsalmon (207 genes) was the most negatively correlated with ABA. MEBrown (2521 genes) had a positive correlation with SA content change, while MEgreen (663 genes) had the highest negative correlation. MEturquoise (4333 genes) was positively correlated with the change in JA content, and its correlation coefficient reached 0.95, which was the highest among all the module factors. The change in the ACC content was similar to that of ABA, and the highest correlation module was also the same; both of these were MEdarkred. From these results, it is suggested that ethylene and ABA may have a synergistic effect in the process of dormancy release.

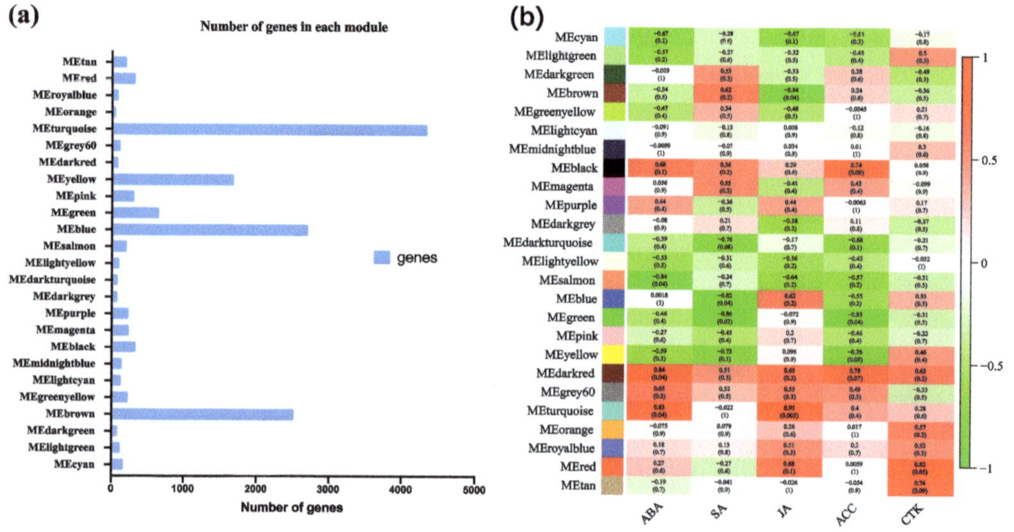

Figure 8. Coexpression module analysis. (**a**) The number of genes in each module. (**b**) Diagram of associations between gene modules and hormone contents. The X axis represents the information of each phenotype value and the Y axis represents the module. Red indicates a positive correlation, green indicates a negative correlation, and a correlation value close to 1 indicates a high correlation. The *p*-value of each significant correlation is shown in parentheses.

The transcript levels of genes in MEdarkred were shown in Figure S3. According to expression patterns, the genes can be classified into two classes. The expression patterns of the class I genes were more similar to the changes of ABA and ACC contents; interestingly, almost all transcription factors were clustered in this class. Table 2 lists the GO classification and KEGG annotation information for transcription factors in MEdarkred, including ethylene response factor 12 (*ERF12*): *VaccdScaff33-process-gene-241.9*, elongation factor G-2: *Vaccdscaff43-augustus-gene-228.38*, ABC transporter F family member 1 ISOFORM X1: *Vaccdscaff-augustus-gene-87.38*, and homeobox-leucine zipper protein REVOLUTA: *Vaccdscaff47-augustus-gene-75.30*. These genes were highly correlated with the changes in ABA and ACC contents and were also associated with ethylene and ABA signal transduction based on KEGG annotation.

Table 2. Information on transcription factors in MEdarkred.

Gene ID	GO	KEGG Annotation
VaccDscaff33-processed-gene-241.9	GO:0050794 regulation of cellular process	ethylene response factor 12
VaccDscaff43-augustus-gene-228.38	GO:0045893 positive regulation of transcription	elongation factor G-2
VaccDscaff8-snap-gene-394.21	GO:0007165 signal transduction | GO:0006351 transcription	two-component response regulator ARR1 isoform X2
VaccDscaff13-augustus-gene-302.19	GO:0042787 protein ubiquitination involved in ubiquitin-dependent protein catabolic process	E3 ubiquitin-protein ligase UPL3
VaccDscaff46-augustus-gene-87.38	GO:0008152 metabolic process	ABC transporter F family member 1 isoform X1
VaccDscaff46-augustus-gene-19.24	GO:0045893 positive regulation of transcription	elongation factor G-2, chloroplastic
VaccDscaff47-augustus-gene-75.30	GO:0048519 negative regulation of biological process	homeobox-leucine zipper protein REVOLUTA
VaccDscaff5-augustus-gene-239.20	—	E3 ubiquitin protein ligase RIE1-like
VaccDscaff703-augustus-gene-0.9	GO:0006355 regulation of transcription, DNA-templated	homeobox-leucine zipper protein HOX11-like

3.10. RT-qPCR Verification

To verify the accuracy of the transcriptome analysis results, 15 DEGs were randomly selected for RT–qPCR analysis (Supplementary Figure S4). Although few genes differed at different sampling points, the expression patterns of these 15 DEGs were basically consistent between the RT–qPCR and RNA-seq results, indicating that the expression data estimated by RNA-seq were reliable.

4. Discussion

Although the transcriptomes of flower buds in endo- and ecodormancy have been widely reported in other species [40–42], few studies have been conducted on the process of floral bud break. Recently, in tea [22] and grape [42], transcriptome studies were performed on different states of flower buds from dormancy to bloom, covering paradormancy, endodormancy, ecodormancy, and bud break with flower buds sampled by month, reflecting the gene expression of different states of buds. Due to the lack of detailed sampling at key dormancy breaking times, the detailed process of gene expression regulation by temperature during bud dormancy release cannot be determined. In this study, samples of flower buds from endodormancy to flowering included two processes: endodormancy release under chilling accumulation and ecodormancy release under warming accumulation. A total of 21 libraries were constructed for RNA-seq analysis, and the ratio of clean reads to the reference genome sequence was 72.04%~79.71%. A total of 12,350 DEGs were found in the different libraries. A large number of transcripts produced in the study were attributed to the deep RNA sequencing of blueberry.

The results showed that the bud break percentage of samples collected on 8 December exceeded 50% first, which was the critical moment for endodormancy release. The number of DEGs in Dec. 8 vs. Nov. 19 reached 8733, which was the largest number of genes among all library comparisons. The comparison between the ecodormancy state and the endodormancy transition stage (Dec. 29 vs. Dec. 8) showed that 538 genes were differentially expressed. During the process of ecodormancy release, there were 5431 DEGs in the initial 6 h of warm treatment and 750 DEGs in 12 h vs. 6 h. However, the number of DEGs did not exceed 700 in the subsequent comparisons. The change trends for the DEGs in the above process are similar to those described in previous reports [6,8]. The metabolism of starch and sucrose and the signal transduction of plant hormones were also enriched. The former regulates plant energy metabolism and the latter regulates endogenous hormone synthesis. These metabolic pathways interact to form a complex dormancy regulatory network [43].

The circadian rhythm is an important regulatory mechanism by which perennial plants adapt to seasonal cyclical changes. Plants regulate the synthesis of hormones, proteins, and sugars related to dormancy by regulating the expression of a series of genes, such as *TOC1*, *PRR*, and *LHY*, and sense the correct time to enter and break dormancy in order to adapt to seasonal and environmental changes. Therefore, it can be understood that a certain diurnal duration and chilling accumulation are necessary for the induction and release of plant dormancy [44]. In the transition from bud endodormancy to ecodormancy, light signal-transduction- and circadian-rhythm-related genes were differentially expressed. In this study, genes related to *PHYA*, *FT*, *COP1*, *LHY*, *ELF3*, and *CO* were enriched during dormancy release. These genes are a major component of the circadian rhythm. The roles of the *CO*/*FT* module in flowering regulation have been well described [45]. Recent studies have found that the *CO*/*FT* module has multiple roles in the regulation of flowering and dormancy of perennial plants [46,47]. CO can directly interact with DELLA and the histone marker reader MRG1/2 to regulate the expression of *FT* [48]. In this study, the expression of *CO* was significantly downregulated during endodormancy release, and the expression of *FT* was significantly upregulated upon the same process, which suggests that the *CO*/*FT* module plays an important role in the dormancy release of blueberries. During the dormancy release, floral buds require significant carbon and energy metabolism to resist the chilling winter conditions [49]. The expression levels of *pckA*, *IDH3*, *SDHA*, and *DLAT* were upregulated during the process of endo- and ecodormancy release, indicating that expressions of these genes are important for energy metabolism in order to synthesize various compounds for bloom [50].

ABA is considered to be a plant hormone that controls dormancy in woody plants [51]. When perennial plants enter the endodormant stage, the increase of ABA activates downstream transcription factors which activate the plant's response to cold stress, making the dormant plants resistant to low temperatures [52]. In pear, the ABA content increased during endodormancy maintenance and decreased during endodormancy release [53], and the results of ABA content change in the present study are in accordance with their findings. In recent years, many genes and signaling pathways activated in response to cold stress have been revealed by transcriptomic analysis. For example, the ICE-CBF-COR signaling pathway is activated by cold resistance [54]. Puig et al. [55] studied the transcriptome changes in peach and revealed that auxin signal transduction may play a role in cold sensitivity and tolerance. RNA-seq analyses of apple plants grown under drought, cold, or acute high-salinity stress revealed five common DEGs involved in multiple pathways and improved comprehensive resistance to stress. Among these are PP2C and ABI5 family members related to ABA signaling [56]. In this study, we found that the expression levels of *VcNCED* and *VcCYP707A* were significantly decreased and increased, respectively, during the process of endodormancy release. The results were consistent with previous research in pear [53]. In some SHB cultivars with low CR, fluctuating warm and cold spells in winter can cause flowers to suffer frost damage. Our previous research-based transcriptome indicated that the exogenous application of ABA could promote the transition of blueberry ("Emerald") flower buds into a deep endodormant state through regulation of the expression of flower bud dormancy-related genes and inhibition of blueberry early flowering [57]. Similar observations have been made in pear [16]. Experiments have proven that exogenous application of ABA can prolong bud dormancy in response to climate change.

JA is a universal phytohormone that plays a key role in stress defense as well as seed germination, plant growth, abscission, and senescence [58]. In recent years, the involvement and regulation of JA in bud dormancy has attracted much attention [22,59]. The present results showed that the contents of JA in blueberry flower buds decreased with the dormancy release; *jasmonate O-methyltransferase* (*Vaccdscaff32-processed-gene-197.0*) is a key gene that is critical for JA synthesis. The expression levels of the gene were consistent with the changes in JA content. It retained low expression levels during the process of dormancy release and was upregulated from ecodormancy to bloom; these results are similar to those of previous research [42]. It has been speculated that JA may play an important role in the process of dormancy release. Ethylene is known to interact with ABA and play a central role in various developmental, stress responses, and seed germination in plants [60,61]. Our experimental results showed that the contents of ABA and ACC decreased in the process of dormancy release in blueberry flower buds, and we found changes in *ERF12*, which is a hormone-related gene, through WGCNA. Changes in expression level of the gene were the most positively correlated with ABA and ACC, similarly to the transcriptome studies of tea. The evidence showed that ERFs were downregulated during endodormancy release [22]. Light, ethylene, and abscisic acid signal transduction pathways consecutively control bud development by setting, modifying, or terminating these processes [41]. ERFs are important transcription factors in ethylene signal transduction, as they regulate downstream genes to complete the ethylene response. Therefore, the relationship and roles of *ERF*, ABA, and ethylene in bud dormancy induction and release need further study.

This study provides a better understanding of the transcriptomic changes that occur from bud endodormancy to flowering. Our analyses using transcriptomic profiles and WGCNA identified modules and candidate genes associated with hormone changes and dormancy release; however, it remains unclear whether the candidate genes may interact with other genes in the same pathway in additive and epistatic ways, which could lead to gene expression profiles' linkage disequilibrium. Further studies are required to confirm this. Moreover, key genes controlling dormancy-related traits need to be predicted and examined further for molecular breeding. Expression genome-wide association studies (eGWAS), genomic prediction (GP), and machine learning (ML) and transgenic experiments have been widely used in investigations of forest trees [62,63]. Expression mapping via eGWAS and genomic prediction of expression profiles are the subjects of our future works. These integrative approaches will be extremely useful for a better comprehension of complex phenological processes in many species.

5. Perspective

Floral bud endodormancy in deciduous fruit trees is an essential process, and it is a biological characteristic of higher plants adapted to seasonal environmental changes through long-term natural selection [64]. SHB is a cultivar group that is better adapted to warm climates than the original northern highbush blueberry (NHB) [65]. The interspecific hybridization and nutrient culture of *Vaccinium* species have played a major role in blueberry cultivation at low latitudes [66–68]. Breeding low-CR varieties is the main goal of SHB breeding. Although many new blueberry varieties with low CR have been bred, their adaptability to the environment is not ideal, especially in the context of global climate change, and early flowering in winter and frost damage often occur. It is imperative to study the mechanism of CR and use it to cultivate more suitable varieties to manipulate chilling requirements and develop improved freezing-tolerant cultivars that can be considered to be long-term solutions. Chilling requirements are a complex quantitative trait. Although recent QTL analyses have provided useful information, the detailed genetic basis of CR traits remains obscure. The mechanisms underlying the formation of CR traits in floral buds need further study using newer genetic approaches such as gene editing, genome-wide association studies, and genome-wide environmental association studies [69]. Identification of the key candidate genes in the dormancy release transcriptome maybe used in genetic breeding efforts to generate new blueberry cultivars with different chilling requirements which can adapt to lower latitudes.

6. Conclusions

Dormancy is an important part of the fruit tree life cycle, and sufficient chilling accumulation is needed to trigger the release of dormancy. In this study, RNA-seq and RT–qPCR were used to systematically study the dormancy release process of blueberry flower buds. By comparing the transcripts of seven dormancy stages, we found some possible pathways related to dormancy release, such as circadian rhythm, plant hormone signal transduction, and so on, which are closely related to the dormancy release process.

By measuring the changes in hormone content in the process of dormancy release, WGCNA was used to obtain the genes related to these changes in hormone contents. The results of this study have important reference value for elucidating the molecular mechanisms of dormancy release in blueberry flower buds. The hormone-related gene modules mined through WGCNA will reveal the mechanisms of the hormones involved in the regulation of bud dormancy release.

Supplementary Materials: The following supporting information can be downloaded at: https://www.mdpi.com/article/10.3390/horticulturae8020176/s1, Additional file 1: Table S1: qRT-PCR primer sequences. Table S2: FPKM values of all genes. Table S3: GO analysis data of each comparison. Table S4: KEGG analysis data of each comparison. Table S5: DEGs in Circadian rhythm-plant. Table S6: DEGs in Citrate cycle (TCA cycle). Table S7: DEGs in Carotenoid biosynthesis. Table S8: DEGs in Plant hormone signal transduction. Additional file 2: Figure S1: Determination of soft threshold. The abscissa represents the soft threshold (β). (**a**) Ordinate corresponds to the index of the scale free network model; (**b**) The average link degree of each soft threshold. Figure S2: Gene cluster dendrograms and module relationships. (**a**) Clustering of genes based on the topological overlap. (**b**) Correlation analysis among 25 modules. Figure S3: RNA-seq validated by RT–qPCR. The histogram and the primary vertical axis show the relative expression values from RT–qPCR, and the line chart and the secondary vertical axis show the FPKM value from RNA-seq. Figure S4: Analysis of changes in the expression levels of genes in MEdarkred. Each column represents different sampling points, and each row represents a gene. The mean expression values were calculated using zero to one. Genes and expression patterns were hierarchically clustered based on the average Pearson's metric. Red indicates high expression, and green indicates low expression.

Author Contributions: Conceptualization, Y.L.; Data curation, Q.Z. and F.L.; Formal analysis, R.L.; Funding acquisition, Y.L. and W.G.; Investigation, Z.Z. and Y.Z. (Yu Zong); Methodology, L.Y. (Linbo Yao), L.Y. (Li Yang) and Y.Z. (Youyin Zhu); Project administration, Y.L.; Resources, W.G.; Software, Y.L.; Supervision, W.C.; Visualization, Y.L.; Writing—original draft, Y.L.; Writing—review & editing, Y.L. and R.M. All authors have read and agreed to the published version of the manuscript.

Funding: This research was funded by Zhejiang Public Welfare Technology Application Research Project (LGN21C150011), and the Project of the Major Program for Science and Technology of Zhejiang Province (2021C02066-9).

Institutional Review Board Statement: Not applicable.

Informed Consent Statement: Not applicable.

Data Availability Statement: In this study, all materials were obtained from Zhejiang Normal University (Jinhua, Zhejiang Province, China), and the samples collection were complied with relevant institutional, national, and international guidelines and legislation. The raw transcriptomic data can be accessed from the NCBI Sequence Read Archive (SRA) platform (http://www.ncbi.nlm.nih.gov/sra/ (accessed on 28 December 2021)) under accession numbers SRR16017035, SRR16017046, SRR16017047 (Nov.19); SRR16017031, SRR16017032, SRR16017033 (Dec.8); SRR16017028, SRR16017029, SRR16017030 (Dec.29); SRR16017027, SRR16017044, SRR16017045 (6 h); SRR16017041, SRR16017042, SRR16017043 (12 h); SRR16017038, SRR16017039, SRR16017040 (18 h), and SRR16017034, SRR16017036, SRR16017037 (24 h).

Conflicts of Interest: The authors declare that they have no competing interest.

References

1. Lang, G.A.; Early, J.D.; Martin, G.C.; Darnell, R.L. Endo-, Para-, and Ecodormancy: Physiological Terminology and Classification for Dormancy Research. *Hortscience* **1987**, *22*, 271–277.
2. Cline, M.G.; Deppong, D.O. The Role of Apical Dominance in Paradormancy of Temperate Woody Plants: A Reappraisal. *J. Plant Physiol.* **1999**, *155*, 350–356. [CrossRef]
3. Qin, Z.; Bai, Y.; Muhammad, S.; Wu, X.; Deng, P.; Wu, J.; An, H.; Wu, L. Divergent roles of FT-like 9 in flowering transition under different day lengths in Brachypodium distachyon. *Nat. Commun.* **2019**, *10*, 812. [CrossRef]
4. Zhuang, W.; Gao, Z.; Wang, L.; Zhong, W.; Ni, Z.; Zhang, Z. Comparative proteomic and transcriptomic approaches to address the active role of GA4 in Japanese apricot flower bud dormancy release. *J. Exp. Bot.* **2013**, *64*, 4953–4966. [CrossRef]

5. Liu, Y.; Ye, N.; Liu, R.; Chen, M.; Zhang, J. H_2O_2 mediates the regulation of ABA catabolism and GA biosynthesis in Arabidopsis seed dormancy and germination. *J. Exp. Bot.* **2010**, *61*, 2979–2990. [CrossRef] [PubMed]
6. Liu, G.; Li, W.; Zheng, P.; Xu, T.; Chen, L.; Liu, D.; Hussain, S.; Teng, Y. Transcriptomic analysis of 'Suli' pear (Pyrus pyrifolia white pear group) buds during the dormancy by RNA-Seq. *BMC Genom.* **2012**, *13*, 700. [CrossRef] [PubMed]
7. Gao, Y.; Yang, Q.; Yan, X.; Wu, X.; Yang, F.; Li, J.; Wei, J.; Ni, J.; Ahmad, M.; Bai, S.; et al. High-quality genome assembly of 'Cuiguan' pear (Pyrus pyrifolia) as a reference genome for identifying regulatory genes and epigenetic modifications responsible for bud dormancy. *Hortic. Res.* **2021**, *8*, 197. [CrossRef]
8. Zhu, Y.; Li, Y.; Xin, D.; Chen, W.; Shao, X.; Wang, Y.; Guo, W. RNA-Seq-based transcriptome analysis of dormant flower buds of Chinese cherry (Prunus pseudocerasus). *Gene* **2015**, *555*, 362–376. [CrossRef]
9. Ionescu, I.A.; López-Ortega, G.; Burow, M.; Bayo-Canha, A.; Junge, A.; Gericke, O.; Møller, B.L.; Sánchez-Pérez, R. Transcriptome and Metabolite Changes during Hydrogen Cyanamide-Induced Floral Bud Break in Sweet Cherry. *Front. Plant Sci.* **2017**, *8*, 1233. [CrossRef]
10. Yu, J.; Conrad, A.O.; Decroocq, V.; Zhebentyayeva, T.; Williams, D.E.; Bennett, D.; Roch, G.; Audergon, J.-M.; Dardick, C.; Liu, Z.; et al. Distinctive Gene Expression Patterns Define Endodormancy to Ecodormancy Transition in Apricot and Peach. *Front. Plant Sci.* **2020**, *11*, 180. [CrossRef]
11. Khalil-Ur-Rehman, M.; Sun, L.; Li, C.-X.; Faheem, M.; Wang, W.; Tao, J.-M. Comparative RNA-seq based transcriptomic analysis of bud dormancy in grape. *BMC Plant Biol.* **2017**, *17*, 18. [CrossRef] [PubMed]
12. Min, Z.; Zhao, X.; Li, R.; Yang, B.; Liu, M.; Fang, Y. Comparative transcriptome analysis provides insight into differentially expressed genes related to bud dormancy in grapevine (*Vitis vinifera*). *Sci. Hortic.* **2017**, *225*, 213–220. [CrossRef]
13. Zhang, H.; Li, H.; Lai, B.; Xia, H.; Wang, H.; Huang, X. Morphological Characterization and Gene Expression Profiling during Bud Development in a Tropical Perennial, Litchi chinensis Sonn. *Front. Plant Sci.* **2016**, *7*, 1517. [CrossRef] [PubMed]
14. Niu, Q.; Li, J.; Cai, D.; Qian, M.; Jia, H.; Bai, S.; Hussain, S.; Liu, G.; Teng, Y.; Zheng, X. Dormancy-associated MADS-box genes and microRNAs jointly control dormancy transition in pear (*Pyrus pyrifoliawhite* pear group) flower bud. *J. Exp. Bot.* **2016**, *67*, 239–257. [CrossRef] [PubMed]
15. Saito, T.; Bai, S.; Imai, T.; Ito, A.; Nakajima, I.; Moriguchi, T. Histone modification and signalling cascade of thedormancy-associated MADS-boxgene, PpMADS13-1, in Japanese pear (*Pyrus pyrifolia*) during endodormancy. *Plant Cell Environ.* **2014**, *38*, 1157–1166. [CrossRef]
16. Yang, Q.; Yang, B.; Li, J.; Wang, Y.; Tao, R.; Yang, F.; Wu, X.; Yan, X.; Ahmad, M.; Shen, J.; et al. ABA-responsive ABRE-BINDING FACTOR3 activates DAM3 expression to promote bud dormancy in Asian pear. *Plant Cell Environ.* **2020**, *43*, 1360–1375. [CrossRef]
17. Wang, Q.; Xu, G.; Zhao, X.; Zhang, Z.; Wang, X.; Liu, X.; Xiao, W.; Fu, X.; Chen, X.; Gao, D.; et al. Transcription factor TCP20 regulates peach bud endodormancy by inhibiting DAM5/DAM6 and interacting with ABF2. *J. Exp. Bot.* **2019**, *71*, 1585–1597. [CrossRef]
18. Singh, R.; Maurya, J.P.; Azeez, A.; Miskolczi, P.; Tylewicz, S.; Stojkovič, K.; Delhomme, N.; Busov, V.; Bhalerao, R.P. A genetic network mediating the control of bud break in hybrid aspen. *Nat. Commun.* **2018**, *9*, 4173. [CrossRef]
19. Lobos, G.A.; Hancock, J.F. Breeding blueberries for a changing global environment: A review. *Front. Plant Sci.* **2015**, *6*, 782. [CrossRef]
20. Bigler, C.; Bugmann, H. Climate-induced shifts in leaf unfolding and frost risk of European trees and shrubs. *Sci. Rep.* **2018**, *8*, 9865. [CrossRef]
21. Legave, J.-M.; Guédon, Y.; Malagi, G.; EL Yaacoubi, A.; Bonhomme, M. Differentiated Responses of Apple Tree Floral Phenology to Global Warming in Contrasting Climatic Regions. *Front. Plant Sci.* **2015**, *6*, 1054. [CrossRef] [PubMed]
22. Hao, X.; Yang, Y.; Yue, C.; Wang, L.; Horvath, D.P.; Wang, X. Comprehensive Transcriptome Analyses Reveal Differential Gene Expression Profiles of Camellia sinensis Axillary Buds at Para-, Endo-, Ecodormancy, and Bud Flush Stages. *Front. Plant Sci.* **2017**, *8*, 553. [CrossRef] [PubMed]
23. Richardson, E.A. A Model for Estimating the Completion of Rest for "Redhaven" and "Elberta" Peach Trees. *HortScience* **1974**, *9*, 331–332.
24. Yooyongwech, S.; Sugaya, S.; Sekozawa, Y.; Gemma, H. Differential adaptation of high- and low-chill dormant peaches in winter through aquaporin gene expression and soluble sugar content. *Plant Cell Rep.* **2009**, *28*, 1709–1715. [CrossRef]
25. Anders, S.; Pyl, P.T.; Huber, W. HTSeq—A Python framework to work with high-throughput sequencing data. *Bioinformatics* **2015**, *31*, 166–169. [CrossRef]
26. Tarazona, S.; García, F.; Ferrer, A.; Dopazo, J.; Conesa, A. NOIseq: A RNA-seq differential expression method robust for sequencing depth biases. *EMBnet. J.* **2012**, *17*, 18. [CrossRef]
27. Love, M.I.; Huber, W.; Anders, S. Moderated Estimation of Fold Change and Dispersion for RNA-Seq Data with DESeq2. *Genome Biol.* **2014**, *15*, 550. [CrossRef]
28. Conesa, A.; Götz, S.; García-Gómez, J.M.; Terol, J.; Talón, M.; Robles, M. Blast2GO: A universal tool for annotation, visualization and analysis in functional genomics research. *Bioinformatics* **2005**, *21*, 3674–3676. [CrossRef] [PubMed]
29. Aparicio, G.; Götz, S.; Conesa, A.; Segrelles, D.; Blanquer, I.; García, J.M.; Hernandez, V.; Robles, M.; Talon, M. Blast2GO goes grid: Developing a grid-enabled prototype for functional genomics analysis. *Stud. Health Technol. Inf.* **2006**, *2006*, 120.
30. Minoru, K.; Michihiro, A.; Susumu, G.; Masahiro, H.; Mika, H.; Masumi, I.; Toshiaki, K.; Shuichi, K.; Shujiro, O.; Toshiaki, T. KEGG for Linking Genomes to Life and the Environment. *Nucleic Acids Res.* **2008**, *36*, 480–484.

31. Ye, J.; Fang, L.; Zheng, H.; Zhang, Y.; Chen, J.; Zhang, Z.; Wang, J.; Li, S.; Li, R.; Bolund, L.; et al. WEGO: A web tool for plotting GO annotations. *Nucleic Acids Res.* **2006**, *34*, W293–W297. [CrossRef] [PubMed]
32. Zhaoyun, C.; Youning, M.; Renxiang, M.; Shasha, Y.; Mingxue, C. Analysis of 17 Cytokinins in Rice by Solid Phase Extraction Purification and Liquid Chromatography-Tandem Mass Spectrometry. *Chin. J. Chromatogr.* **2015**, *33*, 715–721.
33. Zhao, J.; Yu, N.; Ju, M.; Fan, B.; Zhang, Y.; Zhu, E.; Zhang, M.; Zhang, K. ABC transporter OsABCG18 controls the shootward transport of cytokinins and grain yield in rice. *J. Exp. Bot.* **2019**, *70*, 6277–6291. [CrossRef] [PubMed]
34. Šimura, J.; Antoniadi, I.; Široká, J.; Tarkowská, D.; Strnad, M.; Ljung, K.; Novák, O. Plant Hormonomics: Multiple Phytohormone Profiling by Targeted Metabolomics. *Plant Physiol.* **2018**, *177*, 476–489. [CrossRef]
35. Hellens, R.P.; Allan, A.C.; Friel, E.N.; Bolitho, K.; Grafton, K.; Templeton, M.D.; Karunairetnam, S.; Gleave, A.P.; Laing, W.A. Transient expression vectors for functional genomics, quantification of promoter activity and RNA silencing in plants. *Plant Methods* **2005**, *1*, 13. [CrossRef]
36. Langfelder, P.; Horvath, S. WGCNA: An R package for weighted correlation network analysis. *BMC Bioinform.* **2008**, *9*, 559. [CrossRef]
37. Yang, L.; Chen, M.; Cai, K.; Zhang, L.; Zhu, Y.; Ye, Q.; Lü, M.; Liao, F.; Chen, W.; Guo, W. VcFAS, VcSUN and VcOVATE orchestrated the fruit morphogenesis in southern highbush blueberry during the pre-anthesis and fruit development. *Sci. Hortic.* **2018**, *240*, 109–115. [CrossRef]
38. Die, J.V.; Rowland, L.J. Superior Cross-Species Reference Genes: A Blueberry Case Study. *PLoS ONE* **2013**, *8*, e73354. [CrossRef]
39. Yang, L.; Liu, L.; Wang, Z.; Zong, Y.; Yu, L.; Li, Y.; Liao, F.; Chen, M.; Cai, K.; Guo, W. Comparative anatomical and transcriptomic insights into Vaccinium corymbosum flower bud and fruit throughout development. *BMC Plant Biol.* **2021**, *21*, 289. [CrossRef]
40. Bai, S.; Saito, T.; Sakamoto, D.; Ito, A.; Fujii, H.; Moriguchi, T. Transcriptome Analysis of Japanese Pear (Pyrus pyrifolia Nakai) Flower Buds Transitioning Through Endodormancy. *Plant Cell Physiol.* **2013**, *54*, 1132–1151. [CrossRef]
41. Ruttink, T.; Arend, M.; Morreel, K.; Storme, V.; Rombauts, S.; Fromm, J.; Bhalerao, R.; Boerjan, W.; Rohde, A. A Molecular Timetable for Apical Bud Formation and Dormancy Induction in Poplar. *Plant Cell* **2007**, *19*, 2370–2390. [CrossRef] [PubMed]
42. Shangguan, L.; Chen, M.; Fang, X.; Xie, Z.; Gong, P.; Huang, Y.; Wang, Z.; Fang, J. Comparative transcriptome analysis provides insight into regulation pathways and temporal and spatial expression characteristics of grapevine (*Vitis vinifera*) dormant buds in different nodes. *BMC Plant Biol.* **2020**, *20*, 390. [CrossRef] [PubMed]
43. Aguilar-Martínez, J.A.; Poza-Carrión, C.; Cubas, P. Arabidopsis BRANCHED1Acts as an Integrator of Branching Signals within Axillary Buds. *Plant Cell* **2007**, *19*, 458–472. [CrossRef] [PubMed]
44. Cooke, J.E.K.; Eriksson, M.E.; Junttila, O. The dynamic nature of bud dormancy in trees: Environmental control and molecular mechanisms. *Plant Cell Environ.* **2012**, *35*, 1707–1728. [CrossRef] [PubMed]
45. Pin, P.A.; Nilsson, O. The multifaceted roles of FLOWERING LOCUS T in plant development. *Plant Cell Environ.* **2012**, *35*, 1742–1755. [CrossRef]
46. Böhlenius, H.; Huang, T.; Charbonnel-Campaa, L.; Brunner, A.M.; Jansson, S.; Strauss, S.H.; Nilsson, O. CO/FT Regulatory Module Controls Timing of Flowering and Seasonal Growth Cessation in Trees. *Science* **2006**, *312*, 1040–1043. [CrossRef]
47. Xu, F.; Li, T.; Xu, P.-B.; Li, L.; Du, S.-S.; Lian, H.-L.; Yang, H.-Q. DELLA proteins physically interact with CONSTANS to regulate flowering under long days inArabidopsis. *FEBS Lett.* **2016**, *590*, 541–549. [CrossRef]
48. Zhongyuan, B.; Yu, Y.; Zepeng, L.; Yanchao, L.; Wen, J.; Ying, H.; Ai-Wu, D. Regulation of Arabidopsis Flowering by the Histone Mark Readers MRG1/2 via Interaction with CONSTANS to Modulate FT Expression. *PLoS Genet.* **2014**, *10*, e1004617.
49. Ben Mohamed, H.; Vadel, A.M.; Geuns, J.M.; Khemira, H. Biochemical changes in dormant grapevine shoot tissues in response to chilling: Possible role in dormancy release. *Sci. Hortic.* **2010**, *124*, 440–447. [CrossRef]
50. Ben Mohamed, H.; Vadel, A.M.; Geuns, J.M.; Khemira, H. Carbohydrate changes during dormancy release in Superior Seedless grapevine cuttings following hydrogen cyanamide treatment. *Sci. Hortic.* **2012**, *140*, 19–25. [CrossRef]
51. Guak, S.; Fuchigami, L.H. Effects of Applied ABA on Growth Cessation, Bud Dormancy, Cold Acclimation, Leaf Senescence and N Mobilization in Apple Nursery Plants. *J. Hortic. Sci. Biotechnol.* **2001**, *76*, 459–464. [CrossRef]
52. Ma, Y.; Szostkiewicz, I.; Korte, A.; Moes, D.; Yang, Y.; Christmann, A.; Grill, E. Regulators of PP2C Phosphatase Activity Function as Abscisic Acid Sensors. *Science* **2009**, *324*, 1064–1068. [CrossRef] [PubMed]
53. Li, J.; Xu, Y.; Niu, Q.; He, L.; Teng, Y.; Bai, S. Abscisic Acid (ABA) Promotes the Induction and Maintenance of Pear (Pyrus pyrifolia White Pear Group) Flower Bud Endodormancy. *Int. J. Mol. Sci.* **2018**, *19*, 310. [CrossRef] [PubMed]
54. Chinnusamy, V.; Zhu, J.; Zhu, J.-K. Cold stress regulation of gene expression in plants. *Trends Plant Sci.* **2007**, *12*, 444–451. [CrossRef] [PubMed]
55. Puig, C.P.; Dagar, A.; Ibanez, C.M.; Singh, V.; Crisosto, C.H.; Friedman, H.; Lurie, S.; Granell, A. Pre-symptomatic transcriptome changes during cold storage of chilling sensitive and resistant peach cultivars to elucidate chilling injury mechanisms. *BMC Genom.* **2015**, *16*, 245. [CrossRef] [PubMed]
56. Li, X.; Li, M.; Zhou, B.; Yang, Y.; Wei, Q.; Zhang, J. Transcriptome analysis provides insights into the stress response crosstalk in apple (Malus × domestica) subjected to drought, cold and high salinity. *Sci. Rep.* **2019**, *9*, 9071. [CrossRef] [PubMed]
57. An, S.; GAO, Y.; Maidunur, Y.; PAN, Y.; SHAO, W.; ZONG, Y.; CHEN, W.; YANG, L.; GUO, W.; LI, Y. Research on Application Exogenous Abscisic Acid in Inhibiting Early Flowering and Associated Genes Expression Characteristics in Blueberry. *J. Fruit Sci.* **2020**, *38*, 325–334. [CrossRef]

58. Davies, P.J. The Plant Hormones: Their Nature, Occurrence, and Functions. In *Plant Hormones*; Springer: Dordrecht, The Netherlands, 1995.
59. Howe, G.T.; Horvath, D.P.; Dharmawardhana, P.; Priest, H.D.; Mockler, T.C.; Strauss, S.H. Extensive Transcriptome Changes During Natural Onset and Release of Vegetative Bud Dormancy in Populus. *Front. Plant Sci.* **2015**, *6*, 989. [CrossRef]
60. Bleecker, A.B.; Kende, H. Ethylene: A Gaseous Signal Molecule in Plants. *Annu. Rev. Cell Dev. Biol.* **2000**, *16*, 1–18. [CrossRef]
61. Stepanova, A.N.; Alonso, J.M. Alonso Ethylene Signalling and Response Pathway: A Unique Signalling Cascade with a Multitude of Inputs and Outputs. *Physiol. Plant.* **2005**, *123*, 195–206. [CrossRef]
62. Arenas, S.; Cortés, A.J.; Mastretta-Yanes, A.; Jaramillo-Correa, J.P. Evaluating the accuracy of genomic prediction for the management and conservation of relictual natural tree populations. *Tree Genet. Genomes* **2021**, *17*, 12. [CrossRef]
63. Cortés, A.J.; Restrepo-Montoya, M.; Bedoya-Canas, L.E. Modern Strategies to Assess and Breed Forest Tree Adaptation to Changing Climate. *Front. Plant Sci.* **2020**, *11*, 583323. [CrossRef] [PubMed]
64. Samish, M.R. Dormancy in Woody Plants. *Annu. Rev. Plant Physiol.* **1954**, *5*, 183–204. [CrossRef]
65. Nishiyama, S.; Fujikawa, M.; Yamane, H.; Shirasawa, K.; Babiker, E.; Tao, R. Genomic insight into the developmental history of southern highbush blueberry populations. *Heredity* **2020**, *126*, 194–205. [CrossRef] [PubMed]
66. Medina Cano, C.I.; Lobo Arias, M.; Castaño Colorado, Á.A.; Cardona, L.E. Development analysis of Mortiño (Vaccinium meridionale Swart.) plants derived from clonal and sexual propagation. *CTA* **2015**, *16*, 65–77. [CrossRef]
67. Cappai, F.; Garcia, A.; Cullen, R.; Davis, M.; Munoz, P.R. Advancements in Low-Chill Blueberry *Vaccinium corymbosum* L. Tissue Culture Practices. *Plants* **2020**, *9*, 1624. [CrossRef]
68. Frías-Ortega, C.E.; Alejo-Santiago, G.; Bugarín-Montoya, R.; Aburto-González, C.A.; Juárez-Rosete, C.R.; Urbina-Sánchez, E.; Sánchez-Hernández, E. Nutrient Solution Concentration and Its Relationship with Blueberry Production and Quality. *Cienc. Tecnol. Agropecu.* **2020**, *21*, e1296.
69. Yang, Q.; Gao, Y.; Wu, X.; Moriguchi, T.; Bai, S.; Teng, Y. Bud endodormancy in deciduous fruit trees: Advances and prospects. *Hortic. Res.* **2021**, *8*, 139. [CrossRef]

horticulturae

Article

Pathogenic Fungi Diversity of 'CuiXiang' Kiwifruit Black Spot Disease during Storage

Yaming Yang [1], Lijuan Chen [1,2], Chenyu Wang [1], Honghui Peng [1], Weijie Yin [1], Rui Li [1], Cuihua Liu [1], Xiaolin Ren [1,*] and Yuduan Ding [1,*]

1 College of Horticulture, Northwest Agricultural & Forestry University, Xianyang 712100, China; yangym@nwafu.edu.cn (Y.Y.); chenlijuan@nwafu.edu.cn (L.C.); wcy1004@nwafu.edu.cn (C.W.); penghonghui2021@163.com (H.P.); ywj2018050283@nwafu.edu.cn (W.Y.); lirui0410@nwafu.edu.cn (R.L.); liuch@nwafu.edu.cn (C.L.)
2 Institute of Horticulture, Sichuan Academy of Agricultural Sciences, Chengdu 610000, China
* Correspondence: tjw689@126.com (X.R.); dingyuduan@nwafu.edu.cn (Y.D.)

Abstract: Kiwifruit black spot disease has become increasingly widespread in many 'CuiXiang' kiwifruit plantings regions. This research was aimed at the pathogenic microorganisms of black spot of the 'CuiXiang' cultivar. Physiological, morphological and transcriptional characteristics between black spot fruit and healthy fruits were evaluated. Then, it applied a high-throughput internal transcribed spacer (ITS) sequencing to analyze the black spot disease microbial community. The cell structure showed that mycelium was attached to the surface of the kiwifruit through black spot, and that consequently the mitochondria were damaged, starch particles were reduced, and shelf life was shortened. Transcriptome revealed that different genes in kiwifruit with black spot disease were involved in cell wall modification, pathogen perception, and signal transduction. ITS sequencing results described the disease-causing fungi and found that the microbial diversity of black spot-diseased fruit was lower than that of healthy fruit. We predict that candidate pathogenic fungi *Cladosporium cladosporioides*, *Diaporthe phaseolorum*, *Alternaria alternata*, and *Trichothecium roseum* may cause black spot. This study was to explore the pathogenic fungal community of 'CuiXiang' kiwifruit black spot disease and to provide essential information for field prevention.

Keywords: black spot; 'CuiXiang'; kiwifruit; internal transcribed spacer sequencing; transcriptome analysis

Citation: Yang, Y.; Chen, L.; Wang, C.; Peng, H.; Yin, W.; Li, R.; Liu, C.; Ren, X.; Ding, Y. Pathogenic Fungi Diversity of 'CuiXiang' Kiwifruit Black Spot Disease during Storage. *Horticulturae* **2022**, *8*, 13. https://doi.org/10.3390/horticulturae8010013

Academic Editors: Dong Zhang and Libo Xing

Received: 25 November 2021
Accepted: 20 December 2021
Published: 23 December 2021

1. Introduction

The kiwifruit (*Actinidia* spp.) is one of the important horticultural crops that is with rapidly grown in China [1]. China is rich in kiwifruit resources, planting area, and yield to reach the world's demands [2]. 'CuiXiang' (*Actinidia deliciosa* 'CuiXiang') is a mid-early maturing cultivar with excellent fruit quality. 'CuiXiang' is very popular with consumers and is predominantly cultivated in China, particularly in Shaanxi province. 'CuiXiang' is a delicious early-maturing kiwifruit variety that has been selected and bred by Xi'an Kiwifruit Institute and Zhouzhi county's Agricultural Technology Experimental Station, following more than 10 years of breeding work. *A. chinensis* is the paternal progenitor of *A. deliciosa* [3]. Black spot disease (also called blackhead, black mold, or mildew) has intensified on species of *Actinidia deliciasa*, especially 'CuiXiang'. At the early stage (in June in Shaanxi), some small spots appear on the apex (stigma-end) of young fruits and become gradually contiguous. The disease only exhibits on the skin (outer pericarp) and cannot expand into the fruit pulp (mesocarp). The disease can accelerate the fruit's softening and reduce the shelf life. In 2013, black spot disease appeared on 'CuiXiang' fruit [4]. At that time, the disease was regarded as being specific to 'CuiXiang'. However, in 2019, 'XuXiang' fruit was found to have similar symptoms to 'CuiXiang' black spot disease [5]. Black spot disease damages the kiwifruit appearance and reduces the postharvest quality of fruits. It severely affects fruit quality and profitability.

the reads. HISAT2 v2.0.5 was used to build an index of the reference genome and compare it with the reference genome. The gene expression levels were analyzed using the HTSeq v0.9.1 tool and expressed using the FPKM value (fragments per kilobase of transcript per million mapped reads). A total of three biological samples were used in each stage.

In order to identify the differentially expressed genes of kiwifruit with black spot during postharvest, transcriptome analysis was carried out on RNA-seq of fruit samples. The different expression genes (DEG) between healthy and diseased fruit samples were detected. A total of 1143 and 8805 DEGs detected that compared with healthy fruits were harvested at 0 and 5 DAH.

2.5. ITS Sequence and Bioinformatics Analysis of Fungal Communities on Fruit Surface

Healthy and diseased 'CuiXiang' kiwifruits were immersed in 0.5 L of 1% (*w*/*v*) peptone water and 1% (*v*/*v*) Tween-80 was added with a sterile beaker [23]. Then the Erlenmeyer flask was vigorously shaken for approximately 20 min to thoroughly wash the surface of the fruits. The suspension was filtered over a sterile polycarbonate membrane (0.22 μm pore size) held in a sterilized filtration device. The membrane containing the fungal filtrate was placed in a sterile tube and stored at −80 °C until the DNA extraction stage. Then the extracts underwent sequencing analysis (Novogene Bioinformatics Institute, Beijing, China) [24]. The experiment consisted of five separate biological replications from 5 orchards.

ITS sequences were detected by three-generation full-length amplicon sequencing technology. The species and abundance of fungi in 5 pairs of diseased and healthy tissue samples were analyzed. The raw ITS rDNA gene sequencing reads were demultiplexed and quality filtered by CCS (generate highly accurate single-molecule consensus reads) (SMRT Link v7.0) to correct the sequence with the following criteria: (1) the correction parameters were CCS = 3 and the minimum accuracy rate was 0.99 [25], (2) the shortest < 500 and the longest sequences length > 1000 were removed, (3) SSR (simple sequence repeat) filtering and Cutadapt (version 1.14) were used to remove the primers and filter out sequences containing consecutive identical bases > 8, and (4) the result was compared with the full-length annotation database to remove chimera.

We assigned sequences with ≥97% similarity to the same operational taxonomic units (OTUs) in UPARSE (Uparse v7.0.1001) [26] and then selected the sequences with the highest frequency of OTUs as the representative sequence. BLAST [27] in Qiime (Version 1.9.1) and the Unit Database (https://unite.ut.ee/, accessed on 23 March 2020) were used for species annotation analysis, and the community composition of each sample was counted at each classification level: boundary, phylum, outline, order, family, genus, and species. MUSCLE (Version 3.8.31) was used for fast multi-sequence alignment to obtain the systematic occurrence of all OTU representative sequences. Finally, the data of each sample were made uniform. Alpha and beta diversity analyses were based on the data after homogenization. We used Qiime (Version 1.9.1) to calculate Good's coverage indexes. R (Version 2.15.3), LEfSe, and Qiime (Version 1.9.1) were used to analyze the differences in diversity indexes. Unweighted UniFrac distance, principal component analysis, Chao1 index, Shannon index, and Venn diagram analysis were performed for OTU obtained by sequencing of healthy tissue and diseased fruit tissue. Fungal community diversity in healthy and diseased tissues was analyzed at phylum and genus levels.

Text mining work were carried out for search information of the significant fungi. Literature was retrieved from research articles in the NCBI PubMed and Web of Science databases. Information was screened for pathogenic fungi and antagonistic fungi related to plant diseases, and information on reducing their pathogenic association with black spot disease in kiwifruit. Text mining results of the pathogenicity of the significant fungi are listed in Table S1.

2.6. Laboratory Isolation, Purification, Identification, and Inoculation of Pathogenic Fungi

2.6.1. Isolation, Purification, and Identification of Pathogenic Fungus

The fruit surface was disinfected with 75% alcohol for 3–5 s, rinsed with distilled water and dried slightly. The peel at the junction of the fruit disease spots was removed on the very clean workbench and then placed in potato glucose agar medium (PDA), and cultured at 25 °C for 5–7 days. After growing hyphae, a very small number of hyphae at the edge of the colony was removed for separation and purification, and, after purification 5–6 times, samples for sequencing and identification were sent. The sequencing results of the isolated and purified 7 strains were compared with the existing ITS database BLAST.

2.6.2. Inoculation of Pathogenic Fungus

The cultured mycelium was scraped with a sterile knife, diluted and filtered with sterile water. Then the number of spores under the blood cell counting plate were observed. On the day of inoculation, the sample was diluted with sterile water into 1X105 spore suspension/mL for standby. The surface of the healthy fruit was disinfected, and the outer skin layer of the kiwifruit was pricked with an inoculation needle μ L and then the spore suspension was dropped at the small hole and inoculated with sterile water as a control. A layer of polyethylene plastic film (0.01 mm thick) on the back cover was slightly dried to retain moisture. The sample was stored at room temperature (22 ± 2 °C) and relative humidity of 85–90%. After inoculation for 15 days, the incidence of fruit was observed and counted. Duncan's Range Multiple of ANOVA was used for significance test ($p < 0.05$ significant difference).

3. Results and Discussion

3.1. Black Spot Disease Symptom during Postharvest: Accelerating Softening of Fruit, Visioning Hyphae Attached to Fruit Skin and Exhibiting Cell Wall Structure Changes

The kiwifruit stored at room temperature showed that black spot disease accelerated fruit softening (Figure S1). The physiological indexes of kiwifruit diseased fruit (with different degree of disease) and healthy fruit were preliminarily detected under normal temperature storage. Results showed the black spot could significantly affect the firmness of fruit, accelerate ethylene production, and significantly shorten the shelf life. The SSC of black spot kiwifruit was higher than healthy fruit, and the content of water-soluble calcium was significantly lower than that of healthy fruit (Figure S2). The taste of kiwifruit can be affected by a number of components, including SSC and TA [18]. The mineral composition of kiwifruit is an important factor for its nutritional quality, in particular its nutritional properties (sodium/potassium ratio) [28]. Preharvest treatment of Ca-chitosan delayed the loss of firmness in kiwifruit during cold storage [29]. Therefore, the measurement of these physiological characteristics can explain the impact of black spot disease on 'CuiXiang' kiwifruit.

The microscopic aspects illustrated the difference between healthy and diseased fruit at the cell microstructure level. On the surface of diseased kiwifruit skins were scattered black spots (Figure 1A). Paraffin section observation showed that the keratinocytes of the diseased fruit were thicker than those of the healthy fruit, and the epidermal cells of the healthy fruit were arranged more orderly (Figure 1B). The results of transmission electron microscope showed that relatively complete cells, including mitochondria, starch granules and other organelles, could be observed in the healthy pericarp, and a relatively obvious bright–dark–bright zoning structure could be seen, while the mitochondria, starch granules, and other organelles in the diseased pericarp were damaged (Figure 1C). It was also observed by scanning electron microscopy that the diseased peels showed filaments resembling mycelium, with different shapes. Some were a cluster of round ovate shapes, and some were elongated and cylindrical, growing from the epidermal tissue of the peel. The healthy peel had less filamentous material and a smooth surface. Scanning electron microscope results showed that there were hyphae on the surface of the diseased fruit (Figure 1D).

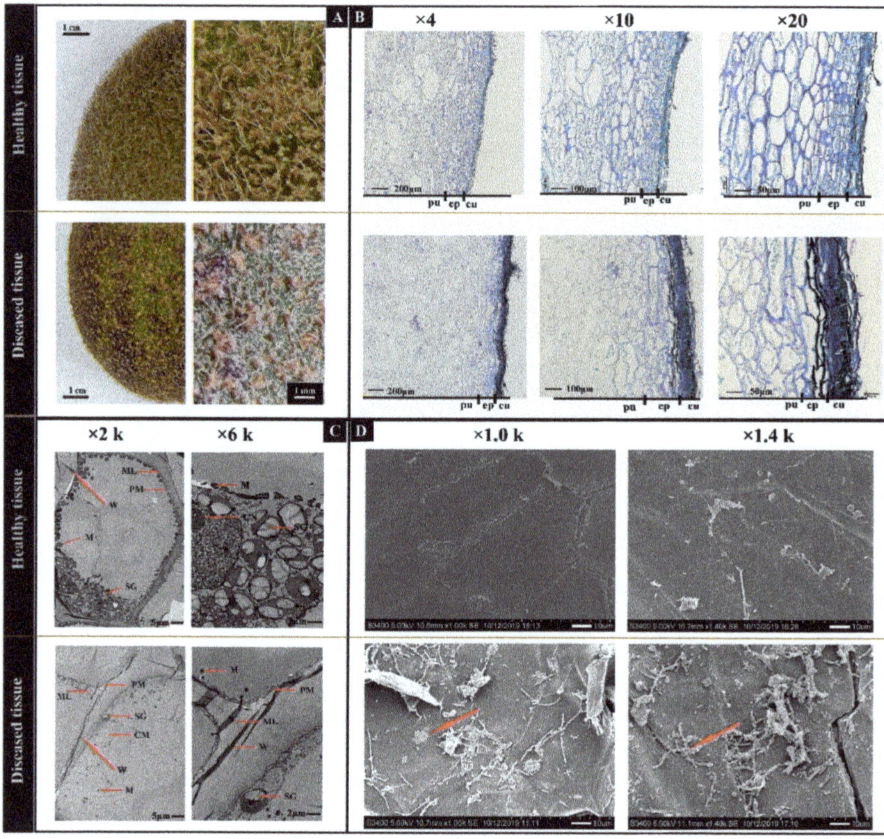

Figure 1. Phenomena of 'CuiXiang' healthy fruit and black spot diseased fruit. (**A**) The surface of diseased and healthy kiwifruit. (**B**) Paraffin sections of outer pericarps of healthy and diseased fruits. The equatorial pericarp of healthy and diseased fruit was fixed, dehydrated, wax-soaked and embedded, sliced, stained and sealed, and finally observed and photographed by a positive fluorescence microscope at Olympus BX51 (cu-cuticle, ep-epidermal cells, pu-pulp cells, SG-starch granule; magnification times: ×4, ×10 and ×20). (**C**) Transmission electron microscope observation of pericarp of healthy and diseased fruits. Note: ML-medium adhesive layer; M-mitochondria; CM-vacuolated mitochondria; PM-plasma membrane; SG-starch grain; W-wrinkle. (**D**) Scanning electron microscope observation of healthy and diseased fruit surface. Healthy and diseased fruit peel were cut into 1.0–1.5 cm^2 pieces and observed under an S-3400N electron microscope (acceleration voltage: 5.00 kv, mm: the distance of the sample from the probe, magnification times: ×1.00 k and ×1.40 k).

In summary, the phenomena of black spot disease symptom of kiwifruit indicated: (1) the thickness of cuticle and epidermal cells on fruit surface will affect its water loss [30], paraffin wax results show that the cuticle of healthy fruit is thinner than that of diseased fruit, which can explain the faster water loss of healthy fruit; (2) the degradation of the cell wall is considered to directly lead to fruit softening [31]; (3) scanning electron microscopy showed that hyphae attached to the host cell wall. The short shelf life of diseased fruit indicates that the pathogen invades the related enzymes that may be released from the host cell wall, resulting in the semi disintegration or complete disintegration of the organelles of the host cells; and (4) mitochondria are an important place for fruit aerobic respiration. The number and integrity of mitochondria affect the process of fruit senescence [32]. In this study, compared with healthy fruit, the mitochondrial structure of diseased fruit was damaged, the number was significantly less, and the number of starch particles decreased and gradually weakened; the cell wall was severely deformed.

3.2. Transcription Influence of Black Spot Disease on Fruits: Cell Wall Modification and Pathogen Signals Genes Were Regulated

A total of 1143 and 8805 DEGs detected that compared with healthy fruits were harvested at 0 and 5 days after harvesting (DAH) (Figure 2A). In this study, DEG involved in cell wall modification was regulated to varying degrees in different storage periods (Figure 2B). Encoding, such as polygalacturonase (PG, Achn144331), xyloglucan endotransglucosylase (XTH, Achn088411), pectate lyase (PL, Achn070291 | Achn315151), and expansin (EXP, Achn382301), were up-regulated. The cell wall is the basic barrier of the plant defense system in response to pathogen attack [33]. Especially during fruit ripening, the disintegration of the cell wall allows pathogens to easily invade [34]. When plants are infected by pathogens, they will detect pathogen signals through pattern recognition receptors (PRRs) [35].

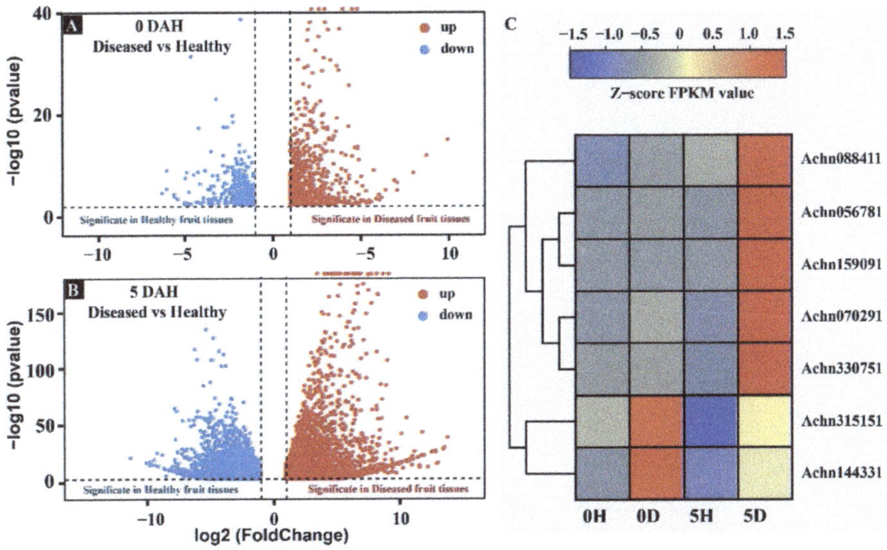

Figure 2. Transcriptome difference of healthy and diseased fruit after harvest. (**A**,**B**) The DEGs between healthy and diseased fruit pulps were assigned with a threshold of | log2 (Fold Change) | > 1 and padj < 0.05. A total of 1143 and 8805 DEGs were detected at 0 and 5 DAH. (**C**) Heatmap of DEGs involved in annotation of cell wall modification and pathogen signals. Encoding genes, such as polygalacturonase (PG, Achn144331), xyloglucan endotransglucosylase (XTH, Achn088411), pectate lyase (PL, Achn070291 | Achn315151), and expansin (EXP, Achn382301), were up-regulated. Some DEGs that involved in pathogen perception and signaling transduction of PRR genes were significantly induced after kiwifruit was infected with black spot disease. They included various types of receptor-like kinases (RLKs) and receptor-like proteins (RLPs), such as LRR receptor-like serine/threonine-protein kinase (Achn056781); receptor-like serine/threonine-protein kinase (Achn330751) and receptor-like protein kinase (Achn159091) were up-regulated.

Some DEGs involved in pathogen perception and signaling transduction of PRR genes were significantly induced after the kiwifruit was infected with black spot disease (Figure 2C). They included various types of receptor-like kinases (RLKs) and receptor-like proteins (RLPs), such as LRR receptor-like serine/threonine-protein kinase (Achn056781);

receptor-like serine/threonine-protein kinase (Achn330751) and receptor-like protein kinase (Achn159091) were up-regulated.

In summary, RNA-seq revealed the transcriptional response of black spot kiwifruit during postharvest storage. Most cell wall modification genes were up-regulated in diseased fruits, suggesting that black spot infection affects shelf life through cell wall related enzymes. At the same time, kiwifruit with black spot disease is also involved in pathogen perception and signal transduction.

3.3. OTU Statistical Analysis of Fungal Community: The Species and Abundance of Microorganisms of Healthy Fruit Tissues Were Higher than That of Diseased Tissues

ITS sequencing was performed to illuminate the fungal environment on fruit surfaces between diseased and health fruits. Sequencing results of fungi in five pairs of samples (from five orchards) showed that there were differences in fungal communities between healthy and diseased fruit surfaces. All 60,095 clean reads were clustered into operational taxonomic units (OTUs) with 97% identity, then were polymerized into 497 OTU sequences. Representing 497 different fungi, homogenization was performed with the least amount of sequence data. Finally, the OTU of the healthy fruit surface was 413, while that of diseased fruit was 271, suggesting that there were more fungal species on healthy fruit surface than infected fruit. The good coverage index showed that the sequences of all samples reached the saturation stage, and the coverage index was ≥97.70% (Table 1).

Table 1. Summary of the ITS sequences of the fungal community samples.

Sample	Clean Reads	Total Bases	Average Length	Goods Coverage	OTU (97%) *
H1	5593	3,822,534	683	0.993	146
H2	6612	4,459,569	674	0.989	155
H3	5899	4,027,333	682	0.988	178
H4	6171	4,252,244	689	0.984	177
H5	5688	3,888,406	683	0.977	170
D1	6089	4,239,139	696	0.989	130
D2	5993	4,142,348	691	0.988	132
D3	6079	4,105,757	675	0.991	138
D4	5899	4,358,927	738	0.99	93
D5	6072	4,186,335	689	0.994	97
Average health	5992.6	4,090,017	682.2	0.9862	165.2
Average disease	6091.8	4,173,338	684.8	0.9854	162

* MUSCLE (Version 3.8.31) was used for fast multi-sequence alignment to obtain the systematic occurrence of all OTU representative sequences. OTUs are used to categorize fungi based on sequence similarity. In metagenomics approaches, OTUs are a cluster of similar sequence variants of the ITS marker gene sequence. Each of these cluster is intended to represent a taxonomic unit of a fungi species or genus depending on the sequence similarity threshold. Typically, OTU clusters are defined by a 97% identity threshold of the ITS gene sequences to distinguish fungi at the genus level.

According to the unweighted UniFrac heatmap and PCoA (principal co-ordinates analysis) map (Figure 3A,B), which reflected the beta diversity of fungal groups, there were significant differences in fungal communities between black spot diseased fruit and healthy fruit. UniFrac is a β-diversity measure that uses phylogenetic information to compare environmental samples. Beta diversity shows the difference between microbial communities from different environments. The main focus is on the difference in taxonomic abundance profiles from different samples. Unweighted UniFrac purely based on sequence distances does not include abundance information. Weighted UniFrac are weighted by relative abundances and includes both sequence and abundance information. UniFrac could couple with standard multivariate statistical techniques including principal coordinates analysis (PCoA), and identifies factors explaining differences among microbial communities. Healthy fruit and susceptible fruit were obviously divided into two different groups, which further explained the differences between healthy fruit and susceptible fruit groups.

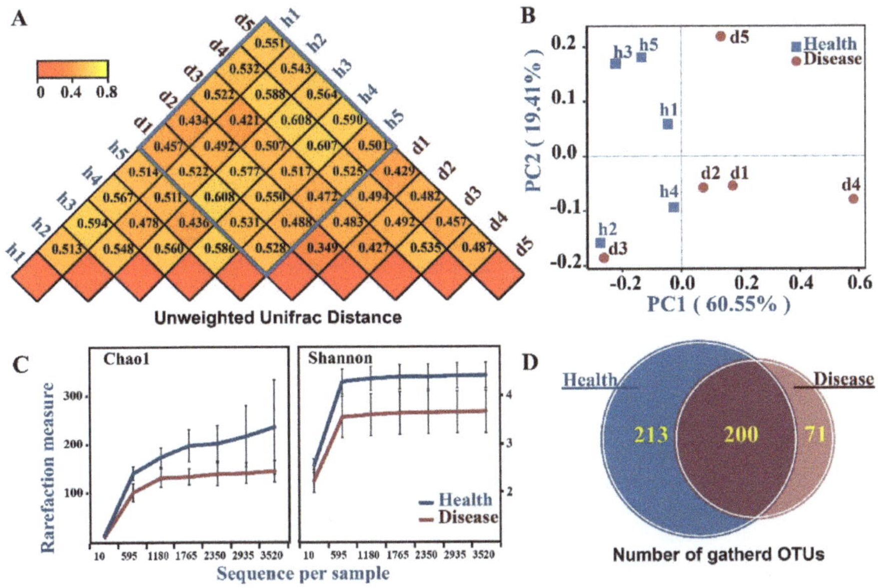

Figure 3. Variance analysis of fungal communities between healthy and diseased tissues. (**A**) Unweighted UniFrac distance of OTUs abundance between healthy and diseased tissues (samples from five orchards). (**B**) PCoA analysis based on UnWeighted UniFrac distance of OTUs abundance. UniFrac is a β-diversity measure that uses phylogenetic information to compare environmental samples. Beta diversity shows the difference between microbial communities from different environments. The main focus is on the difference in taxonomic abundance profiles from different samples. Unweighted UniFrac purely based on sequence distances does not include abundance information. Weighted UniFrac are weighted by relative abundances and includes both sequence and abundance information. UniFrac could couple with standard multivariate statistical techniques, including principal coordinates analysis (PCoA), and identifies factors explaining differences among microbial communities. (**C**) Alpha diversity analysis of samples. Alpha diversity summarizes the structure of an ecological community with respect to its richness (number of taxonomic groups) and evenness (distribution of abundances of the groups). The Chao1 index is a qualitative measure of alpha diversity which includes species richness. It returns an estimate of species richness based on a vector or matrix of abundance data. The Shannon index summarizes the diversity in the population while assuming all species are represented in a sample and that they are randomly sampled. The Shannon index increases as both the richness and evenness of the community increase. (**D**) Venn figure of OTU count in each group. There are 200 OTUs that are common to healthy fruits and diseased fruits, 213 OTUs solely in healthy fruits, and 71 OTUs only in diseased fruits. We focused on OTUs unique to healthy or diseased fruit.

According to alpha diversity index and Venn diagram (Figure 3C,D), which reflect fungal diversity, Chao1 index and Shannon index of healthy fruit were higher than those of diseased fruit, suggesting that the species and abundance of microorganisms on the healthy fruit surface were higher than that on the diseased fruit surface. The Chao1 index is a qualitative measure of alpha diversity which includes species richness. It returns an estimate of species richness based on a vector or matrix of abundance data. The Shannon index summarizes the diversity in the population while assuming all species are represented in a sample and that they are randomly sampled. The Shannon index increases as both the richness and evenness of the community increase. High diversity of microbial species contributes to maintaining a stable microbial environment on fruit surface [36,37]. Once that balance is broken, the number of pathogens that cause black spot increases dramatically.

There were 200 common OTUs on the surface of infected and healthy fruit, 213 OTUs were unique to the surface of healthy fruit, and 71 OTUs were unique to the surface of infected fruit. The OTU statistical analysis indicated that dominant and pathogenetic species existed in fungal communities of the diseased samples.

3.4. Prediction and Analysis of Pathogenic Fungi

Species annotation and relative abundance between disease and health tissues indicated candidate pathogenic fungi. Species annotation at phylum level show that fungi in healthy fruits are qualitatively similar to those in diseased fruits (Figure 4A). The relative abundance of Ascomycota in healthy fruit was higher than that in diseased fruit. The second class is unclassified fungi, comprising 28.20% of diseased fruit, which was higher than 21.60% of healthy fruit. The microbiome of healthy fruits mainly consists of Basidiomycota, Chytridiomycota, Monoblepharomycota, and Entomophthoromycota, and its relative abundance is also higher than that of black spot fruits. At the genus level, 35 genera with the greatest difference in fungal abundance in healthy tissues were noted (Figure 4B). Fungi belonging to *Tilletiopsis*, *Fusarium*, *Golubevia* and *Sarocladium* had high abundance in diseased fruits. Healthy fruit surface enriched to *Acremonium*, *Farysia*, *Cladosporium*, *Moesziomyces*, *Aureobasidium*, *Symmetrospora*, *Rhizophlyctis*, *Kondoa*, *Filobasidium*, *Erythrobasidium*, *Ciliophora*, *Gibbcrclla*, *Acaromyces*, *Hanseniaspora*, *Phoma*, *Fusariclla*, *Plectosphaerella*, *Vishniacozyma*, and *Gibellulopsis*.

Biomarkers with statistically significant differences between diseased and healthy fruit were searched by LDA Effect Size (LEfSe) (Figure 5A,B). At different levels of classification, fungi with significant differences between healthy and diseased fruit included C-Exobasidiomycetes, F-Golubeviaceae, O-Golubeviales, G-*Golubevia*, and *Golubevia* sp. (Figure 5C). These fungi may play important roles in diseased fruit.

Text mining work was carried out to search for the pathogenicity of the significant fungi (Table S1). Results showed that a variety of pathogenic and antibiological fungi had been reported related to plant pathogens, e.g., plant leaf spots and fruit rot. Previously, *Diaporthe eres* (sexual) or *Phomopsis* (monogenesis) [6], *Cladosporium* [7], *Didymella glomerate* [8], *Nigrospora oryzae* [9], *Alternaria alternate* [10], and *Pseudocercospora* actinidiae [11] have been reported to be associated with kiwifruit black disease. On the other hand, there were antagonistic species of pathogenic fungi with high abundance in healthy tissues, such as *Hanseniaspora uvarum*, *Acremonium sclerotigenum*, *Aureobasidium pullulans*, and *Acaromyces ingoldii*. These three species are known to play a role in bioantagonism [35,38–40]. *Acremonium sclerotigenum* is positive for the production of β-1-3 glucanase, cellulase, and compounds, which are involved in biocontrol mechanisms [38].

3.5. Identification of Pathogenic Fungus

A total of seven strains were isolated and purified from diseased fruit surface (Figure 6; Table S2). They were sequenced and BLAST annotated. Results showed that fungi strains of number (No.) 4, No. 7, and No. 8 were annotated as *Diaporte ecualyptorum*, *Paraphaeosphaeria michotii* and *Fusarium nygamai*, respectively. Strains No. 1, No. 2, No.5 and No. 6 belong to neopestalotiopsis, Fusarium, Epicoccum and Alternaria. A total of seven kinds of fungi, distilled water and mixed fungal solution were inoculated into healthy 'CuiXiang' kiwifruit at 22 ± 2 °C. After 15 days, it can be observed that the part inoculated with sterile water has no rot or disease spots. The fruit spots inoculated with No. 1, No. 4, No. 5, No. 6, and No. 7 fungi showed leather shape, hard and no pathological changes in the pulp. The pulp of the fruit inoculated with strain No.2 appeared white, and the white hyphae on the peel of strain No. 2 were similar to strain No. 8. White hyphae also grew on the peel inoculated with mixed fungal solution, and the pulp rotted. It can be seen from Table S2 that the fruits inoculated with fungi show lesions to a certain extent, and the lesion diameter is significantly different under the inoculation of different strains, but it is not the disease of black spot. In conclusion, there was no symptom of black spot disease after inoculation with the seven isolated strains.

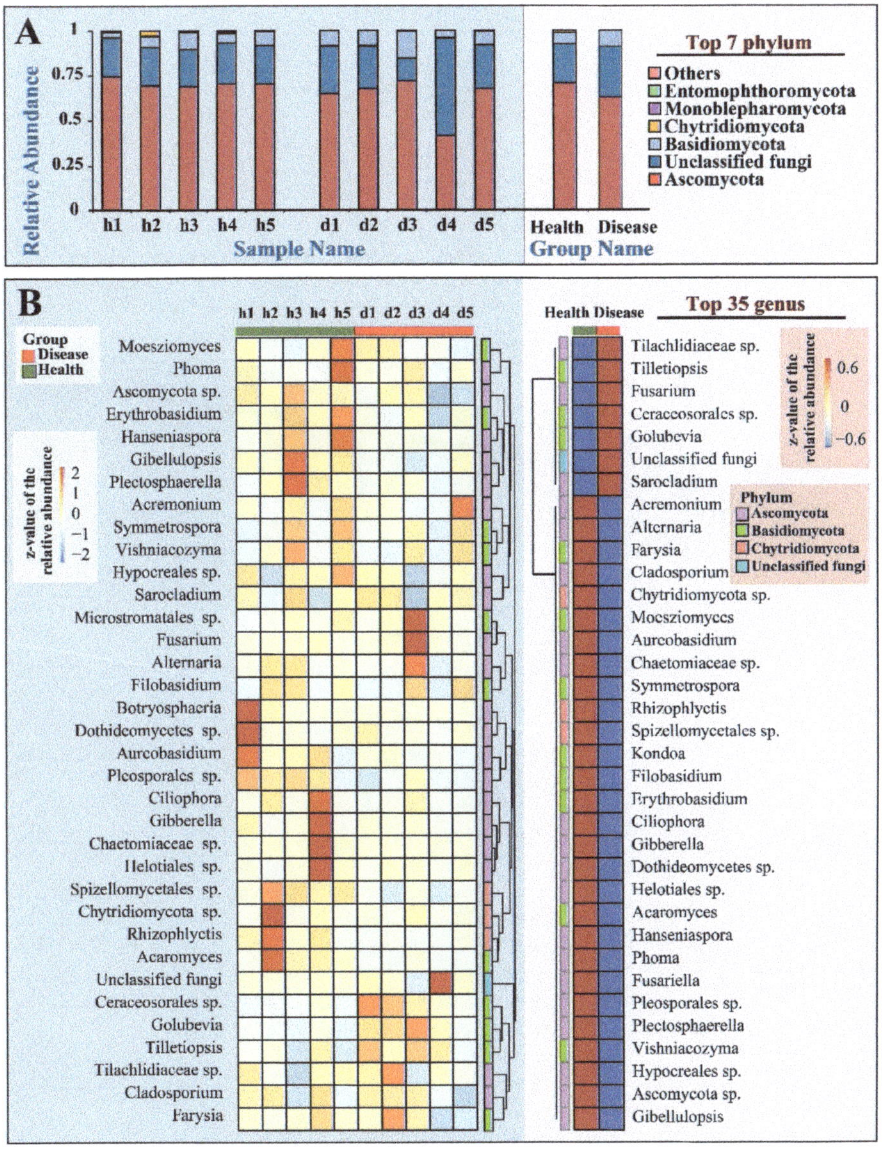

Figure 4. Diversity of fungal communities of healthy and diseased tissues at the phylum (**A**) and genus (**B**) level. (**A**) A histogram is illustrating diversity of fungal communities at phylum level. It shows relative abundance of top 7 phylum. (**B**) Histograms are illustrating diversity of fungal communities at genus level. It shows relative abundance of top 35 genera. The corresponding value of heat map is the z-value of the relative abundance. Class others represent the sum of the relative abundance of all OTUs except those listed in the figure.

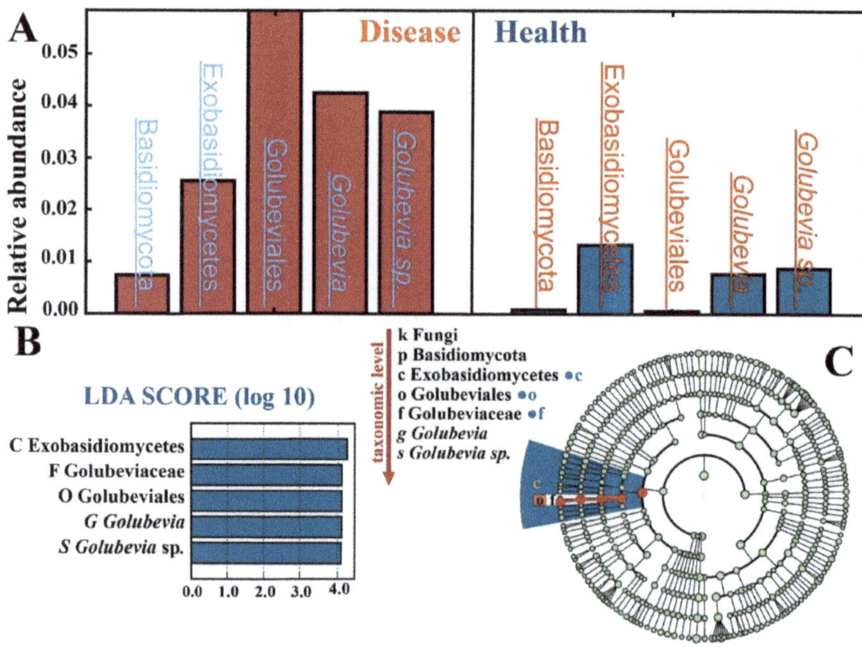

Figure 5. Biomarkers with statistically significant differences between healthy and diseased fruit searched by LDA Effect Size (LEfSe). (**A**). Comparison of relative abundance of biomarkers between groups. (**B**). A histogram of the distribution of LDA values. The LDA value distribution histogram shows the species whose LDA Score is greater than the set value, that is the biomarker with statistically significant differences between groups. The length of the histogram represents the influence of different species (LDA Score). (**C**). Cladogram of the biomarkers. Biomarkers with statistically significant differences between diseased and healthy tissue groups were searched with LDA Effect Size (LEfSe). LEfSe statistical results included a histogram of the distribution of LDA values, an evolutionary bifurcation (phylogenetic distribution), and a comparison of the abundance of biomarkers with statistically significant differences between diseased and healthy tissues groups. In the branching diagram, the circles radiating from the inside out represent the taxonomic level from the phylum to the species. Each small circle at a different classification level represents a classification at that level, and the diameter of the small circle is proportional to the relative abundance size. All species without significant differences were uniformly colored green, and the biomarker for differential species was stained in the following groups. The red nodes represent the microbial groups that play an important role in the pathogenesis group.

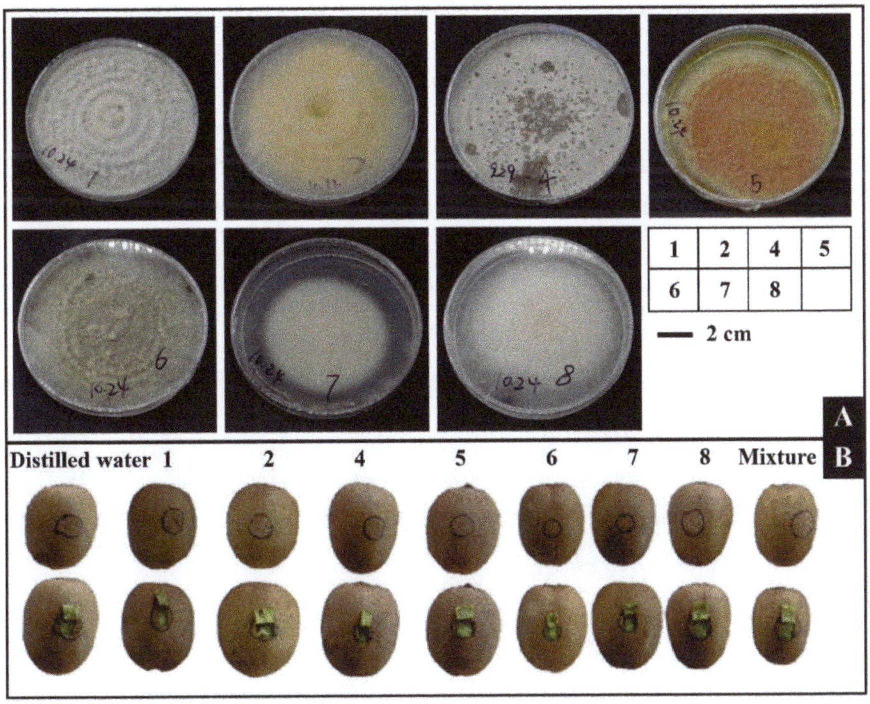

Figure 6. Isolation and inoculation of pathogenic fungus. (**A**) Isolation and purification of pathogenic fungi. (**B**) Observation on inoculation of pathogen from 'CuiXiang' fruit.

4. Conclusions

Black spot disease on kiwifruit is affecting the quality and profitability of postharvest production in Shaanxi province. We applied a high-throughput ITS sequencing method to the analysis of the black spot disease microbial community. The purpose of this study was to explore pathogenic fungal species of 'CuiXiang' kiwifruit black spot disease and to provide a basis for field prevention and treatment. The present study facilitated the illustration of disease mechanisms. In conclusion:

(1) Physiological, morphological and transcriptional characteristics between black spot fruit and healthy fruits were evaluated, and it was found that black spot can affect the ethylene production of healthy kiwifruit, which results in the firmness decreasing rapidly and the shelf life shortening. The results of the electron microscope show that black spots of kiwifruit were mainly distributed on the fruit surface, surface attached hyphae, and the number of starch granules and mitochondria decreased in kiwifruit with black spot disease, and transcriptome results showed that differential genes were involved in cell wall modification, pathogen perception, and signal transduction.

(2) There was a significant difference in fungal communities between healthy and diseased tissues. We analyzed key fungal species with differences between diseased and healthy fruits and predicted candidate pathogenic fungi included *Cladosporium cladosporioides*, *Diaporthe phaseolorum*, *Alternaria alternata*, and *Trichothecium roseum*, which have been reported to cause plant disease. Some species that were relatively abundant in healthy tissue have biological control effects in agricultural production, such as *Hanseniaspora uvarum*, *Acremonium sclerotigenum*, *Acaromyces ingoldii*, and *Aureobasidium pullulans*.

(3) A total of seven pathogens were isolated from harvested kiwifruit with disease infection. The results showed that single pathogen and mixed pathogen failed to show

black spots on the surface of the kiwifruit, which also showed that the formation of black spot disease was a complex process and needed more in-depth research.

This study described the microstructure of 'CuiXiang' kiwifruit black spot disease and identified potential pathogenic microorganisms by describing fungal diversity between healthy and diseased tissues. This study made a preliminary study on kiwifruit black spot, which can provide a reference for further research in the future.

Supplementary Materials: The following are available online at https://www.mdpi.com/article/10.3390/horticulturae8010013/s1. Figure S1: Inspection of the surface of 'CuiXiang' kiwifruit with different condition of black spot disease. Figure S2: Physicochemical properties of different severity of 'CuiXiang' kiwifruit black spot disease during storage. Table S1: Text mining results of pathogenicity of the significant fungi. Table S2: The lesion diameter of 'CuiXiang' kiwifruit inoculated with 7 kinds of fungi respectively.

Author Contributions: Data curation, L.C.; formal analysis, Y.Y. and L.C.; funding acquisition, Y.D.; investigation, C.W. and R.L.; methodology, Y.Y., H.P., C.W. and C.L.; project administration, W.Y.; resources, C.W.; software, H.P.; supervision, Y.D. and X.R.; validation, W.Y.; visualization, W.Y. and R.L.; writing—original draft, Y.Y. and L.C.; writing—review and editing, Y.Y. and L.C. All authors have read and agreed to the published version of the manuscript.

Funding: This work was supported by Primary Research and Development Plan of Ningxia Hui Autonomous Region (2021BBF0214); Introduction of Domestic Doctoral Programs in Shaanxi Province (16–18); and the Chinese Universities Scientific Fund (2452019018).

Institutional Review Board Statement: Not applicable.

Informed Consent Statement: Not applicable.

Data Availability Statement: The datasets presented in this study can be found in online repository of the NCBI Sequence Read Archive: https://www.ncbi.nlm.nih.gov/bioproject/PRJNA673939/ (accessed on 19 December 2021).

Conflicts of Interest: The authors declare that there are no conflict of interest regarding the publication of this article.

References

1. Qi, X.; Guo, D.; Wang, R.; Zhong, Y.; Fang, J. Development status and suggestions on Chinese ki-wifruit industry. *J. Fruit Sci.* **2020**, *37*, 754–763.
2. Liu, Q.; Guo, Y.; He, P. Current Situation, Problems and Countermeasures of Kiwifruit Industry in China. *Guizhou Agric. Sci.* **2020**, *48*, 69–73.
3. Hirsch, A.-M.; Longeon, A.; Guyot, M. Fraxin and esculin: Two coumarins specific to *Actinidia chinensis* and *A. deliciosa* (kiwifruit). *Biochem. Syst. Ecol.* **2002**, *30*, 55–60. [CrossRef]
4. Jin, P. Formation reason and prevention and control measures of black spot in Cuixiang Kiwifruit fruit. *Northwest Hortic. (Fruit Trees)* **2015**, *5*, 26–28.
5. Yang, T.; Meng, J.; Chen, C. Kiwifruit 'CuiXiang' black spot prevention and control. *J. Shanxi Fruit Trees* **2019**, *3*, 84–86.
6. Fu, B.; Wang, J.; Ren, P.; Li, Y.; Zhao, J.; Jin, P.; Zhang, F. Identification of the pathogen causing black spot on kiwifruit in Shaanxi Province. *Acta Phytopathol. Sin.* **2020**, *50*, 112–116.
7. Wang, X.; Zhao, L.; Wang, B.; Li, L.; Zhang, Y. Prevention and control technology of black spot of Cuixiang ki-wifruit. *Shaanxi Agric. Sci.* **2016**, *62*, 125–126.
8. Pan, H.; Chen, M.Y.; Deng, L.; Wang, Z.P.; Li, L.; Zhong, C.H. First Report of *Didymella glomerata* Causing Black Spot Disease of Kiwifruit in China. *Plant Dis.* **2018**, *102*, 2654. [CrossRef]
9. Li, L.; Pan, H.; Chen, M.Y.; Zhang, S.J.; Zhong, C.H. First Report of *Nigrospora oryzae* Causing Brown/Black Spot Disease of Kiwifruit in China. *Plant Dis.* **2018**, *102*, 243. [CrossRef]
10. Kwon, J.-H.; Cheon, M.-G.; Kim, J.-W.; Kwack, Y.-B. Black Rot of Kiwifruit Caused by *Alternaria alternata* in Korea. *Plant Pathol. J.* **2011**, *27*, 298. [CrossRef]
11. Kikuhara, K.; Nakashima, C. Sooty spot of kiwifruit caused by Pseudocercospora actinidiae Deighton. *J. Gen. Plant Pathol.* **2008**, *74*, 185–187. [CrossRef]
12. Zhang, Z.; Guo, C.; Wang, Y.; Zhen, X. Identification and ITS Sequence Analysis of Gerbera Root Rot Pathogen. *Plant Pathol. Bull.* **2005**, *5*, 392–396.
13. Yuan, Y.; Li, J.; Lin, S.; Jia, H.; Pan, Y.; Luo, H.; Deng, M. Study on microbial diversity and function prediction of Pekinensis bean juice based on 16S rDNA high-throughput sequencing technology. *Food Ind. Sci. Technol.* **2020**, *41*, 95–100.

14. Janisiewicz, W.J.; Ii, W.M.J.; Peter, K.A.; Kurtzman, C.P.; Buyer, J. Yeasts associated with plums and their potential for controlling brown rot after harvest. *Yeast* **2014**, *31*, 207–218. [CrossRef]
15. Abdelfattah, A.; Wisniewski, M.; Droby, S.; Schena, L. Spatial and compositional variation in the fungal communities of organic and conventionally grown apple fruit at the consumer point-of-purchase. *Hortic. Res.* **2016**, *3*, 16047. [CrossRef]
16. Volschenk, Q.; Du Plessis, E.M.; Duvenage, F.J.; Korsten, L. Effect of postharvest practices on the culturable filamentous fungi and yeast microbiota associated with the pear carpoplane. *Postharvest Biol. Technol.* **2016**, *118*, 87–95. [CrossRef]
17. Li, L.; Pan, H.; Chen, M.; Zhang, S.; Zhong, C. Isolation and identification of pathogenic fungi causing postharvest fruit rot of kiwifruit (*Actinidia chinensis*) in China. *J. Phytopathol.* **2017**, *165*, 782–790. [CrossRef]
18. Barboni, T.; Cannac, M.; Chiaramonti, N. Effect of cold storage and ozone treatment on physicochemical parameters, soluble sugars and organic acids in *Actinidia deliciosa*. *Food Chem.* **2010**, *121*, 946–951. [CrossRef]
19. Park, Y.S.; Polovka, M.; Suhaj, M.; Ham, K.-S.; Kang, S.G.; Park, Y.-K.; Arancibia-Avila, P.; Toledo, F.; Sánchez, M.R.; Gorinstein, S. The postharvest performance of kiwi fruit after long cold storage. *Eur. Food Res. Technol.* **2015**, *241*, 601–613. [CrossRef]
20. Shahkoomahally, S.; Chaparro, J.X.; Beckman, T.G.; Sarkhosh, A. Influence of Rootstocks on Leaf Mineral Content in the Subtropical Peach cv. UFSun. *Am. Soc. Hortic. Sci.* **2020**, *55*, 496–502. [CrossRef]
21. da Silva, J.S.; Lavorante, A.F.; Paim, A.P.S.; da Silva, M.J. Microwave-assisted digestion employing diluted nitric acid for mineral determi nation in rice by ICP OES. *Food Chem.* **2020**, *319*, 8. [CrossRef]
22. Rahman, M.U.; Ma, Q.; Ahmad, B.; Hanif, M.; Zhang, Y. Histochemical and Microscopic Studies Pre-dict that Grapevine Genotype "Ju mei gui" is Highly Resistant against *Botrytis cinerea*. *Pathogens* **2020**, *9*, 253. [CrossRef]
23. Lopez Velasco, G.; Tydings, H.A.; Boyer, R.R.; Falkinham, J.O., III; Ponder, M.A. Characterization of interactions between Escherichia coli O157:H7 with epiphytic bacteria in vitro and on spinach leaf surfaces—ScienceDirect. *Int. J. Food Microbiol.* **2012**, *153*, 351–357. [CrossRef] [PubMed]
24. Zhang, L.; Wang, S. Bacterial community diversity on in-shell walnut surfaces from six representative provinces in China. *Sci. Rep.* **2017**, *7*, 10054. [CrossRef]
25. Edgar, R.C.; Haas, B.J.; Clemente, J.C.; Quince, C. UCHIME improves sensitivity and speed of chimera detection. *Bioinformatics* **2011**, *27*, 2194–2200. [CrossRef]
26. Edgar, R.C. UPARSE: Highly accurate OTU sequences from microbial amplicon reads. *Nat. Methods* **2013**, *10*, 996–998. [CrossRef] [PubMed]
27. White, J.R.; Nagarajan, N.; Pop, M. Statistical Methods for Detecting Differentially Abundant Features in Clinical Metagenomic Samples. *PLoS Comput. Biol.* **2009**, *5*, e1000352. [CrossRef]
28. Santoni, F.; Paolini, J.; Barboni, T.; Costa, J. Relationships between the leaf and fruit mineral compositions of *Actinidia deliciosa* var. Hayward according to nitrogen and potassium fertilization. *Food Chem.* **2013**, *147*, 269–271. [CrossRef] [PubMed]
29. Kumarihami, H.; Cha, G.H.; Kim, J.G.; Kim, H.U.; Lee, M.; Kwack, Y.B.; Cho, J.G.; Kim, J. Effect of Preharvest Ca-chitosan Application on Postharvest Quality of 'Garmrok' Kiwifruit during Cold Storage. *Hortic. Sci. Technol.* **2020**, *38*, 239–248.
30. Veraverbeke, E.; Verboven, P.; van Oostveldt, P.; Nicolai, B. Prediction of moisture loss across the cuticle of apple (Malus syl-vestris subsp. mitis (Wallr.)) during storage: Part 2. Model simulations and practical applications. *Postharvest Biol. Technol.* **2003**, *30*, 75–88. [CrossRef]
31. Gilbert, H.J. The Biochemistry and Structural Biology of Plant Cell Wall Deconstruction. *Plant Physiol.* **2010**, *153*, 444–455. [CrossRef]
32. Zhang, B.; Chen, K.; Bowen, J.; Allan, A.; Espley, R.; Karunairetnam, S.; Ferguson, I. Differential expression within the LOX gene family in ripening kiwifruit. *J. Exp. Bot.* **2006**, *57*, 3825–3836. [CrossRef]
33. Wang, Y.; Xiong, G.; He, Z.; Yan, M.; Zou, M.; Jiang, J. Transcriptome analysis of *Actinidia chinensis* in response to Bot-ryosphaeria dothidea infection. *PLoS ONE* **2020**, *15*, e0227303.
34. Haile, Z.M.; Guzman, N.D.; Grace, E.; Moretto, M.; Sonego, P.; Engelen, K.; Zoli, L.; Moser, C.; Baraldi, E. Transcriptome Profiles of Strawberry (*Fragaria vesca*) Fruit Interacting with *Botrytis cinerea* at Different Ripening Stages. *Front. Plant Sci.* **2019**, *10*, 1131. [CrossRef]
35. Liu, H.; Guo, J.; Cheng, Y.; Liu, P.; Long, C.; Deng, B. Inhibitory activity of tea polyphenol and *Hanseniaspora uvarum* against *Botrytis cinerea* infections. *Lett. Appl. Microbiol.* **2010**, *51*, 258–263. [CrossRef] [PubMed]
36. Parish, M.; Beuchat, L.; Suslow, T.; Harris, L.; Garrett, E.; Farber, J.; Busta, F. Methods to Reduce/Eliminate Pathogens from Fresh and Fresh-Cut Produce. *Compr. Rev. Food Sci. Food Saf.* **2003**, *2*, 161–173. [CrossRef]
37. Chen, J.; Yan, R.; Hu, Y.; Zhang, N.; Hu, H. Compositional shifts in the fungal diversity of garlic scapes during postharvest transportation and cold storage. *LWT* **2019**, *115*, 108453. [CrossRef]
38. Ghule, S.B.; Sawant, I.S.; Sawant, S.D.; Saha, S.; Devarumath, R.M. Isolation and identification of three new mycoparasites of *Erysiphe necator* for biological control of grapevine powdery mildew. *Australas. Plant Pathol.* **2019**, *48*, 351–367. [CrossRef]
39. Liu, H.M.; Guo, J.H.; Cheng, Y.J.; Li, L.; Pu, L.; Wang, B.Q.; Deng, B.X.; Long, C.A. Control of gray mold of grape by Han-seniaspora uvarum and its effects on postharvest quality parameters. *Ann. Microbiol.* **2010**, *60*, 31–35. [CrossRef]
40. Di Francesco, A.; Di Foggia, M.; Corbetta, M.; Baldo, D.; Ratti, C.; Baraldi, E. Biocontrol Activity and Plant Growth Promotion Exerted by *Aureobasidium pullulans* Strains. *J. Plant Growth Regul.* **2020**, *40*, 1233–1244. [CrossRef]

horticulturae

MDPI

Article

Genome-Wide Identification and Bioinformatics Analysis of Auxin Response Factor Genes in Highbush Blueberry

Yu Zong [1,2,†], Lili Gu [1,†], Zhuli Shen [1], Haiting Kang [1], Yongqiang Li [1,2], Fanglei Liao [1,2], Lishan Xu [1,2,*] and Weidong Guo [1,2,*]

1 College of Chemistry and Life Sciences, Zhejiang Normal University, Jinhua 321004, China; yzong@zjnu.cn (Y.Z.); liligu7@163.com (L.G.); szl_77@163.com (Z.S.); haitingkang@163.com (H.K.); lyq@zjnu.cn (Y.L.); fangleiliao@zjnu.cn (F.L.)

2 Zhejiang Provincial Key Laboratory of Plant Biotechnology, Jinhua 321004, China

* Correspondence: xls@zjnu.cn (L.X.); gwd@zjnu.cn (W.G.)

† These authors contributed equally to this work.

Abstract: Auxin response factors (ARFs) are a transcription factor family that regulates the expression of auxin phase-responsive genes. Here, we performed a genome-wide investigation of the tetraploid blueberry (*Vaccinium corymbosum* cv. 'Draper') genome sequence. Physical and chemical properties, phylogenetic evolution, gene structure, conservative motifs, chromosome location, and cis-acting elements of blueberry *ARF* genes were comprehensively evaluated. A total of 70 blueberry *ARF* genes (*VcARF*) were found in its genome, which could be divided into six subfamilies. *VcARF* genes were unevenly distributed on 40 chromosomes and were observed to encode protein sequences ranging in length from 162 to 1117 amino acids. Their exon numbers range from 2 to 22. *VcARF* promoter regions contain multiple functional domains associated with light signaling, aerobic metabolism, plant hormones, stress, and cell cycle regulation. More family members of *VcARF* genes were discovered in blueberry than in previously studied plants, likely because of the occurrence of whole-genome duplication and/or tandem duplication. *VcARF* expression patterns were analyzed at different stages of fruit development, and *VcARF3*, *VcARF4*, *VcARF14*, *VcARF37*, and *VcARF52* were observed to play important roles. *VcARF3* and *VcARF4* appeared to function as repressors, while *VcARF14* acted as an essential factor in fruit firmness differences between firm and soft flesh cultivars.

Keywords: genome wide; blueberry; auxin response factor; fruit firmness

Citation: Zong, Y.; Gu, L.; Shen, Z.; Kang, H.; Li, Y.; Liao, F.; Xu, L.; Guo, W. Genome-Wide Identification and Bioinformatics Analysis of Auxin Response Factor Genes in Highbush Blueberry. *Horticulturae* **2021**, *7*, 403. https://doi.org/10.3390/horticulturae7100403

Academic Editor: Dong Zhang

Received: 3 September 2021
Accepted: 7 October 2021
Published: 15 October 2021

1. Introduction

The ubiquitous involvement of auxin in almost all aspects of plant development and in plant responses to the environment, abiotic stress, and growth tropisms underlines its importance [1]. The initiation and regulation of these processes are mostly accomplished by the expression and regulation of auxin-related genes [2], which have received considerable research attention. These genes include the Skp1–Cullin–F-Box protein complex, containing the transport inhibitor response 1 protein (SCF^{TIR1}) auxin receptor and its related auxin signaling F-box protein receptor family members, and two families of partially redundant proteins: the auxin response factors (ARFs) and their cognate auxin/indole-3-acetic acid (Aux/IAA) repressors [3–5]. ARFs activate or inhibit the expression of auxin response genes by binding to auxin-responsive elements (AuxREs) in the promoter region and recruit Aux/IAA to perform their functions.

As transcription factors, ARFs possess a modular structure and consist of three major domains: an amino-terminal DNA-binding domain (DBD), a middle domain, and a carboxy-terminal PB1 (Phox and Bem1) domain contained within a region previously known as domain III/IV [2,6,7]. The B3-type DBD is flanked by dimerization domains (DDs), at least in ARF1 and ARF5, and also contains a Tudor-like ancillary domain. The DBD recognizes and binds to the TGTCTC element (AuxRE) in the promoter region of auxin-responsive genes [8]. The DD mediates the dimerization of ARF1 and ARF5, which is essential for ARF5 cooperative binding to target DNA [9]. The function of the Tudor-like ancillary domain is unknown, but it might be involved in an interaction with the DD. The middle domain of ARFs either activates or represses transcription [10,11]; repressor domains are enriched in proline, serine, and threonine, and activator domains are enriched in glutamine. This basis has been used to identify and classify ARFs from many plant species through genome-wide analyses.

Most information concerning ARF function derives from both forward genetic analysis and reverse genetic phenotypic analysis in Arabidopsis, which showed that ARF functional redundancy is universal [12,13]. The Arabidopsis genome contains 22 full-length ARF genes and one incomplete ARF23 [14]. This is thought to be a pseudogene because it contains a stop codon in the DBD, leading to a partially encoded protein [15]. AtARF1 mutations do not themselves confer phenotypes but can enhance the phenotypic traits of *AtARF2* mutations [16]. *AtARF2* mutations result in late flowering leaf senescence, and an increased seed number. *ARF3/ARF4* plays an important role in the development of reproductive and vegetative tissues, and *AtARF3* also has a function in floral meristem determination, floral organ pattern formation, and pistil development [17,18]. *AtARF6* and *AtARF8* encode a pair of functionally redundant transcription factors that regulate the pistil and stamen development of immature flowers. The stamen filaments of *arf6* and *arf8* double mutants are shorter than those of the wildtype and show delayed anther dehiscence, leading to female sterility [19]. Although almost no phenotypic change was detected between *AtARF7* or *AtARF19* single-mutant and wildtype plants, *AtARF7* and *AtARF19* double mutants showed inhibited adventitious root formation, a reduced number of lateral roots, and no leaf cell enlargement, indicating that the two genes are complementary in function [20].

ARFs were also reported to be involved in the fruit ripening process, with most fruit development studies being carried out in tomato. Of the 21 *ARF* tomato genes [21], *SlARF2* encodes the main regulatory factor that controls fruit ripening through ethylene signals and biosynthesis [22], and *SlARF4* functions in the accumulation of fruit chlorophyll and affects early fruit development by regulating glucose metabolism [23]. *SlARF9* regulates cell division at an early stage of fruit development [24], while *SlARF10* controls the formation of chlorophyll, starch synthesis, and sugar accumulation in fruits [25]. However, the response of most *ARF* genes to environmental and hormonal signals has been poorly studied, and there are still unidentified key factors in the *ARF* expression regulatory network. The temporal and spatial expression of *ARF* genes is also unclear.

Following advances in sequencing technology, plant genome sequencing is now more affordable. Thus, several *ARF* plant gene families have been identified at the whole-genome level, including, but not limited, to rice (*Oryza sativa*) [14], maize (*Zea mays*) [26], tomato (*Solanum lycopersicum*) [21], apple (*Malus domestica*) [27], sweet orange (*Citrus sinensis*) [28], grape (*Vitis vinifera*) [29], longan (*Dimocarpus longan*) [30], and strawberry (*Fragaria vesca*) [31]. However, the genome-wide identification of *ARF* genes has not been performed in blueberry (*Vaccinium* spp.). In comparison with fruit trees, the genome resources of blueberry are limited. The first whole-genome sequence of the blueberry cultivar 'W8520' was released by Bian et al. in 2014 [32], but its assembly quality and usability were limited. Although Gupta et al. [33] released another assembly based on the genome sequencing of the blueberry cultivar 'O'Neal' in 2014, 27.35% of genome sequences contained gaps and mistakes. More recently, Colle and colleagues published a high-quality haplotype-phased blueberry genome [34] that was assembled from the highbush blueberry cultivar 'Draper'. This genome has enabled several molecular and bioinformatics studies of blueberry to be conducted and allows the opportunity to identify transcription factor genes throughout the genome.

In the current study, genome-wide *VcARF*s were identified from the highbush blueberry 'Draper' genome. By analyzing the number of family members, gene structure, and amino acid sequences, we clarified the structural and evolutionary characteristics of the blueberry *ARF* gene family. The expression patterns of different *VcARF* genes were also estimated from transcriptomic data, and pivotal *VcARF*s that strongly correlate with blueberry fruit development were screened. Moreover, the transcript abundance of key *VcARF* genes was validated in firm flesh cultivar 'Star' and soft flesh cultivar 'O'Neal' at a range of developmental stages. Our findings provide a reference for the genome-wide identification of transcription factor genes and a study on the regulation of *VcARF* genes in blueberry fruit firmness.

2. Materials and Methods

2.1. Plant Materials

The fruits of two southern highbush blueberry cultivars 'Star' (firm flesh) and 'O'Neal' (soft flesh) were collected at four development stages following the sampling strategy of Zifkin et al. [35]. For each cultivar, 50 healthy fruits for every stage were sampled from different fruit setting positions of three plants. A total of 30 fruits of similar sizes at each fruit stage were screened [35]. The fruits were immediately frozen in liquid nitrogen after collection and stored at −80 °C until required.

2.2. RNA Isolation and Reverse Transcription

Total RNA was extracted using a modified CTAB method as described by Chang et al. [36]. The quality and concentration of blueberry RNA were determined using agarose gel electrophoresis and spectrophotometric analyses, respectively. cDNA was synthesized from 1 μg RNA using the PrimeScript™ RT reagent Kit with gDNA Eraser (Takara Biotechnology Co., Ltd., Dalian, China) following the manufacturer's instructions. Synthesized 1st-strand cDNAs were diluted 10-fold and stored at −80 °C for further use.

2.3. Identification of the ARF Gene Family in Blueberry

A local BLAST database for genome and amino acid sequences of the highbush blueberry 'Draper' was constructed using BLAST-2.9.0+ [37]. *A. thaliana* ARF peptide sequences were downloaded from the plant transcription factor database at Peking University (Beijing, China) (http://planttfdb.gao-lab.org/, accessed on 1 April 2020). Potential blueberry ARF transcription factors were retrieved from deduced amino acid sequences using BLASTp in the BLAST toolkit [37]. Query hits with an e-value $\leq 1e-20$ were kept for further analyses. Profile hidden Markov models of the ARF conserved domain (PF06507) were downloaded from the Pfam website (http://pfam.xfam.org/, accessed on 1 April 2020) [38]. HMMER3.3 [39] was used to search for homologous ARF proteins in blueberry based on the ARF conserved domain. Other parameters of HMMER were set as default. Candidate ARF genes obtained from BLASTp and HMMER results were merged, and redundant sequences were removed. Non-redundant results were submitted to the NCBI conserved domain database (https://www.ncbi.nlm.nih.gov/cdd, accessed on 3 April 2020) for domain analysis, and the sequences with ARF annotation results were retained for sequential analyses. Prediction and analysis of physicochemical properties of all ARF protein sequences in blueberry were made using the Expasy website (https://web.expasy.org/protparam/, accessed on 4 April 2020).

2.4. Analyses of Amino Acid Sequences and VcARF Gene Structure

MAFFT v7.453 [40] and IQ-TREE1.6.12 [41] were used to align protein sequences of *Arabidopsis* and blueberry and to construct a phylogeny tree based on the maximum likelihood method, respectively. Ultrafast bootstrapping was carried out 1000 times to test the accuracy of the phylogeny tree. Other parameters of IQ-TREE were set as default. Identified *VcARF* sequences were obtained from the 'Draper' genome using a customized Perl script and aligned with MAFFT v7.453. A phylogenetic tree was built with IQ-TREE1.6.12. Gene structural analysis was performed using the gene structure display server online tool (http://gsds.gao-lab.org/index.php, accessed on 6 April 2020). The structure of coding regions, introns, and non-coding regions of all selected *ARF* genes in the blueberry genome were plotted, and the phylogenetic relationship of *VcARF* genes was drawn.

2.5. Chromosome Location of VcARFs and Conservative Motif Analysis of VcARF Proteins

Genomic location information of *VcARF* genes was extracted from annotation general feature format files using Shell script. *VcARF* genes were plotted onto the chromosomes with the chromPlot R package [42]. MEME (https://meme-suite.org/meme/doc/meme.html, accessed on 5 April 2020) was used to analyze the conserved motifs of all VcARF protein sequences. The number of motifs was set as 15, and other parameters were kept as default.

2.6. Identification of Cis Elements on VcARF Promoters

A Perl script was used to retrieve 2 kbp sequences upstream of the transcription start site of *VcARF* genes. Cis elements in the promoter regions were predicted by PlantCARE (http://bioinformatics.psb.ugent.be/webtools/plantcare/html/, accessed on 6 April 2020). The categories and number of those cis elements were calculated and drawn using Microsoft Excel 2013 (Microsoft, Redmond, WA, USA).

2.7. Expression Pattern of VcARF Genes during Fruit Development

The gene expression profile of different plant organs was obtained from a study by Colle et al. (2019) [34]. A heat map was generated by ggplot2 [43] based on the fragments per kilobase of exon per million mapped fragments (FPKM) values. The expression level of *VcARF* transcripts in fruits of 'Star' and 'O'Neal' cultivars at four development stages was determined by quantitative PCR on an ABI StepOne Plus^{TM} Real-Time fluorescence qPCR system (Applied Biosystems Co., Ltd., Beijing, China). Specific primers (Table 1) were designed using Primer Premier 5 software (PREMIER Biosoft International, Palo Alto, CA, USA). The experiments were conducted on three biological replicates, and the results were normalized by using the *VcGAPDH* gene. Gene expression data were analyzed with a relative quantification method ($2^{-\Delta\Delta C_t}$) [44]. Statistical analyses were carried out using SPSS software v.18.0 (IBM, Armonk, NY, USA).

Table 1. *VcARF* genes and their primers used for quantification PCR in the current study.

Gene Name	Forward Primer Sequence (5′-3′)	Reverse Primer Sequence (5′-3′)
VcARF3	GTGCTGGACCCCTTGTTACT	GGCAGCTGTTGATCCAATCC
VcARF4	GCTGGACCCCTTGTTACTCTT	ATTGAACGAAGGCAGCTGTTG
VcARF14	TATGGCGGGACCGTAACAAC	ACTGAGTCACCGGTAAAGAGC
VcARF37	CCGTAACAACCACCCCGATT	GGGGAAGTAGACAACGTGGG
VcARF52	GCACCAGATCACCCGATTCC	TACCAAGGGCAATCCCCTGC
VcGAPDH	TGAGAAAGAATACAAGCCAGAT	CAGGCAACACCTTACCAA

3. Results

3.1. Identification of the Blueberry VcARF Gene Family

We predicted a total of 70 *VcARF* genes in the blueberry genome and named the genes *VcARF1–70*, according to their physical location on the chromosome. *VcARF* gene sequence lengths were highly varied, ranging from 1831 bp (*VcARF47*) to 17,346 bp (*VcARF25*). VcARF protein sequences ranged from 162 (VcARF18) to 1117 (VcARF50) amino acids, with molecular weights of 18.7–124.7 kDa and isoelectric points of 4.76–9.34 (Table 2).

Table 2. *VcARF* gene family in the genome of *Vaccinium corymbosum c.v.* 'Draper'.

Gene Name	Genome Position		Gene Length	Protein Length	Molecular Weight	Isoelectric Point
	Start	End				
VcARF1	18,439,481	18,443,830	4349	692	76,350.16	6.38
VcARF2	17,675,083	17,684,250	9167	573	62,768.41	5.91
VcARF3	35,513,223	35,522,595	9372	710	78,858.21	5.78
VcARF4	5,089,913	5,099,455	9542	706	78,334.61	6.11
VcARF5	17,374,973	17,379,982	5009	692	76,311.12	6.32
VcARF6	26,259,124	26,264,169	5045	707	78,336.45	6.23
VcARF7	23,759,116	23,764,049	4933	692	76,321.11	6.3
VcARF8	16,724,921	16,733,537	8616	709	78,059.18	7.05
VcARF9	22,380,725	22,385,479	4754	692	76,294.09	6.41
VcARF10	27,366,853	27,375,393	8540	889	98,168.12	6.20
VcARF11	30,200,121	30,204,632	4511	683	75,418.63	7.52
VcARF12	30,535,118	30,544,916	9798	782	86,669.81	6.21
VcARF13	31,908,720	31,918,385	9665	1073	118,643.05	6.08
VcARF14	31,971,246	31,975,574	4328	325	36,417.71	5.79
VcARF15	5,638,023	5,646,179	8156	896	99,747.22	5.46
VcARF16	2,964,750	2,975,312	10,562	818	91,211.21	5.88
VcARF17	39,198,442	39,206,540	8098	1092	21,781.74	6.26
VcARF18	7,176,872	7,179,011	2139	162	18,729.58	5.70
VcARF19	13,331,772	13,340,852	9080	574	62,818.47	5.91
VcARF20	30,285,241	30,295,980	10,739	880	98,106.11	6.94
VcARF21	25,800,314	25,808,896	8582	889	98,196.17	6.26
VcARF22	28,595,844	28,600,407	4563	683	75,510.72	7.52
VcARF23	28,958,634	28,968,521	9887	782	86,602.73	6.16
VcARF24	31,429,406	31,436,228	6822	827	92,363.56	6.21
VcARF25	27,786,604	27,803,950	17,346	260	28,558.81	5.67
VcARF26	30,557,305	30,565,768	8463	884	97,559.4	6.09
VcARF27	33,308,214	33,312,630	4416	683	75,381.6	7.85
VcARF28	33,590,293	33,600,143	9850	771	85,367.26	6.09
VcARF29	36,291,973	36,299,899	7926	747	84,171.56	7.97
VcARF30	36,316,721	36,323,616	6895	510	58,026.11	9.34
VcARF31	36,347,477	36,352,792	5315	572	63,904.6	5.55
VcARF32	17,724,006	17,728,602	4596	674	74,482.07	5.98
VcARF33	35,733,011	35,740,217	7206	623	69,462.96	7.82
VcARF34	34,950,200	34,957,557	7357	896	99,694.07	5.46
VcARF35	35,564,675	35,571,979	7304	896	99,804.31	5.46
VcARF36	4,968,620	4,978,140	9520	870	96,855.96	5.81
VcARF37	7,369,136	7,383,171	14,035	993	112,009.2	6.18
VcARF38	12,687,653	12,696,034	8381	884	97,599.42	6.12
VcARF39	9,683,280	9,692,882	9602	782	86,682.81	6.21
VcARF40	9,951,498	9,956,041	4543	683	75,383.64	7.85
VcARF41	23,235,922	23,244,210	8288	574	62,848.5	5.91
VcARF42	29,522,528	29,525,843	3315	267	30,862.32	5.21
VcARF43	35,261,834	35,272,570	10,736	880	98,106.11	6.94
VcARF44	17,822,463	17,827,109	4646	674	74,375.88	6.02
VcARF45	27,973,739	27,982,227	8488	828	92,414.65	6.21
VcARF46	1,273,882	1,279,100	5218	565	63,200.73	5.55
VcARF47	1,307,489	1,309,320	1831	402	45,446.23	8.97
VcARF48	17,166,955	17,171,684	4729	674	74,338.78	6.06
VcARF49	28,538,840	28,549,308	10,468	773	86,055.44	5.83
VcARF50	295,030	303,309	8279	1117	124,713.39	6.64
VcARF51	17,097,454	17,102,070	4616	674	74,269.76	6.06
VcARF52	6,479,810	6,486,650	6840	822	91,863	6.14
VcARF53	11,275,837	11,283,809	7972	745	82,051.86	6.89
VcARF54	28,655,149	28,663,653	8504	667	73,970.28	8.61
VcARF55	14,535,659	14,540,819	5160	707	78,310.36	6.23
VcARF56	18,962,251	18,971,430	9179	405	46,288.47	4.76
VcARF57	16,109,186	16,113,795	4609	707	78,280.34	6.23
VcARF58	16,725,386	16,730,572	5186	707	78,306.42	6.23
VcARF59	21,377,071	21,385,465	8394	972	107,956.17	5.98
VcARF60	21,432,199	21,439,229	7030	580	65,502.63	5.99
VcARF61	26,970,680	26,980,225	9545	870	96,855.96	5.81
VcARF62	27,717,127	27,724,764	7637	636	70,995.13	8.82
VcARF63	19,890,774	19,899,622	8848	901	99,470.38	6.15
VcARF64	19,942,680	19,949,685	7005	580	65,561.79	6.08
VcARF65	25,169,999	25,180,560	10,561	827	92,046.12	5.75
VcARF66	3,134,356	3,144,615	10,259	816	90,885.94	5.82
VcARF67	1,635,026	1,639,892	4866	540	60,252.66	8.29
VcARF68	16,996,236	17,002,796	6560	213	24,700	5.6
VcARF69	19,316,087	19,323,149	7062	623	69,462.01	7.82
VcARF70	18,207,611	18,214,839	7228	623	69,508.02	7.84

3.2. Phylogenetic Analysis of VcARF Amino Acids and Structure of VcARF Genes

To explore the evolutionary relationships of ARF proteins among *Arabidopsis* and blueberry, we constructed a phylogenetic tree based on the alignments of 22 *Arabidopsis* ARFs (AtARFs) and 70 blueberry ARFs (VcARFs). All ARF proteins could be divided into six groups, from clade I to clade VI, according to the classification pattern of *Arabidopsis*. Both *Arabidopsis* and blueberry ARFs were present in each group. Clade VI (Figure 1, green branches) contained most ARFs, including 12 AtARF and 17 VcARF members. Clade IV (Figure 1, orange branches) contained 23 ARFs, of which only three were AtARFs. Clade III (Figure 1, blue branches) included the smallest number of ARF transcription factors, with only three VcARFs and one AtARF (AtARF5).

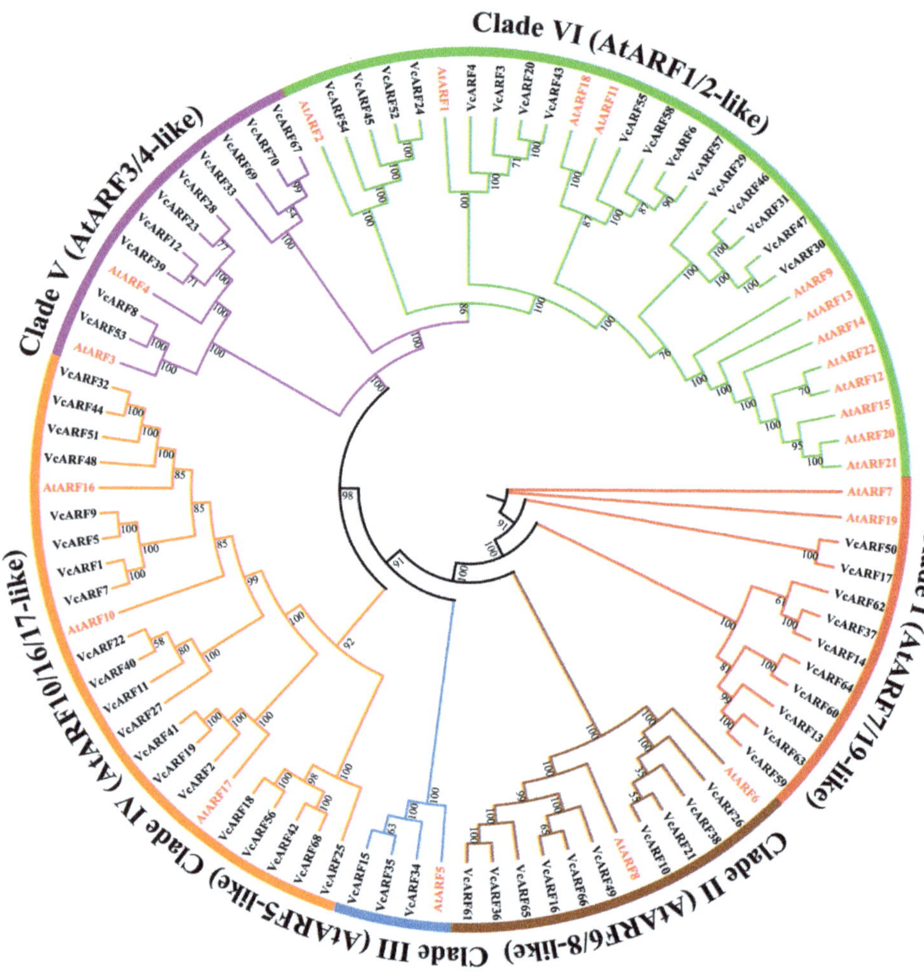

Figure 1. Phylogenetic relationships and classification of ARF proteins from *Arabidopsis* and blueberry. Numbers on the nodes indicate bootstrap values.

Gene structural analysis provided further evidence to support the phylogenetic topology groupings of multigene families. To gain further insights into the structural diversity of blueberry *ARF* genes, we analyzed the exon/intron organization of full-length cDNAs with corresponding genomic DNA sequences of individual *ARF* genes (Figure 2). Similar ARF classification patterns to the phylogenetic tree were observed, with *VcARFs* clustering into six groups according to their gene structure. Most closely related *VcARFs* within the same groups shared similar gene structures in terms of either intron numbers or exon lengths. Taking clade IV as an example, most genes in this group had two to four exons, with the exception of *VcARF18*, *VcARF25*, *VcARF42*, *VcARF56*, and *VcARF68*. These genes contained no untranslated regions or longer introns. The gene structure appeared to be more variable in clades I, II, and VI, which had the largest number of exon/intron structural variants.

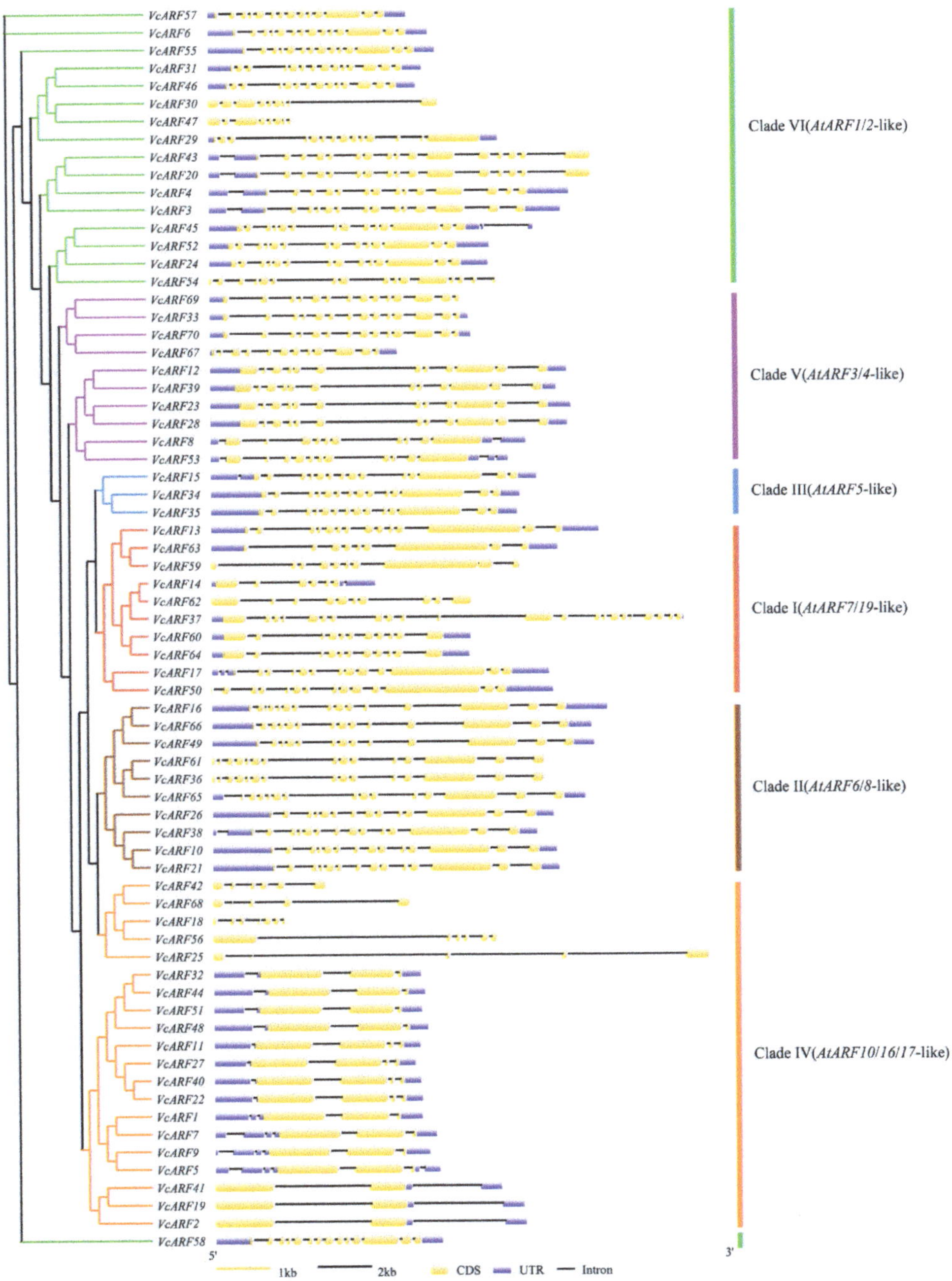

Figure 2. Intron and exon distribution of *VcARF* genes.

3.3. Chromosome Location and Conservative Motif Analysis

Next, we linked 70 *VcARF* genes onto a chromosome-scale genome assembly of the tetraploid highbush blueberry genome. All *VcARF* genes were mapped onto 40 linkage groups (Figure 3). The distribution of *VcARFs* on each chromosome was uneven, ranging from one to four. No *VcARF* was mapped onto chromosomes 4, 7, 16, 31, 32, 36, 45, or 48. Chromosomes 23 and 40 each had four *VcARFs*. *VcARFs* clustered in adjacent regions on chromosomes containing more than two *VcARFs*.

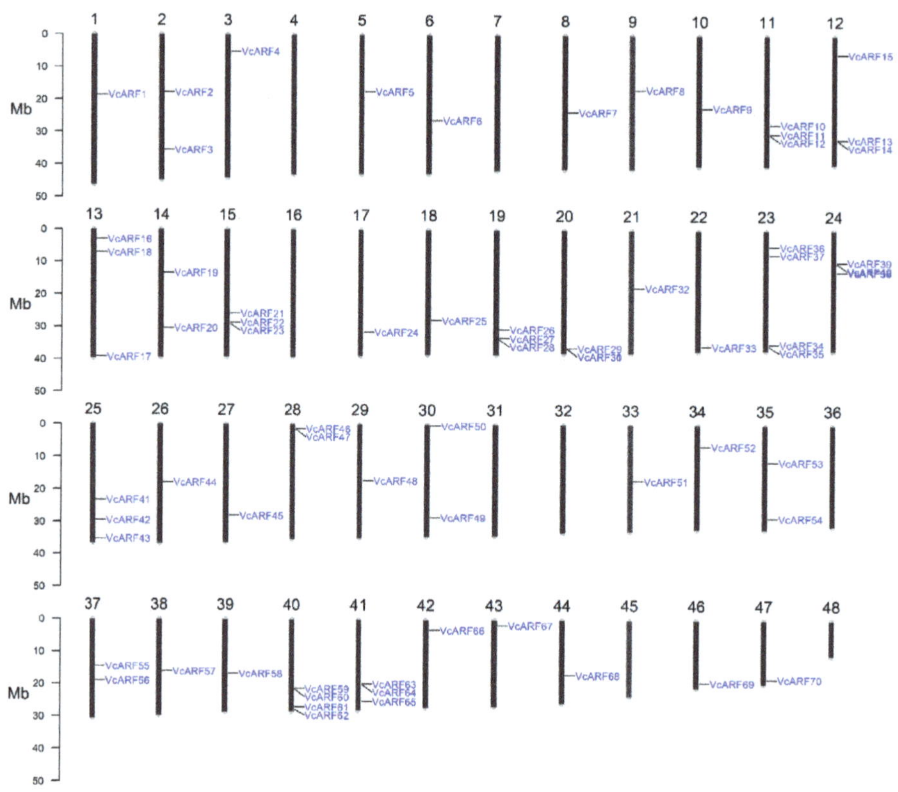

Figure 3. The chromosomal location of blueberry *VcARF* genes.

Conservative motifs of the blueberry VcARF protein were analyzed using the MEME online database. Identified motifs varied in length from 11–41 amino acid residues. A total of 14 conserved motifs were identified from VcARF members in clade II (Figure 4, brown branches) and clade III (Figure 4, blue branches), with VcARF49 as an exception. The same number of motifs was identified in clades II and III, with just a difference in their order. Neither motif 3 nor motif 13 were identified in VcARF49. Some conservative motifs were absent from clades I (Figure 4, red branches), IV (Figure 4, orange branches), and VI (Figure 4, green branches), especially for VcARF18, VcARF25, VcARF42, VcARF56, and VcARF68, in which only three conserved motifs were identified. The position and number of conserved motifs among VcARF members in clade I were highly varied, showing the most motif differences among all groups. Only five conserved motifs were found in VcARF62, which is less than that of other VcARFs in this group.

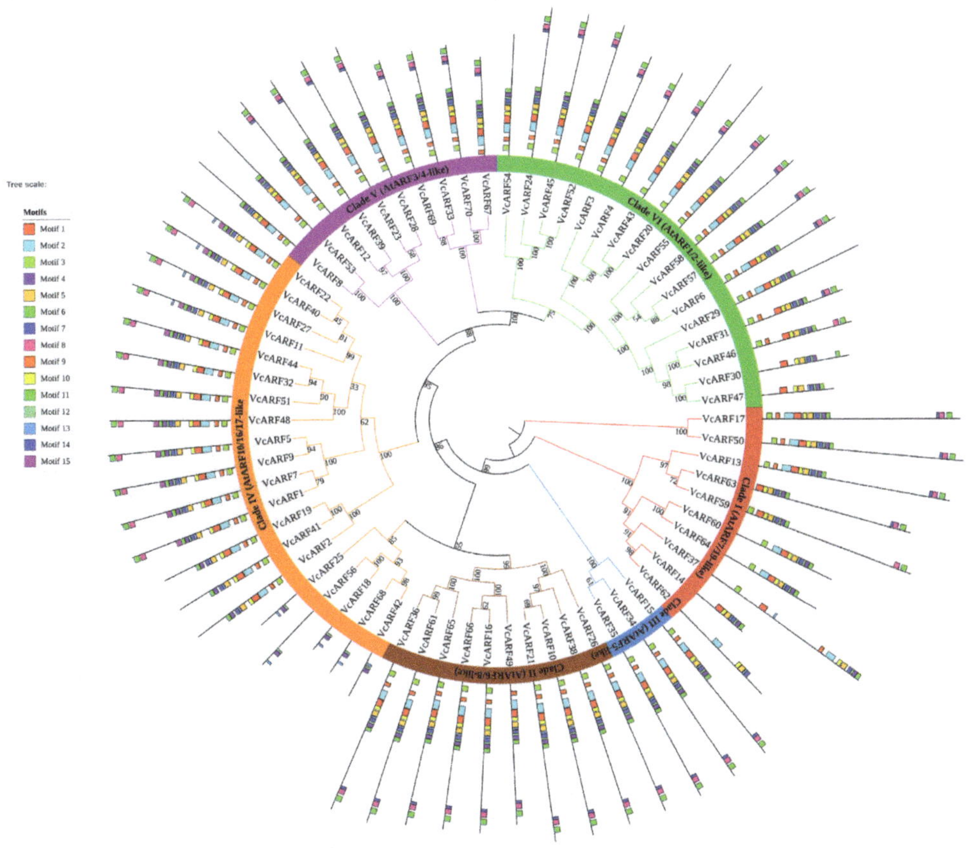

Figure 4. Conserved motif analysis of VcARF proteins in blueberry.

3.4. Prediction of Cis-Acting Elements in Promoters of Gene Family Members

To identify potential cis-acting elements of *VcARFs*, we analyzed 2 kbp promoter regions using bioinformatics analysis. This indicated that most cis elements of *VcARF* promoters belong to the responsive elements of plant hormones and environmental factors. In addition to two core elements (a TATA box and CAAT box), light responsive elements and MYB binding sites were identified in all *VcARF* genes (Figure 5). Oxygen-related metabolism and methyl jasmonate-responsive elements accounted for 88.6% and 75.7%, respectively, of *VcARF* cis-acting elements. Gibberellins and auxin-related responsive elements were also common in *VcARF* promoter regions, with 52.8% of *VcARF* genes containing the latter. Only *VcARF62* and *VcARF67* contained wound-responsive elements. Few cis elements related to development, i.e., cell cycle regulation, circadian control, and endosperm growth, were found in *VcARF* promoter regions. However, four cell cycle regulation and two circadian control responsive elements were found in the promoters of *VcARF15* and *VcARF63*, respectively.

3.5. Expression Pattern of VcARF Genes at Three Development Stages of 'Draper' Fruits

FPKM values of transcripts in different tissues of blueberry 'Draper' were retrieved from Colle et al. [34]. *VcARF* FPKM values at green mature (grnfrt), pink mature (pinkfrt), and mature (ripe) stages were extracted and used to construct a heat map (Figure 6). Dramatic differences were found among the expression patterns of those *VcARF* genes. A total of 58 *VcARF* genes showed low expression and few changes in expression during the fruit ripening process. Genes that clustered into the same clade in Figures 1 and 2 displayed similar expression patterns. *VcARF4* and *VcARF52* had the highest expression level in green mature fruit, showing a transcript abundance that decreased with fruit ripening, leading to an obvious downregulation during the loss of fruit firmness. Other genes such as *VcARF14* and *VcARF37* in clade I (*AtARF7/19*-like), *VcARF16* and *VcARF66* in clade II (*AtARF6/8*-like), and *VcARF3* in clade VI (*AtARF1/2*-like) had higher transcript levels in green mature fruit compared with the other two stages.

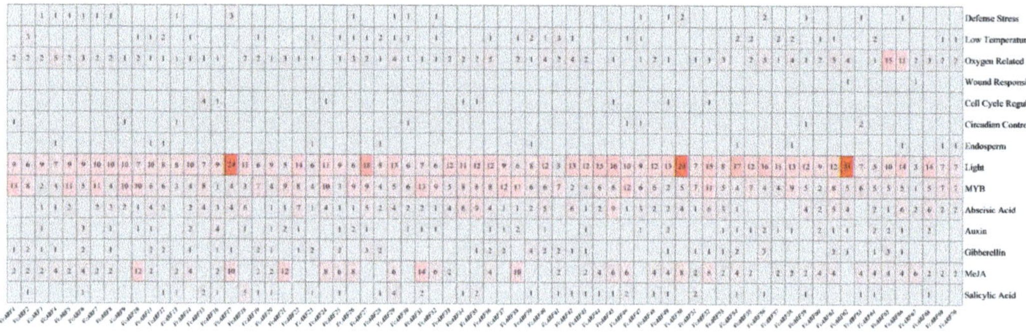

Figure 5. Types and numbers of cis-acting elements in the promoter regions of *VcARF* genes in blueberry.

3.6. Gene Expression during 'Star' and 'O'Neal' Fruit Development and Ripening

Because the expression level of *VcARF* genes varied according to different maturation stages, we speculated that some genes had an effect on fruit firmness. *VcARF* genes showing significant expression differences in the ripening process of blueberry 'Draper' were therefore selected to have their expression profiles validated at four stages of development of firm and soft flesh blueberries. Hence, the expression patterns of *VcARF3*, *VcARF4*, *VcARF14*, *VcARF37*, and *VcARF52* were evaluated in the fruit of 'Star' and 'O'Neal' using qPCR (Figure 7). Moderate changes were observed between the expression profiles of these five *VcARF* genes in 'Draper' transcriptomic data and qPCR results. The expression patterns of *VcARF3* and *VcARF37* in 'Star' and 'O'Neal' were similar to those in 'Draper'. The expression of *VcARF4* and *VcARF14* in firm flesh cultivar 'Star' decreased sharply from stage S5, which showed similar FPKM value changes to those observed in 'Draper'. However, the expression levels of these two genes were almost unchanged in the soft flesh cultivar 'O'Neal', particularly for *VcARF14*. By contrast, a slight change in the expression level of *VcARF52* was seen in 'Star' compared with 'O'Neal'.

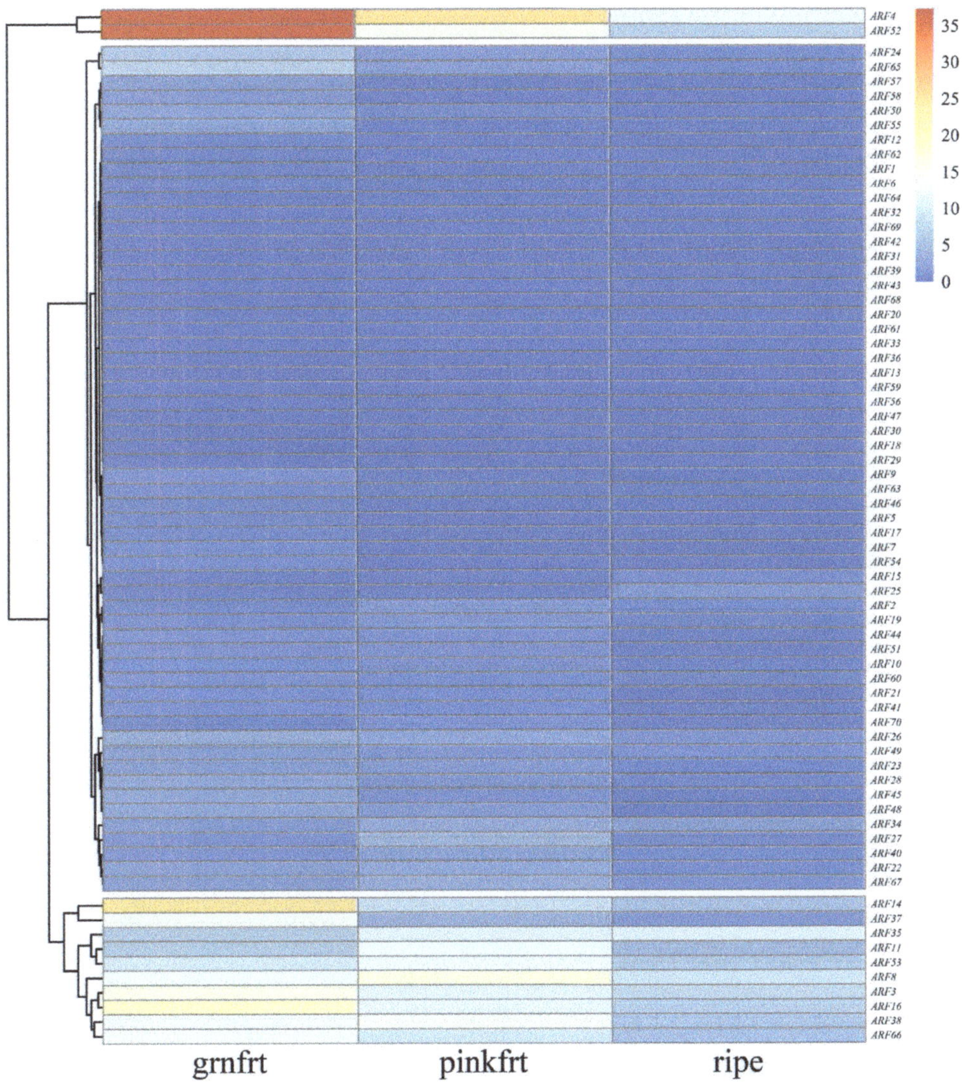

Figure 6. Heat map of RNA-seq data for *ARF* genes during fruit development.

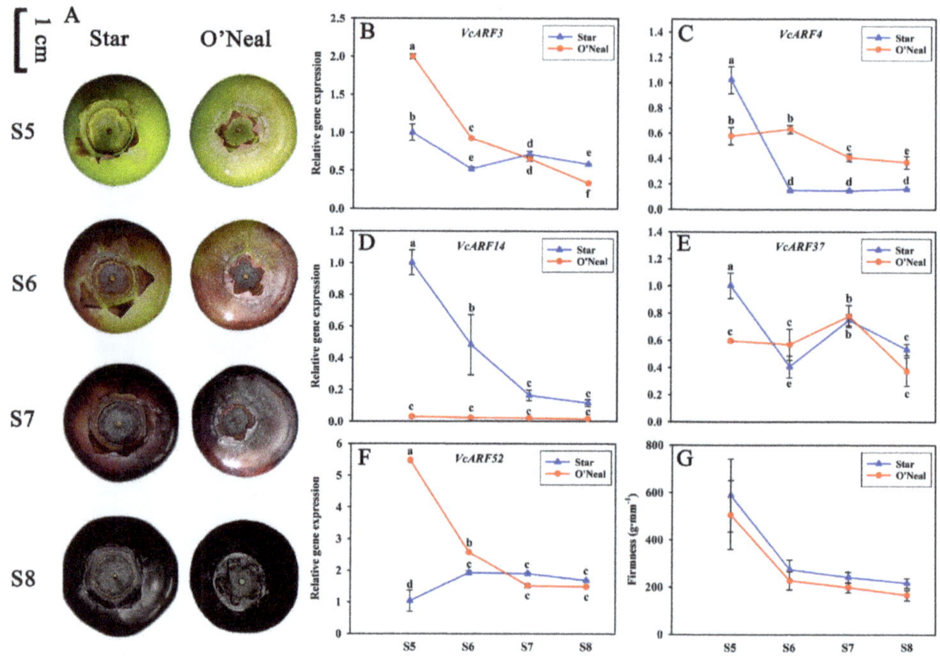

Figure 7. (**A**) Fruits of 'Star' and 'O'Neal' at different stages of development. (**B–F**) Relative expression of five *VcARF* genes at four development stages of 'Star' and 'O'Neal'. (**G**) Fruit firmness of 'Star' and 'O'Neal' at different stages of development.

4. Discussion

4.1. Expanded VcARF Family Members and High Sequence Divergence in Blueberry

ARF gene families have been identified in many sequenced plant genomes, with 22 *ARFs* identified in *A. thaliana*, 25 in *O. sativa* [14], 22 in *S. lycopersicum* [21], and 17 in *V. vinifera* [29]. In the current study, we identified 70 *VcARF* genes in a tetraploid blueberry genome, suggesting that the blueberry *ARF* gene family is expanded compared with other genomes. Long evolutionary periods typically lead to the presence of multiple members of a specific gene family [14,45]. Moreover, a recent whole-genome duplication event has been reported in the blueberry genome, in which tandem duplications may have contributed to metabolic diversity or gene functionalization [34]. Gene duplication has been shown to have a non-negligible effect on the formation of gene families [46], so we speculate that whole-genome duplication and tandem duplication are the main contributing events for the expansion of the blueberry *ARF* gene family.

Phylogenetic analysis indicated that many blueberry *VcARF* family members share high levels of similarity, while the adjacent chromosome locations of some *VcARF* genes provide further support for tandem duplication. Duplicated *ARF* genes exhibited different expression patterns, probably because of the lack of intense evolutionary selection pressure and the need for diversification [47,48]. Previous studies demonstrated that phylogenetically close *ARFs* may also be genetically linked, while phylogenetically distant ARFs may not be [48]. However, we found no correlation between phylogenetic distance and genetic link in blueberry *ARFs*. Because ARFs are transcription factors that regulate the expression of auxin response genes, those *VcARF* genes would be expected to contain AuxREs in their promoter regions. However, only 37 *VcARF* genes contained AuxREs in their promoter regions, of which 12 had more than two AuxREs. This indicates that the underlying mechanism for auxin inducibility of *VcARF* genes needs to be further elucidated in blueberry.

4.2. The Potential Contribution of VcARF Genes to Firmness Divergence between Firm Flesh and Soft Flesh Blueberries

Previous studies on the ripening process of tomato [21], grape [29], and apple [27] suggested that *ARF* genes were involved in the regulation of fruit development, especially with respect to floral organ and fruit development. *AtARF1* appeared to function as a transcriptional repressor in planta [16], while *ARF1* repressed auxin-induced gene expression in transient assays [11], and *arf1* mutations increased the transcription of *Aux/IAA* genes in *Arabidopsis* flowers [16]. Our phylogenetic analysis results showed that VcARF3 and VcARF4 had the highest amino acid similarity to AtARF1. Moreover, transcript abundances of *VcARF3* and *VcARF4* decreased greatly at stage S5 and changed slightly during blueberry fruit ripening, suggesting that *VcARF3* and *VcARF4* function during the early stages of fruit maturation. This indicates a similarity in the roles of *VcARF3* and *VcARF4* with that of *AtARF1*. We detected the significant downregulation of *VcARF3* expression during ripening in the 'O'Neal' cultivar, and a significant early expression decline of *VcARF4* in 'Star'. Therefore, both *VcARF3* and *VcARF4* have a likely repressor function during blueberry fruit ripening, with *VcARF3* having a dominant effect on fruit firmness.

VcARF14 and *VcARF37* are both phylogenetically related to *Arabidopsis AtARF7* and *AtARF19*. Because of the reported high level of similarity between ARF7 and ARF19 proteins, the expression of one ARF allows for functional compensation for the loss of the other in *arf7* and *arf19* single mutants [12]. In tomato, *SlARF7* acts as a negative regulator of fruit set until pollination and fertilization have taken place and moderates the auxin and gibberellin response during fruit growth. RNA interference-silenced *SlARF7* expression leads to parthenocarpic fruit growth, indicating that SlARF7 acts as a negative regulator of both fruit set and fruit development [24]. In our study, the expression of *VcARF14* in 'Star' showed significant differences with that in 'O'Neal', while an overall low transcript abundance and almost no expression changes were observed for *VcARF14* during fruit ripening in 'O'Neal'. *VcARF37* showed similar expression patterns between 'Star' and 'O'Neal'. Based on the protein sequence similarity between VcARF14 and AtARF7, we speculate that VcARF14 may also be a negative regulator of fruit ripening. The fruit firmness of 'Star' and 'O'Neal' decreased sharply from stage S5, but that of 'Star' was significantly greater than that of 'O'Neal' at all four stages, especially at stage S5 and S6. The higher expression levels of *VcARF14* in 'Star' compared with 'O'Neal' may account for this difference in firmness, but further studies should investigate the regulatory mechanisms of *VcARF14* during fruit softening.

VcARF52 was identified as an AtARF1/2-like transcription factor in clade VI (Figure 1) and has a similar sequence to that of *Arabidopsis* ARF2. ARF2 is most similar to ARF1, with these two proteins having both distinct and overlapping functions in *A. thaliana* [16]. ARF2 was reported to regulate leaf senescence and floral organ abscission independently of the ethylene and cytokinin response pathways [16]. ARF2A is a recognized auxin signaling component that may interconnect signals of ethylene and additional hormones to co-ordinate and initiate the complex ripening process of tomato [22]. Over-expressing *ARF2A* in tomato resulted in blotchy ripening in which certain fruit regions reddened and showed accelerated ripening [22]. This suggested that *SlARF2A* has a positive impact on tomato ripening. However, we found that the expression of *VcARF52* in 'O'Neal' decreased with fruit ripening, while its expression in 'Star' did not significantly change. Therefore, *VcARF52* probably accelerated the softening process at an early stage of blueberry fruit development but had little effect at later stages. Although ARF2 is thought to regulate fruit firmness by affecting fruit development and ripening, phylogenetic studies indicate that ARF1 and ARF2 diverged prior to the monocot–dicot split [49,50] so would have had ample time to evolve distinct biochemical activities. Thus, *VcARF52* may not conform to the canonical auxin response model.

Taken together, our findings suggest that *ARF* genes play essential roles in the ripening process of blueberry fruit. We propose that future studies should focus on *VcARF3*, *VcARF4*,

VcARF14, and *VcARF52* to elucidate their function in determining differences in fruit firmness between firm and soft flesh cultivars.

Author Contributions: Conceptualization, Y.Z., L.X. and W.G.; data curation, L.G. and H.K.; formal analysis, L.G., Z.S. and Y.Z.; funding acquisition; L.X. and W.G.; writing—original draft, Y.Z. and L.G.; writing—review and editing, Y.L. and F.L. All authors have read and agreed to the published version of the manuscript.

Funding: This research was supported by the Key Research Project of Science Technology Department of Zhejiang Province, grant number 2021C02066-9; the Key Research and Development Program of Zhejiang Science and Technology Department, grant number 2018C02007; and the Natural Science Foundation of Zhejiang Province, grant number LQ16C150001.

Institutional Review Board Statement: Not applicable.

Informed Consent Statement: Not applicable.

Data Availability Statement: Not applicable.

Conflicts of Interest: The authors declare no conflict of interest.

References

1. Chandler, J.W. Auxin response factors. *Plant Cell Environ.* **2016**, *39*, 1014–1028. [CrossRef] [PubMed]
2. Guilfoyle, T.J.; Hagen, G. Auxin response factors. *Curr. Opin. Plant Biol.* **2007**, *10*, 453–460. [CrossRef] [PubMed]
3. Chapman, E.J.; Estelle, M. Mechanism of auxin-regulated gene expression in plants. *Annu. Rev. Genet.* **2009**, *43*, 265–285. [CrossRef]
4. Wang, R.; Zhang, Y.; Kieffer, M.; Yu, H.; Kepinski, S.; Estelle, M. HSP90 regulates temperature–dependent seedling growth in Arabidopsis by stabilizing the auxin co-receptor F-box protein TIR1. *Nat. Commun.* **2016**, *7*, 10269. [CrossRef] [PubMed]
5. Salehin, M.; Bagchi, R.; Estelle, M. SCFTIR1/AFB-Based Auxin Perception: Mechanism and Role in Plant Growth and Development. *Plant Cell* **2015**, *27*, 9–19. [CrossRef]
6. Guilfoyle, T.J.; Hagen, G. Getting a grasp on domain III/IV responsible for Auxin Response Factor–IAA protein interactions. *Plant Sci.* **2012**, *190*, 82–88. [CrossRef] [PubMed]
7. Wright, R.C.; Nemhauser, J.L. New tangles in the auxin signaling web. *F1000Prime Rep.* **2015**, *7*, 19. [CrossRef]
8. Li, S.; Xie, Z.; Hu, C.; Zhang, J. A review of Auxin Response Factors (ARFs) in plants. *Front. Plant Sci.* **2016**, *7*, 47. [CrossRef]
9. Boer, D.R.; Freire-Rios, A.; van den Berg, W.A.M.; Saaki, T.; Manfield, I.; Kepinski, S.; López-Vidrieo, I.; Franco-Zorrilla, J.M.; de Vries, S.C.; Solano, R.; et al. Structural basis for DNA binding specificity by the *a*uxin-dependent ARF transcription factors. *Cell* **2014**, *156*, 577–589. [CrossRef]
10. Tiwari, S.B.; Hagen, G.; Guilfoyle, T. The Roles of Auxin Response Factor Domains in Auxin-Responsive Transcription. *Plant Cell* **2003**, *15*, 533–543. [CrossRef]
11. Ulmasov, T.; Hagen, G.; Guilfoyle, T.J. Activation and repression of transcription by auxin-response factors. *Proc. Natl. Acad. Sci. USA* **1999**, *96*, 5844–5849. [CrossRef]
12. Okushima, Y.; Overvoorde, P.J.; Arima, K.; Alonso, J.M.; Chan, A.; Chang, C.; Ecker, J.R.; Hughes, B.; Lui, A.; Nguyen, D.; et al. Functional genomic analysis of the AUXIN RESPONSE FACTOR gene family members in *Arabidopsis thaliana*: Unique and overlapping functions of *ARF7* and *ARF19*. *Plant Cell* **2005**, *17*, 444–463. [CrossRef]
13. Hardtke, C.; Ckurshumova, W.; Vidaurre, D.P.; Singh, S.A.; Stamatiou, G.; Tiwari, S.B.; Hagen, G.; Guilfoyle, T.J.; Berleth, T. Overlapping and non-redundant functions of the *Arabidopsis* auxin response factors *MONOPTEROS* and *NONPHOTOTROPIC HYPOCOTYL 4*. *Development* **2004**, *131*, 1089–1100. [CrossRef] [PubMed]
14. Wang, D.; Pei, K.; Fu, Y.; Sun, Z.; Li, S.; Liu, H.; Tang, K.; Han, B.; Tao, Y. Genome-wide analysis of the auxin response factors (ARF) gene family in rice (*Oryza sativa*). *Gene* **2007**, *394*, 13–24. [CrossRef] [PubMed]
15. Guilfoyle, T.J.; Hagen, G. Auxin response factors. *J. Plant Growth Regul.* **2001**, *20*, 281–291. [CrossRef]
16. Ellis, C.M.; Nagpal, P.; Young, J.C.; Hagen, G.; Guilfoyle, T.J.; Reed, J.W. AUXIN RESPONSE FACTOR1 and AUXIN RESPONSE FACTOR2 regulate senescence and floral organ abscission in *Arabidopsis thaliana*. *Development* **2005**, *132*, 4563–4574. [CrossRef] [PubMed]
17. Sessions, A.; Nemhauser, J.L.; McColl, A.; Roe, J.L.; Feldmann, K.A.; Zambryski, P.C. ETTIN patterns the Arabidopsis floral meristem and reproductive organs. *Development* **1997**, *124*, 4481–4491. [CrossRef] [PubMed]
18. Pekker, I.; Alvarez, J.P.; Eshed, Y. Auxin Response Factors mediate *Arabidopsis* organ asymmetry via modulation of KANADI activity. *Plant Cell* **2005**, *17*, 2899–2910. [CrossRef]
19. Wu, M.; Tian, Q.; Reed, J. Arabidopsis microRNA167 controls patterns of ARF6 and ARF8 expression, and regulates both female and male reproduction. *Development* **2006**, *133*, 4211–4218. [CrossRef]
20. Wilmoth, J.C.; Wang, S.; Tiwari, S.B.; Joshi, A.D.; Hagen, G.; Guilfoyle, T.J.; Alonso, J.M.; Ecker, J.R.; Reed, J.W. NPH4/ARF7 and ARF19 promote leaf expansion and auxin-induced lateral root formation. *Plant J.* **2005**, *43*, 118–130. [CrossRef]

21. Wu, J.; Wang, F.; Cheng, L.; Kong, F.; Peng, Z.; Liu, S.; Yu, X.; Lu, G. Identification, isolation and expression analysis of auxin response factor (ARF) genes in *Solanum lycopersicum*. *Plant Cell Rep.* **2011**, *30*, 2059–2073. [CrossRef]
22. Breitel, D.A.; Chappell-Maor, L.; Meir, S.; Panizel, I.; Puig, C.P.; Hao, Y.; Yifhar, T.; Yasuor, H.; Zouine, M.; Bouzayen, M.; et al. AUXIN RESPONSE FACTOR 2 intersects hormonal signals in the regulation of tomato fruit ripening. *PLoS Genet.* **2016**, *12*, e1005903. [CrossRef]
23. Sagar, M.; Chervin, C.; Mila, I.; Hao, Y.; Roustan, J.-P.; Benichou, M.; Gibon, Y.; Biais, B.; Maury, P.; Latché, A.; et al. SlARF4, an Auxin Response Factor involved in the control of sugar metabolism during tomato fruit development. *Plant Physiol.* **2013**, *161*, 1362–1374. [CrossRef]
24. de Jong, M.; Wolters-Arts, M.; Feron RMariani, C.; Vriezen, W.H. The *Solanum lycopersicum* auxin response factor 7 (*SlARF7*) regulates auxin signaling during tomato fruit set and development. *Plant J.* **2009**, *57*, 160–170. [CrossRef]
25. Yuan, Y.; Mei, L.; Wu, M.; Wei, W.; Shan, W.; Gong, Z.; Zhang, Q.; Yang, F.; Yan, F.; Luo, Y.; et al. SlARF10, an auxin response factor, is involved in chlorophyll and sugar accumulation during tomato fruit development. *J. Exp. Bot.* **2018**, *69*, 5507–5518. [CrossRef]
26. Xing, H.; Pudake, R.N.; Guo, G.; Xing, G.; Hu, Z.; Zhang, Y.; Sun, Q.; Ni, Z. Genome-wide identification and expression profiling of auxin response factor (ARF) gene family in maize. *BMC Genom.* **2011**, *12*, 178. [CrossRef] [PubMed]
27. Luo, X.; Sun, M.; Xu, R.; Shu, H.; Wang, J.; Zhang, S. Genome wide identification and expression analysis of the ARF gene family in apple. *J. Genet.* **2014**, *93*, 785–797. [CrossRef]
28. Li, S.; Ouyang, W.; Hou, X.; Xie, L.; Hu, C.; Zhang, J. Genome-wide identification, isolation and expression analysis of auxin response factor (ARF) gene family in sweet orange (*Citrus sinensis*). *Front. Plant Sci.* **2015**, *6*, 119. [CrossRef]
29. Wan, S.; Li, W.; Zhu, Y.; Liu, Z.; Huang, W.; Zhan, J. Genome-wide identification, characterization and expression analysis of the auxin response factor gene family in *Vitis vinifera*. *Plant Cell Rep.* **2014**, *33*, 1365–1375. [CrossRef] [PubMed]
30. Peng, Y.; Fang, T.; Zhang, Y.; Zhang, M.; Zeng, L. Genome-Wide identification and sxpression analysis of Auxin Response Factor (ARF) gene family in Longan (*Dimocarpus longan* L.). *Plants* **2020**, *9*, 221. [CrossRef] [PubMed]
31. Wang, S.; Shi, F.; Dong, X.; Li, Y.; Zhang, Z.; Li, H. Genome-wide identification and expression analysis of auxin response factor (ARF) gene family in strawberry (*Fragaria vesca*). *J. Integr. Agric.* **2019**, *18*, 1587–1603. [CrossRef]
32. Bian, Y.; Ballington, J.; Raja, A.; Brouwer, C.; Reid, R.; Burke, M.; Wang, X.; Rowland, L.J.; Bassil, N.; Brown, A. Patterns of simple sequence repeats in cultivated blueberries (*Vaccinium* section *Cyanococcus* spp.) and their use in revealing genetic diversity and population structure. *Mol. Breed.* **2014**, *34*, 675–689. [CrossRef]
33. Gupta, V.; Estrada, A.D.; Blakley, I.; Reid, R.; Patel, K.; Meyer, M.D.; Andersen, S.U.; Brown, A.F.; Lila, M.A.; Loraine, A.E. RNA-Seq analysis and annotation of a draft blueberry genome assembly identifies candidate genes involved in fruit ripening, biosynthesis of bioactive compounds, and stage-specific alternative splicing. *GigaScience* **2015**, *4*, 5. [CrossRef]
34. Colle, M.; Leisner, C.P.; Wai, C.M.; Ou, S.; Bird, K.A.; Wang, J.; Wisecaver, J.H.; Yocca, A.E.; I Alger, E.; Tang, H.; et al. Haplotype-phased genome and evolution of phytonutrient pathways of tetraploid blueberry. *GigaScience* **2019**, *8*, giz012. [CrossRef]
35. Zifkin, M.; Jin, A.; Ozga, J.; Zaharia, L.I.; Schernthaner, J.P.; Gesell, A.; Abrams, S.R.; Kennedy, J.A.; Constabel, C.P. Gene expression and metabolite profiling of developing highbush blueberry fruit indicates transcriptional regulation of flavonoid metabolism and activation of abscisic acid metabolism. *Plant Physiol.* **2011**, *158*, 200–224. [CrossRef]
36. Chang, S.; Puryear, J.; Cairney, J. A simple and efficient method for isolating RNA from pine trees. *Plant Mol. Biol. Rep.* **1993**, *11*, 113–116. [CrossRef]
37. Altschul, S.F.; Gish, W.; Miller, W.; Myers, E.W.; Lipman, D.J. Basic local alignment search tool. *J. Mol. Biol.* **1990**, *215*, 403–410. [CrossRef]
38. Mistry, J.; Chuguransky, S.; Williams, L.; Qureshi, M.; Salazar, G.A.; Sonnhammer, E.L.L.; Tosatto, S.C.; Paladin, L.; Raj, S.; Richardson, L.J.; et al. Pfam: The protein families database in 2021. *Nucleic Acids Res.* **2020**, *49*, D412–D419. [CrossRef] [PubMed]
39. Finn, R.D.; Clements, J.; Eddy, S.R. HMMER web server: Interactive sequence similarity searching. *Nucleic Acids Res.* **2011**, *39*, W29–W37. [CrossRef]
40. Katoh, K.; Standley, D.M. MAFFT: Multiple sequence alignment software version 7: Improvements in performance and usability. *Mol. Biol. Evol.* **2013**, *30*, 772–780. [CrossRef]
41. Nguyen, L.; Schmidt, H.; Von Haeseler, A.; Minh, B.Q. IQ-TREE: A fast and effective stochastic algorithm for estimating maximum-likelihood phylogenies. *Mol. Biol. Evol.* **2014**, *32*, 268–274. [CrossRef]
42. Anand, L.; Lopez, C.M.R. ChromoMap: An R package for interactive visualization and annotation of chromosomes. *bioRxiv* **2020**, 605600. [CrossRef]
43. Wickham, H. ggplot2: Elegant graphics for data analysis. *J. R. Stat. Soc. Ser. A* **2011**, *174*, 245–246. [CrossRef]
44. Livak, K.J.; Schmittgen, T.D. Analysis of relative gene expression data using real-time quantitative PCR and the $2^{-\Delta\Delta C}{}_T$ method. *Methods* **2001**, *25*, 402–408. [CrossRef] [PubMed]
45. Danilevskaya, O.N.; Meng, X.; Hou, Z.; Ananiev, E.V.; Simmons, C.R. A genomic and expression compendium of the expanded PEBP gene family from maize. *Plant Physiol.* **2007**, *146*, 250–264. [CrossRef]
46. Wang, Y.; Tan, X.; Paterson, A.H. Different patterns of gene structure divergence following gene duplication in Arabidopsis. *BMC Genom.* **2013**, *14*, 652. [CrossRef] [PubMed]
47. Lynch, M.; Force, A. The probability of duplicate gene preservation by subfunctionalization. *Genetics* **2000**, *154*, 459–473. [CrossRef]

48. Kumar, R.; Tyagi, A.K.; Sharma, A.K. Genome-wide analysis of auxin response factor (ARF) gene family from tomato and analysis of their role in flower and fruit development. *Mol. Genet. Genom.* **2011**, *285*, 245–260. [CrossRef]
49. Remington, D.L.; Vision, T.J.; Guilfoyle, T.J.; Reed, J. Contrasting modes of diversification in the *Aux/IAA* and *ARF* gene families. *Plant Physiol.* **2004**, *135*, 1738–1752. [CrossRef]
50. Sato, A.; Yamamoto, K.T. Overexpression of the non-canonical *Aux/IAA* genes causes auxin-related aberrant phenotypes in Arabidopsis. *Physiol. Plant.* **2008**, *133*, 397–405. [CrossRef]

horticulturae

MDPI

Article

Genome-Wide Identification of the 1-Aminocyclopropane-1-carboxylic Acid Synthase (*ACS*) Genes and Their Possible Role in Sand Pear (*Pyrus pyrifolia*) Fruit Ripening

Jing-Guo Zhang [1,2], Wei Du [2], Jing Fan [2], Xiao-Ping Yang [2], Qi-Liang Chen [2], Ying Liu [1], Hong-Ju Hu [2,*] and Zheng-Rong Luo [1,*]

1 Key Laboratory of Horticultural Plant Biology, Huazhong Agricultural University, Wuhan 430070, China; zhjg@webmail.hzau.edu.cn (J.-G.Z.); liuying1991@webmail.hzau.edu.cn (Y.L.)

2 Research Institute of Fruit and Tea, Hubei Academy of Agricultural Science, Wuhan 430064, China; duwei528529@126.com (W.D.); fanjing2013pear@163.com (J.F.); yangxiaoping1981@163.com (X.-P.Y.); cql99@sina.com (Q.-L.C.)

* Correspondence: hongjuhu@sina.com (H.-J.H.); luozhr@mail.hzau.edu.cn (Z.-R.L.)

Abstract: Ethylene production is negatively associated with storage life in sand pear (*Pyrus pyrifolia* Nakai), particularly at the time of fruit harvest. 1-Aminocyclopropane-1-carboxylic acid synthase (ACS) is the rate-limiting enzyme in ethylene biosynthesis and is considered to be important for fruit storage life. However, the candidate *ACS* genes and their roles in sand pear remain unclear. The present study identified 13 *ACS* genes from the sand pear genome. Phylogenetic analysis categorized these ACS genes into four subgroups (type I, type II, type III and putative AAT), and indicated a close relationship between sand pear and Chinese white pear (*P. bretschneideri*). According to the RNA-seq data and qRT-PCR analysis, *PpyACS1*, *PpyACS2*, *PpyACS3*, *PpyACS8*, *PpyACS9*, *PpyACS12* and *PpyACS13* were differently expressed in climacteric and non-climacteric-type pear fruits, 'Ni-nomiyahakuri' and 'Eli No.2', respectively, during fruit ripening. In addition, the expressions of *PpyACS2*, *PpyACS8*, *PpyACS12* and *PpyACS13* were found to be associated with system 1 of ethylene production, while *PpyACS1*, *PpyACS3*, and *PpyACS9* were found to be associated with system 2, indicating that these *ACS* genes have different roles in ethylene biosynthesis during fruit development. Overall, our study provides fundamental knowledge on the characteristics of the *ACS* gene family in sand pear, in addition to their possible roles in fruit ripening.

Keywords: sand pear; fruit ripening; ACS gene family; ethylene biosynthesis; climacteric; non-climacteric

Citation: Zhang, J.-G.; Du, W.; Fan, J.; Yang, X.-P.; Chen, Q.-L.; Liu, Y.; Hu, H.-J.; Luo, Z.-R. Genome-Wide Identification of the 1-Aminocyclopropane-1-carboxylic Acid Synthase (*ACS*) Genes and Their Possible Role in Sand Pear (*Pyrus pyrifolia*) Fruit Ripening. *Horticulturae* **2021**, *7*, 401. https://doi.org/10.3390/horticulturae7100401

Academic Editor: Dong Zhang

Received: 13 September 2021
Accepted: 13 October 2021
Published: 14 October 2021

Publisher's Note: MDPI stays neutral with regard to jurisdictional claims in published maps and institutional affiliations.

1. Introduction

Ethylene is a phytohormone that participates in many events of plant growth and development, like organ abscission, stress responses, fruit ripening and senescence [1,2]. Generally, ethylene biosynthesis is divided into two systems, named system 1 and system 2. System 1allows basal ethylene production in vegetative tissues and unripe fruit. System 2 involves the burst of ethylene production during the ripening of climacteric fruit. Moreover, system 1 is regulated in an auto-inhibitory manner, while system 2 is regulated in an autocatalytic manner. Fruit ripening behavior is generally characterized as being of a climacteric or a non-climacteric type, depending on whether or not the process is associated with a peak of ethylene production and a rise in respiration rate [3–5].

Ethylene is synthesized from *S*-adenosyl-L-methionine (SAM) via 1-aminocyclopropane-1-carboxylic acid (ACC). The two major processes in ethylene biosynthesis are the metabolic conversion of SAM to ACC by ACC synthase (EC 4.4.1.14) and secondly the conversion of ACC to ethylene by ACC oxidase (EC 1.14.17.4) [6]. In plant species, ACS is the rate-limiting enzyme in ethylene synthesis and is encoded by a widely diverse multi-gene family [7]. Ethylene biosynthesis is a complicated process involving numerous ACS genes.

The variable expression of ACS genes may be implicated in the synthesis of ethylene at a specific developmental stage. For example, there are 14 ACS genes in tomato (*Solanum lycopersicum*), among them *SlACS1A*, *SlACS2*, *SlACS4*, *SlACS11* and *SlACS12* showed different expression profiles during fruit ripening [8]. Three ACS genes (*MdACS1*, *MdACS3a*, and *MdACS6*) in apple have been extensively examined, and each of them was found to express distinctly [9–11]. *MdACS3a* was found to be expressed abundantly one month before fruit ripening, and its alleles are crucial in controlling fruit shelf-life, whereas *MdACS1* was shown to be responsible for the burst of ethylene production in system 2. *MdACS6* was expressed in the early stages of fruit development, before *MdACS3a* and *MdACS1* expression began [11].Among six *Pp-ACSs* isolated from the peach genome, *Pp-ACS1* and *Pp-ACS4* showed ripening-related increased expression during fruit development and ripening [12].

The sand pear (*Pyrus pyrifolia* Nakai) is widely distributed in southern China, where the ancestral pear species originated, and where a large collection of germplasm, including wild genotypes, landraces, and improved cultivars, exists [13,14]. Sand pear exhibits both climacteric and non-climacteric behavior depending on the cultivar or genotype, making it a suitable material for studying the fundamental basis of the various ripening behaviors [4]. In this study, the genome-wide analyses of ACS genes were conducted across five Rosaceae and two non-Rosaceae species. A total of 13 *PpyACSs* were identified in sand pear fruit and RNA-seq analysis showed the great variations between their expression levels. To further understand the role of *ACS* in fruit ripening, the predominantly expressed six ACS genes were selected and their expression profiles were compared in the fruits 'Ninomiyahakuri' and 'Eli No. 2' at different ripening stages. The findings of this study could help researchers better understand the molecular mechanism of fruit ripening in sand pear.

2. Materials and Methods

2.1. Plant Materials and Treatments

Two sand pear cultivars ('Ninomiyahakuri' and 'Eli No. 2') grown in Wuhan City, Hubei Province, P.R. China, were used in this study (Figure S1). Interestingly, both cultivars have different ripening behavior. Comparatively, fruits of 'Ninomiyahakuri' yield more ethylene during fruit ripening, whereas the fruits of 'Eli No. 2' yield less ethylene. The fruits of 'Ninomiyahakuri' and 'Eli No. 2' were collected 30 days before harvest (30 DBH; commercial maturity = 20 July 2017) and were kept at room temperature (24 °C) for 20 days. For lab analysis, fruits were sampled after an interval of 5 days after harvest (1, 5, 10, 15, 20 DAH). Five fruits were sampled at each sampling point for determination of ethylene production. Then the fruits were sliced, frozen in liquid nitrogen and stored at −80 °C for RNA extraction.

For spatial gene expression analysis, young fruit, mature fruit, flowers, leaf and stem were collected from 'Eli No. 2' pear trees. All samples were processed and stored as discussed above, for further analysis.

2.2. Measurements of Ethylene Production

The ethylene production was estimated with a gas chromatograph (Agilent 7890A, Santa Clara, CA, USA) fitted with a flame ionization detector. Briefly, fruits were placed in an airtight container (2.0 dm^3) equipped with septa and maintained at 24 °C for 1 h. Afterwards, 1 cm^3 of gas was sampled through the headspace of the container by a syringe.

2.3. RNA Isolation and RNA-Seq

Total RNA was extracted with TRIzol reagent (Invitrogen, Carlsbad, CA, USA) following the manufacturer's instructions. RNA concentration was measured by NanoDrop 2000 (Thermo, Waltham, MA, USA) and RNA integrity was evaluated by the RNA Nano 6000 Assay Kit using the Agilent Bioanalyzer 2100 (Agilent Technologies, Santa Clara, CA, USA).

Transcriptome libraries were constructed from samples of 'Ninomiyahakuri' and 'Eli No. 2' collected at three ripening stages: 1 (pre-ripening, 30 days before harvest), 2 (at harvest) and 3 (fruit senescence, 10 days after harvest). 1 μg RNA per sample was used for RNA-seq, which was carried out by Beijing Biomarker Technology Co., Ltd. (Beijing, China). In short, RNA-seq libraries were constructed using NEBNext®Ultra™ RNA Library Prep Kit for Illumina® (NEB, Ipswich, MA, USA) according to the manufacturer's protocol [15]. The cDNA library samples were sequenced by the Illumina Hi-Seq 2000 Sequencer. The transcriptome was aligned to the sand pear reference genome using Top Hat and the read count for each gene was obtained by the Cufflinks software. Quantification of gene expression levels were measured in fragments per kilobase of transcript per million fragments mapped (FPKM), and the DEGSeq R package (1.10.1) was used to analyze the differential expression of the pairwise comparisons [16].

2.4. Genome-Wide Identification of ACS Genes

In order to recognize the ACS gene family in *P. pyrifolia* and other four Rosaceae species (*P. bretschneideri*, *P. communis*, *Malus domestica* and *Prunus persica*), we searched the Genome Database for Rosaceae by BLASTP and BLASTN, using AtACSs sequences as queries [17,18]. The domain composition of ACSs were predicted by SMART (http://smart.embl-heidelberg.de/, accessed on 10 March 2021) and Pfam (http://pfam.san ger.ac.uk/, accessed on 10 March 2021). All of the identified ACS genes were analyzed using the NCBI's conserved domain database (https://www.ncbi.nlm.nih.gov/Structure/cdd/wrpsb.cgi, accessed on 12 March 2021) to determine the domains they possess. Multiple sequence alignment of ACS proteins was performed with Clustal X, the redundant or incomplete sequences were then removed. Additionally, ACS genes were named according to the guideline developed for Rosaceae family members [19].

The amino acid length and chromosome information were obtained from the genomic file. Protein molecular weights (Da) and isoelectric points (pI) were calculated by the online EXPASY server (http://web.expasy.org/protparam/, accessed on 12 March 2021). Gene motif analysis was performed with MEME v5.0.4 (http://meme-suite.org/tools/meme, accessed on 13 March 2021). Gene structure, motifs and chromosome location were presented using TBtools [20]. The Plant-mPLoc (http://www.csbio.sjtu.edu.cn/cgibin/PlantmPLoc.cgi, accessed on 15 March 2021) was performed to predict the subcellular localization of the ACS proteins. *Cis*-acting regulatory elements in the 2.0 kb of 5′-UTR of *PpyACS* were identified by the Plant-CARE database [21].

2.5. Phylogenetic Analysis

To identify orthologous genes, OrthoVenn2 was used to search against seven species genomes (Supplementary Table S1) using PpyACS as queries [22,23]. ACS amino acid sequences were aligned by using the MUSCLE program, and phylogenetic trees were built with the neighbor-joining method and 1000 bootstrap iterations in MEGAv7.0 software [24].

2.6. Quantitative Real-Time PCR (qRT-PCR) Analysis

The primers were designed by the Primer 6.0 program and listed in Table S2. Synthesis of the first-strand cDNA was conducted with PrimeScript™ RT reagent Kit with gDNA Eraser (TaKaRa, Shiga, Japan). cDNA was diluted 2-fold, and used as a template for qPCR. The PCR mix (50 mm^3) contained: 1.0 mm^3 Ex Taq and 25.0 mm^3 2X Ex Taq Buffer (both Takara), 1.0 mm^3 dNTP Mix, 1.5 mm^3 of each primer (10X), 5.0 mm^3 cDNA, and 15.0 mm^3 PCR-Grade Water. The melt curve was assessed from 65 to 95 °C, with 0.5 °C increments. Relative levels of gene expression were calculated with the $2^{-\Delta\Delta Ct}$ method [25]. qRT-PCR gene expression was analyzed by CFX Connect™ (BIO-RAD, Hercules, CA, USA) using the PP2A gene as an internal control [26]. All analyses were achieved by three independent biological replicates.

3. Results

3.1. ACS Gene Sequence Identification

In this study, 56 ACS genes from pear were identified and it included 13 genes from apple and 8 from peach, 10 genes from Chinese white pear, 12 genes from European pear as well as 13 genes from sand pear. The detailed information of gene IDs, genomic positions, coding region lengths and translated protein sequence about pear ACS genes were performed in Table S3. Lineage-specific whole-genome duplication (WGD) was found to have occurred in the ancestor of pear and apple [27], which may have resulted in almost double the number of ACS genes in these four species than in peach. The numbers of ACS genes identified in three species of *Pyrus* spp. have slight differences that may be due to their genome assembly quality.

Based on the presence or absence of C-terminal phosphorylation motifs, ACS proteins are classified into three categories (type I, type II, and type III) [28]. Type I ACSs have a C-terminal segment with phosphorylation sites for both mitogen-activated protein kinases (MAPKs) and calcium-dependent protein kinases (CDPKs) (CDPKs). Type II isozymes only have the CDPK target site, but type III isozymes lack both the MAPK and the CDPK sites. There is also another ACS-like homolog, AtACS10 and AtACS12, which are thought to be amino acid transferases lacking ACS activity and are known as putative AAT [29].

All identified ACS genes contained a "1-Aminocyclopropane-1-carboxylic acid synthase" domain but belong to the different domain families: PLN02450 superfamily or PLN02450 exist in all types of ACS except type III, which specifically possess PLN2607. Interestingly, two type I ACS homologous genes (*AtACS1* and *AtACS2*) in Arabidopsis have a PLN02376 domain, which is not present in other investigated plants (Table S3).

In sand pear, all PprACSs contained the seven conserved regions of ACS (Figure 1) [23] and were located in 8 chromosomes unevenly: chromosomes 15 and 2 resided three genes, chromosomes 1 had two genes, and chromosomes 4, 6, 7, 8 and 14 only showed one gene (Figure S2). Subcellular prediction results showed that all PpyACS proteins were localized in the chloroplast, except for type II ACS, which were cytoplasm-localized proteins. The length of PpyACS-encoded protein sequences ranged from 445 (PpyACS11) to 611 (PpyACS13) amino acids. The predicted molecular weights and pI ranged from 49.84 kDa (PpyACS11) to 67.55 kDa (PpyACS13) and from 5.63 (PpyACS10) to 8.76 (PpyACS12), respectively (Table 1).

Table 1. Genetic bioinformation of all identified *ACS* genes in *Pyrus pyrifolia*.

Gene Name	Gene ID	Chromosome	Protein Length (aa)	No. of Exons	Molecular Weight (kDa)	Isoelectric Point (pI)	Type	Subcellular Prediction
PpyACS1	Ppy15g2510.1	Chr15:20022979..20024784+	473	4	53.24	6.47	Type II	Cytoplasm
PpyACS5	Ppy02g1684.1	Chr02:14156236..14158531+	473	4	55.54	6.9	Type II	Cytoplasm
PpyACS6	Ppy08g0607.1	Chr08:4476046..4478458-	488	5	59.57	6.69	Type II	Cytoplasm
PpyACS7	Ppy15g0540.1	Chr15:3580798..3589288-	502	5	54.67	7.06	Type II	Cytoplasm
PpyACS3	Ppy01g0738.1	Chr01:12651750..12654529-	495	4	54.86	7.96	Type I	Chloroplast
PpyACS8	Ppy07g1563.1	Chr07:23041756..23045681+	529	5	53.18	8.53	Type I	Chloroplast
PpyACS9	Ppy06g0809.1	Chr06:12079561..12081861-	487	4	56.82	7.56	Type I	Chloroplast
PpyACS4	Ppy14g0933.1	Chr14:11862299..11864545-	487	4	55.15	6.64	Type I	Chloroplast
PpyACS2	Ppy15g1745.1	Chr15:12377889..12379594-	446	3	50.1	5.65	Type III	Chloroplast
PpyACS10	Ppy02g0565.1	Chr02:3901240..3902919-	447	3	50.31	5.63	Type III	Chloroplast
PpyACS11	Ppy02g0566.1	Chr02:3918274..3920013-	445	3	49.84	5.84	Type III	Chloroplast
PpyACS12	Ppy04g0675.1	Chr04:6740597..6743142+	538	4	58.89	8.76	Putative AAT	Chloroplast
PpyACS13	Ppy01g1669.1	Chr01:20075291..20079590-	611	6	67.55	8.3	Putative AAT	Chloroplast

The exon-intron organizations and motifs of all PpyACS genes were examined in *P. pyrifolia*. As shown in Figure 2, 16 conserved motif sequences were detected (Table S4). All PpyACS contained motif 1~10. Motif 11 and 12 were only found in type I ACS, whereas motif 15 was unique to putative AAT genes. Motif 13 was found in type I and type III ACS. In four type II ACS genes, motif 14 was present in PpyACS1 and PpyACS5, whereas motif 16 only existed in PpyACS1 and PpyACS5. In general, the motifs are quite conserved within each type. In addition, the presence of a varying number and length of introns among the PpyACSs contribute to significant variations in gene length. The number of introns per PpyACSs also varied from two to five: the type III PpyACSs were characterized by two introns and the other PpyACSs displayed three to four introns except for PpyACS13, which contained five introns.

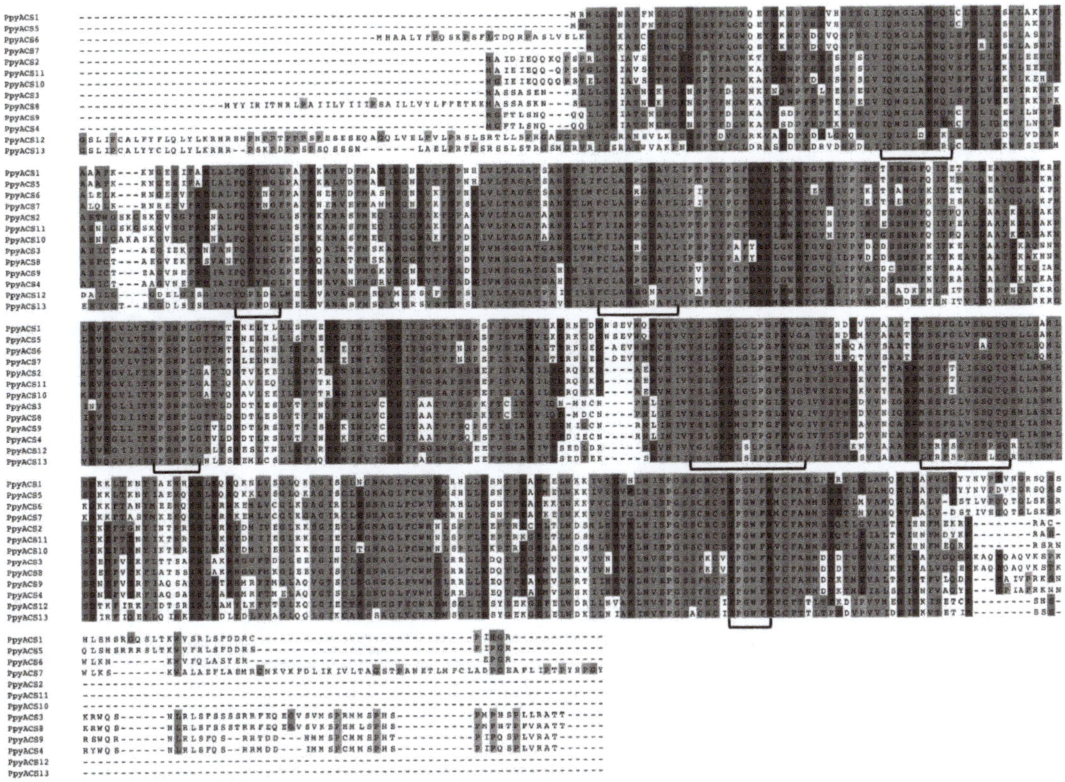

Figure 1. Amino acid sequence alignment of the PpyACSs. The seven conserved domains of the ACS isozymes are marked as red boxes. The seven highly conserved regions among all ACC synthases are underlined.

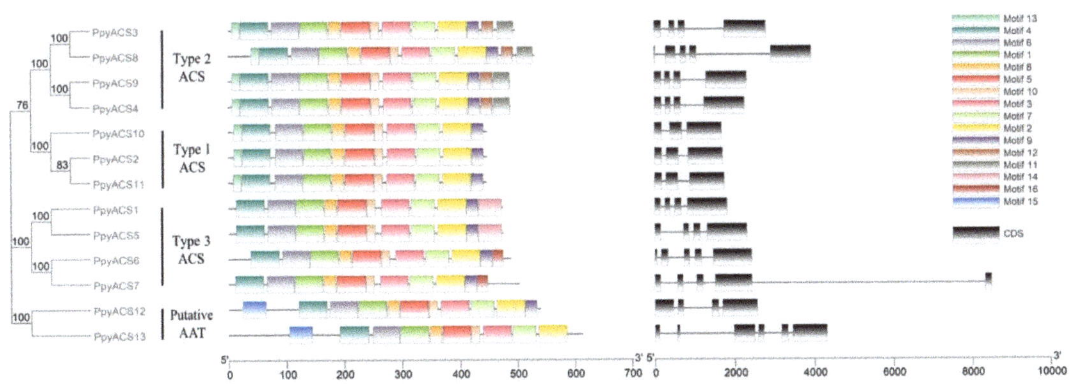

Figure 2. The phylogenetic relationship, conserved motifs and CDS structure of PpyACS proteins. A phylogenetic tree of the PpyACS protein family (**left side**) was constructed by MEGA7 using the Neighbor-Joining (NJ) method and 1000 bootstraps. The distribution of 16 conserved motifs across PpyACS protein (**middle side**) was predicted by the MEME (Multiple Em for Motif Elicitation) tools. The CDS structure of PpyACS genes were shown on the (**right side**). The sequences and lengths of motifs among PpyACS proteins were presented in Table S4.

To further understand the transcriptional regulation and potential functions of *Ppy*ACS genes, the upstream promoter sequences of *PpyACSs* (2000 bp upstream of the start codon) were isolated and predicted (Figure 3). A number of cis-elements implicated in hormone and stress responses were found in the upstream sequences of these *PpyACSs*. MeJA responsiveness elements (TGACG-and CGTCA-motif) were present in all *PpyACSs* except *PpyACS13*. ABRE, ABA-responsive element was identified in most of the *PpyACSs*, except *PpyACS4*, *PpyACS11* and *PpyACS12*. Three elements involved in gibberellin response, namely GARE-motif, TATC-box and P-box, were identified in all *PpyACSs*, except *PpyACS7*. Auxin-responsive elements (AuxRR-core and TGA-element) were found in most of the *PpyACSs*, except *PpyACS6*, *PpyACS7*, *PpyACS8* and *PpyACS11*. We also noticed that cis-elements associated with abiotic stress response were widely distributed within the promoters of *PpyACSs*. For example, an element essential for the anaerobic induction, named ARE, was identified in the *PpyACSs*, except *PpyACS6* and *PpyACS12*. A low temperature-responsive element (LTR) was identified in *PpyACS1*, *PpyACS3*, *PpyACS9* and *PpyACS10*, whereas TC-rich repeats involved in defense response were observed in *PpyACS1* and *PpyACS2*. Meanwhile, a MYB transcription factor binding site (MBS) involved in drought stress response was present in four members, including *PpyACS2*, *PpyACS7*, *PpyACS10* and *PpyACS11*. In addition, a wound-responsive element (WUN motif) was found in *PpyACS5*, *PpyACS9* and *PpyACS11*.

3.2. Phylogenetic Analysis of Putative ACS Genes

A total of 88 ACS proteins from seven species were phylogenetically categorized into four subgroups that perfectly fit their types (Figure 4). Type II included the highest number of ACS genes (27), followed by type I, which contained 25 ACS genes. This result provides further evidence that type I and type II contains a greater percentage of ACS genes to other types. Interestingly, each species has two members belonging to putative AAT that were presumed without ACS activity. In addition, from species phylogeny based on ACS orthologs genes (Figure S3, Table S5), a close relationship of ACS genes was observed between *Pyrus* and *Malus*, and the closest relationship was detected between *P. pyrifolia* and *P. bretschneideri* genes.

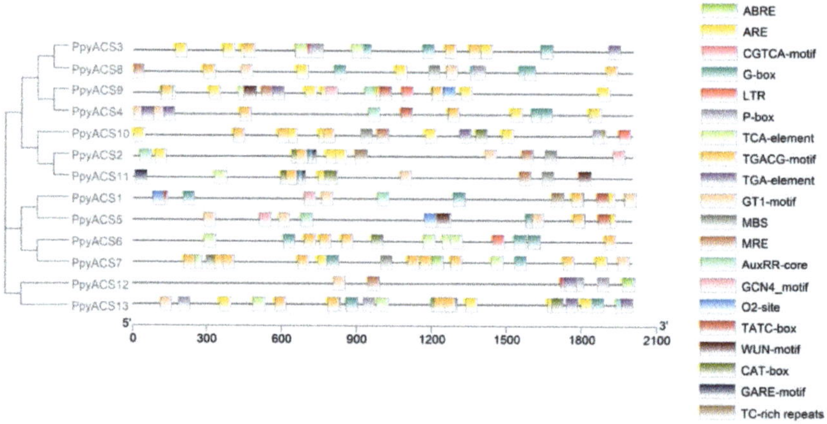

Figure 3. Cis-elements analysis of *PpyACS* genes. Note: *ABRE*, abscisic acid-responsive element; *ARE*, element essential for the anaerobic induction; *LTR*, low-temperature responsiveness; *MBS*, MYB binding site involved in drought-inducibility; *MRE*, MYB binding site involved in light responsiveness; *TCA-element*, salicylic acid-responsive element; *TGA-element*, auxin-responsive element; *TC-rich repeats*, defense and stress responsiveness; *TGACG-* and *CGTCA-motif*, MeJA responsiveness; *GARE-motif*, gibberellin-responsive element; *TATC-box*, gibberellin-responsive element; *WUN-motif*, wound-responsive element; *P-box*, gibberellin-responsive element; *AuxRR-core*, auxin-responsive element.

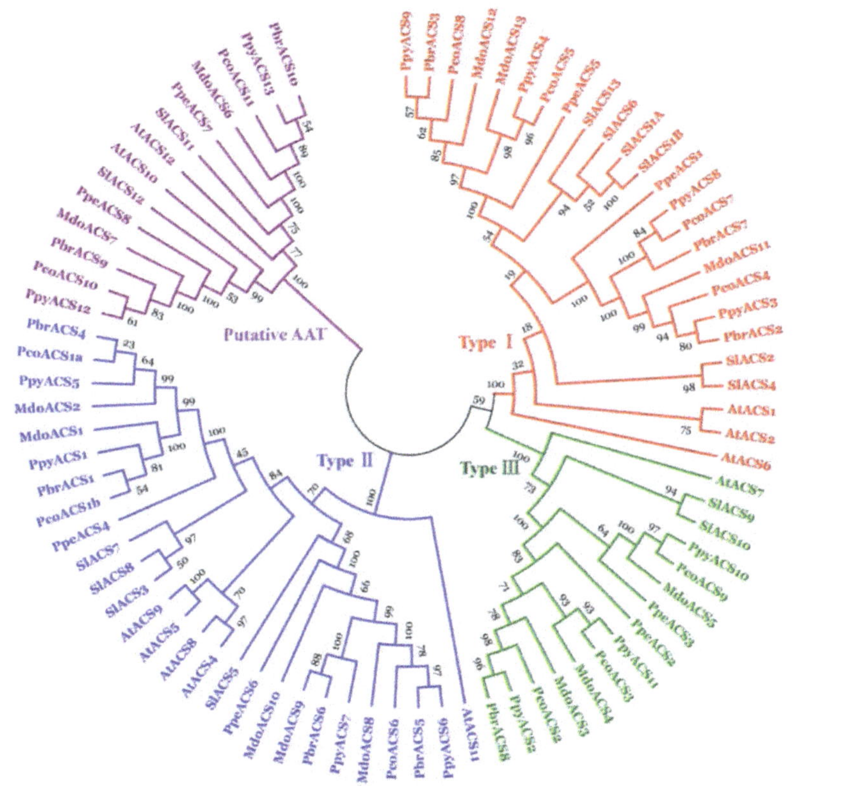

Figure 4. Phylogenetic analysis of ACS proteins from five Rosaceae species (*Pyrus pyrifolia*, *Pyrus bretschneideri*, *Pyrus communis*, *Malus domestica* and *Prunus persica*), and two non-Rosaceae species (*Arabidopsis thaliana* and *Solanum lycopersicum*). The phylogenetic tree was constructed using MEGA7.0 software by the neighbor-joining method. Different colors indicate different subfamilies of ACS.

3.3. Expression of PpyACSs in Different Organs

The qRT-PCR analysis showed that *Ppy*ACS genes were differentially expressed in different tissues. Among 13 *Ppy*ACS genes, *PpyACS1*, *PpyACS2*, *PpyACS3* and *PpyACS9* were expressed highly in mature fruits. The other nine *PpyACS* genes were expressed in the remaining studied tissues/organs, such as *PpyACS4* and *PpyACS5*, which were highly expressed in the leaf, while *PpyACS5*, *PpyACS7*, *PpyACS8*, *PpyACS10*, *PpyACS12* and *PpyACS13* in flowers (Figure 5).

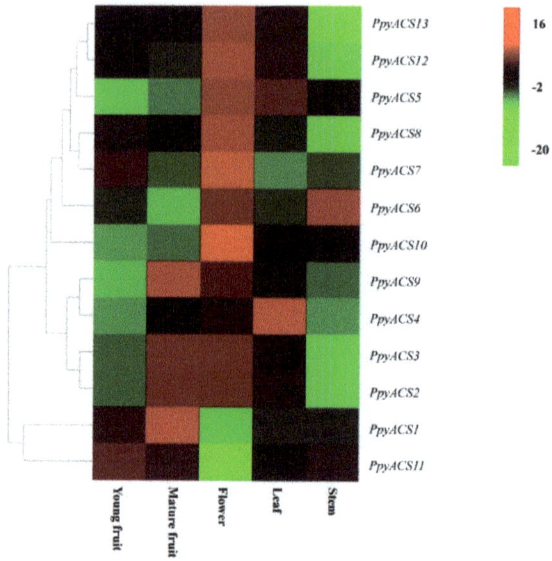

Figure 5. Heat map of the expression level of *PpyACSs* in different organs. The tissues used for the expression profiling are indicated at the bottom of each column. A cluster dendrogram is shown on the left. ACSs are divided into 3 major groups based on their expression. The color key at the top right corner represents the Z-score values transformed from log2-based expression values obtained by qRT-PCR.

3.4. RNA-Seq of Sand Pear Fruit and Identification of PpyACSs

In this study, six transcriptome libraries were constructed and sequenced for cv. 'Ninomiyahakuri' and 'Eli No. 2' in three ripening stages, including 1 (pre-ripening, 30 days before harvest), 2 (at harvest) and 3 (fruit senescence, 10 days after harvest). Over 89.2% of clean reads were uniquely mapped to the sand pear genome [30]. Expression of 13 *PpyACSs* in the pear genome was detected. The transcript abundance of *PpyACSs* varied greatly among members in different climacteric type sand pears, with that of one gene, *PpyACS1*, accounting for the most of total abundance in 'Ninomiyahakuri' (climacteric type) but none in 'Eli No. 2' (non-climacteric type). And six genes (*PpyACS2, PpyACS3, PpyACS8, PpyACS9, PpyACS12* and *PpyACS13*) were expressed during the fruit maturation in both cultivars but in different FPKM (Fragments per kilobase of transcript per million fragments mapped) (Figure 6). Based on these results, seven genes were selected for further analysis.

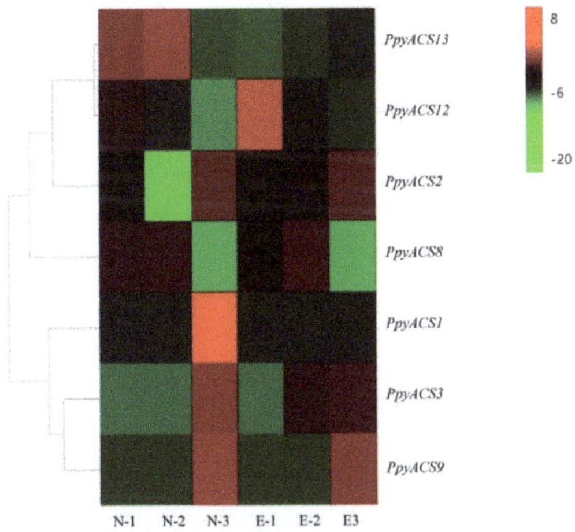

Figure 6. Heat map of the expression level of *PpyACSs* in fruit ripening. 'Ninomiyahakuri' and 'Eli No. 2' in three ripening stages, including 1 (pre-ripening, 30 days before harvest), 2 (at harvest) and 3 (fruit senescence, 10 days after harvest). A cluster dendrogram is shown on the left. ACSs are divided into two major groups based on their expression. The color key at the top right corner represents the Z-score values transformed from log2-based expression values obtained by RNA-seq (Table S6).

3.5. Changes in Internal Ethylene Concentration in Flesh during Fruit Ripening

To further elucidate the relationship between the expression pattern of the seven *PpyACS* genes and internal ethylene concentration, dynamic changes in internal ethylene concentration in the fruits of two cultivars were determined. As shown in Figure 7, there was a rapid increase in internal ethylene concentration and a large amount of ethylene was produced in 'Ninomiyahakuri' fruits that showed ripening behavior typical to climacteric fruits. In contrast, almost undetectable ethylene production was observed in 'Eli No. 2' fruits during postharvest.

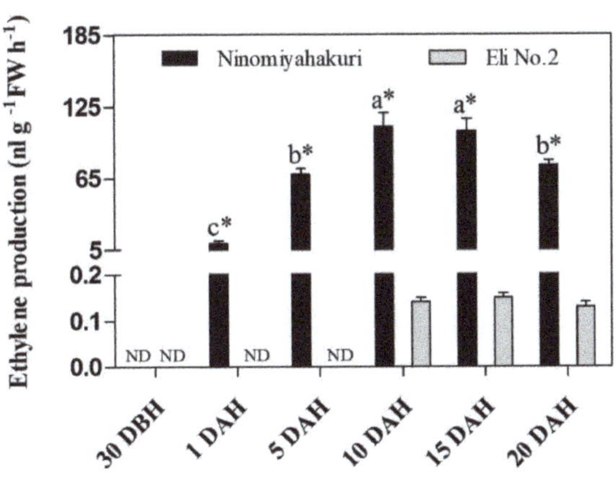

Figure 7. The ethylene production of fruits in ripening period of 'Ninomiyahakuri' and 'Eli No. 2' fruits. DBH: day before harvest. DAH: day after harvest. ND: None detected. '*' in each graph indicates a significant difference between 'Ninomiyahakuri' and 'Eli No. 2' cultivar (*t*-test, $n = 3$, $p < 0.05$). Different lower-case letters indicate significant differences among time points in the same cultivar (Duncan-test, $n = 3$, $p < 0.05$).

3.6. Expression of PpyACSs during Fruit Ripening

To validate expression profiles of the ACS genes isolated from transcriptome analysis, the seven *PpyACSs* were selected for qRT-PCR testing in fruits of cv. 'Ninomiyahakuri' and 'Eli No. 2' (Figure 8). *ACS1* was dramatically induced by ripening in 'Ninomiyahakuri' and showed the highest expression during postharvest, but was undetectable throughout ripening in 'Eli No. 2', which means *PpyACS1* is the dominant gene that regulates ripening-associated ethylene production in climacteric fruit. The expression of *PpyACS2* was also strongly increased during storage, not only in climacteric fruits but also in non-climacteric varieties, although the peak was approximately 2-fold lower than that in climacteric fruits. It is worth noting that three out of four type I genes have shown distinct expression patterns in fruit ripening, such as *PpyACS3* and *PpyACS9*, which were highly expressed after harvest in climacteric fruits, but the expression level of *PpyACS8* was higher at the pre-harvest stage than at the post-harvest stage in both climacteric and non-climacteric fruits. During storage, the expression level of *PpyACS8* decreased first and then increased. Interestingly, *PpyACS12* and *PpyACS13*, two ACS homologues of *AtACS10* and *AtACS12* which are presumed as amino acid transferases without ACS activity, were highly expressed at 30 days before commercial harvest and the expression level decreased slightly after harvest.

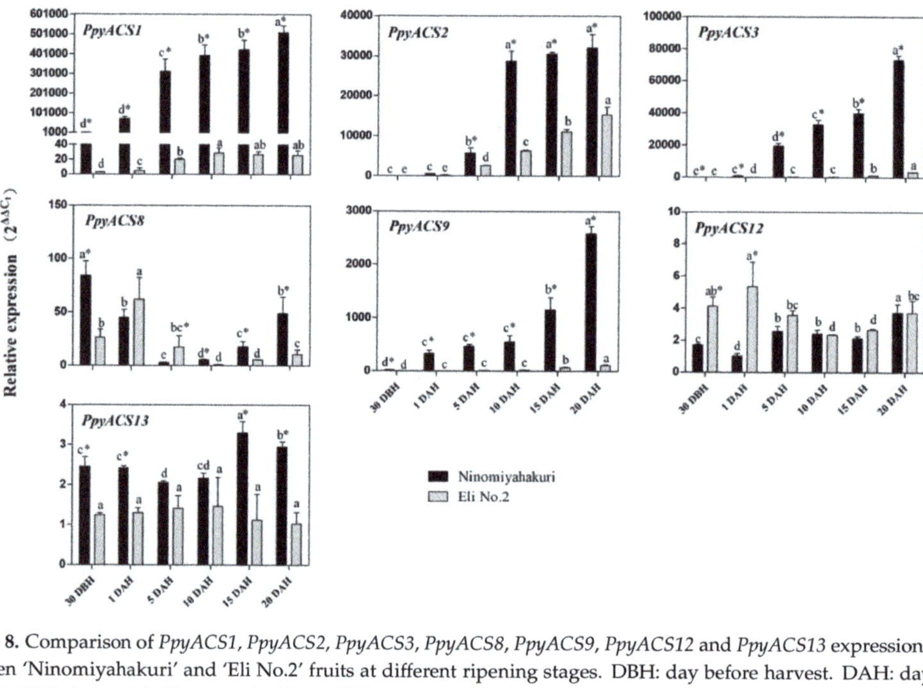

Figure 8. Comparison of *PpyACS1*, *PpyACS2*, *PpyACS3*, *PpyACS8*, *PpyACS9*, *PpyACS12* and *PpyACS13* expression levels between 'Ninomiyahakuri' and 'Eli No.2' fruits at different ripening stages. DBH: day before harvest. DAH: day after harvest. '*' Each graph indicates a significant difference between 'Ninomiyahakuri' and 'Eli No. 2' cultivar (t-test, $n = 3$, $p < 0.05$). Different lower-case letters indicate significant differences between time points in the same cultivar (Duncan-test, $n = 3$, $p < 0.05$).

4. Discussion

ACS is encoded by a multigene family with unique spatial and temporal expression patterns [8,12,23,31–33]. Although two ACSs have been found in sand pear [4], there are likely to be unidentified ACS genes in the pear genome, and their relationship with ripening behavior is unknown. Currently, the full-length genome of sand pear (*P. pyrifolia*) has been sequenced, providing valuable gene resources for further functional genomics study [30]. In this study, a total of 13 ACS genes were found in the genome of sand pear, and the number is the same as that identified in *P. ussuriensis* [34]. A comprehensive overview of the ACS gene family was undertaken, including the gene structures, conserved motifs, chromosome locations, phylogeny, and *cis*-elements in the promoter sequence, as well as expression patterns.

It was noticed that the number of species in which the fruits of different varieties exhibit both climacteric and non-climacteric behavior and the classification of fruits into climacteric and non-climacteric categories are an over-simplification [35]. Consistent with previous reports [4,36], substantial differences in ethylene production during fruit ripening was observed between two sand pear cultivars, which indicated that sand pear shows climacteric as well as non-climacteric behavior, depending on the genotype. Previous research showed that ethylene production in climacteric sand pears is attributed to *PpyACS1* and *PpyACS2* expression during the ripening stage [4], but the mechanism of fruit ripening in non-climacteric fruit is unclear.

In this study, genome-wide expression analysis based on RNA-seq data and qRT-PCR was carried out with seven genes identified as the most strongly expressed genes during fruit ripening, *PpyACS1*, *PpyACS2*, *PpyACS3*, *PpyACS8*, *PpyACS9*, *PpyACS12* and *PpyACS13* (Figure 6, Table 2). *PpyACS1* was pre-dominantly expressed only in climacteric fruits after harvest and in accordance with the change in system 2 ripening-associated ethylene production, which indicated that *PpyACS1* may play key roles in the determination of ethylene production of climacteric fruit ripening. The same result was verified in previous studies [4,31,37,38]. Otherwise, as a type II gene, *PpyACS1* was probably under the regulation of ETO1 (ETHYLENE OVERPRODUCER 1) [28], which needs further research.

Table 2. Summary of *PpyACSs* gene expression during sand pear fruit development and ripening.

Fruit Type	Main ACSs Involved in Ethylene Production	
	Immature Stage	**Mature Stage**
Climacteric	PpyACS8, PpyACS12, PpyACS13	PpyACS1, PpyACS2, PpyACS3, PpyACS9
Non-climacteric	PpyACS8, PpyACS12, PpyACS13	PpyACS2

Type III ACS proteins lack both MAPK and CDPK sites for post-translational regulation and may be of minor importance in fruit ripening [12]. However, a type IIIACS gene, *PPACS2* (*PpyACS2*), appears to be only expressed in the ripening of sand pear cultivars producing high or moderate levels of ethylene [4]. Surprisingly, our results showed that *PpyACS2* was expressed highly, even in non-climacteric fruit after harvest, which indicated that *PpyACS2* was responsible for system 1 ethylene biosynthesis in fruit ripening, similar to its ortholog gene, *MdACS3*, in apple.

Several investigations have revealed that type I functional divergence was the prevailing trend in the evolution of ACS subclades in plants [39]. In tomato fruit, it is well established that *SlACS6* and *SlACS1A* expression is associated with system 1 of ethylene production, while *SlACS4* and *SlACS2* are associated with system 2 [8,40]. Interestingly, all these four ACSs belong to type I (Figure 4) [41]. However, no type I ACS genes were expressed in apple fruit [33]. In the present study, three type I genes were newly identified—*PpyACS3*, *PpyACS8* and *PpyACS9*—and were involved in ethylene biosynthesis during fruit ripening. *PpyACS8* was highly expressed in the pre-ripening stage and is mostly responsible for the auto-inhibitory low levels of system 1, while *PpyACS3* and *PpyACS9* were highly expressed after fruit ripening and are associated with the burst of ethylene production in system 2. These results suggest that the type I ACS genes probably obtained unique and redundant functions during evolution.

It has been suggested that plant ACS genes evolved from ACS-like genes, which come from AATase genes [39]. The number and sequence of ACS-like genes are highly conserved in each species identified, which supported the ACS originate evolution of ACS-like genes. ACS-like genes, such as *AtACS10* and *AtACS12*, encode aminotransferases and without ACS activity, they were not generally considered to be authentic ACS genes. However, ACS-like genes in tomato, *SlACS11* and *SlACS12* display a significant up-regulation during fruit ripening. Li et al. (2015) have reported that ethylene in young apple fruits is mainly produced by catalysis by *MdACS6*, an ACS-like gene that was mistaken for a type III gene [11,33]. In our data, the expression levels of *PpyACS12* and *PpyACS13* were low but expressed constitutively throughout the experimental period in both cultivars. Therefore, ACS-like gene expression patterns in tomato, apple and sand pear suggested that ACS-like genes are also responsible for the system 1 ethylene biosynthesis.

Analysis of the pear genome has revealed numerous duplicated genes (paralogs) forming large synteny blocks, covering major portions of chromosomes [27]. Interestingly, all PpyACS genes are distributed on chromosome pairs with partial synteny blocks rather than on chromosome pairs with more than 90% syntenic relations. For instance, *PpyACS1* (*Ppy15g2510.1*) and *PpyACS5* (*Ppy02g1684.1*) were paralogs, and have sequence similarities of more than 94%, located in the chromosome pairs with partial synteny blocks, Chr02 (upper) and Chr15 (middle-upper). The result indicated that *PpyACS* gene expansion resulted from the WGD event. Interestingly, both *PpyACS1* and *PpyACS2*, two key ACS genes that operate in system1 and system 2 ethylene biosynthesis in fruit ripening, respectively, were located in the middle of chromosome 15. These findings implied that this region is most important in the regulation of fruit ripening and therefore needs to be focused on in future investigations.

The localization of proteins in different organelles may be associated with their function [42,43]. Previous reports have indicated that ACS is localized in the cytoplasm [44–46]. Interestingly, subcellular localization predictions show that only type II ACS is located in the cytoplasm and other types of ACS in chloroplasts. Further subcellular localization testing is required to confirm this.

5. Conclusions

In conclusion, the genome sequencing data showed a total of 13 ACS genes in sand pear, nine of which were novel members. Seven of these ACS genes were found to be involved in sand pear fruit ripening, while four of them were found to be involved in system 1 ethylene biosynthesis, indicating that they play diverse roles in ethylene biosynthesis systems. This study also highlighted the need to include additional ACS genes in studies on ethylene production in sand pear fruits.

Supplementary Materials: The following are available online at https://www.mdpi.com/article/10.3390/horticulturae7100401/s1, Figure S1: Ripe fruit of 'Ninomiyahakuri' and 'Eli No. 2' pears, Figure S2: Distribution of ACS genes on the *Pyrus pyrifolia* chromosomes, Figure S3: Species tree of five Rosaceae species and two reference species. Table S1: ACS homologous genes identified in five Rosaceae species and two non-Rosaceae plant species, Table S2: Primers used for qRT-PCR analysis, Table S3: Detailed information of all ACS family genes identified in the seven species, Table S4: Sequences and lengths of motifs among PpyACS proteins, Table S5: ACS orthologs genes identified in the seven species, Table S6: Different expression of ACS genes in sand pear fruit ripening.

Author Contributions: Conceptualization, J.-G.Z. and Z.-R.L.; methodology, J.-G.Z., Y.L. and J.F.; formal analysis, J.-G.Z. and W.D.; investigation, J.-G.Z., H.-J.H., X.-P.Y. and Q.-L.C., writing—original draft preparation, J.-G.Z.; writing—review and editing, J.-G.Z., H.-J.H., W.D. and Z.-R.L.; supervision, J.F. and X.-P.Y.; project administration, J.-G.Z.; funding acquisition, H.-J.H. and Z.-R.L. All authors have read and agreed to the published version of the manuscript.

Funding: This research was funded by the China Agriculture Research System of MOF and MARA (CARS-28) and the Major Program of Hubei Agricultural Science and Technology Innovation Center (2020-620-000-002-05).

Institutional Review Board Statement: Not applicable.

Informed Consent Statement: Not applicable.

Acknowledgments: The authors thank Syed Bilal Hussain (University of Florida, USA) for peer review of the manuscript and for providing valuable comments and suggestions. We are also grateful for the patience and help of the editors and the reviewers.

Conflicts of Interest: The authors declare no conflict of interest.

References

1. Li, S.; Chen, K.; Grierson, D. A critical evaluation of the role of ethylene and MADS transcription factors in the network controlling fleshy fruit ripening. *New Phytol.* **2019**, *221*, 1724–1741. [CrossRef] [PubMed]
2. Dubois, M.; Van den Broeck, L.; Inzé, D. The pivotal role of ethylene in plant growth. *Trends Plant Sci.* **2018**, *23*, 311–323. [CrossRef]
3. Liu, Y.; Tang, M.; Liu, M.; Su, D.; Chen, J.; Gao, Y.; Bouzayen, M.; Li, Z. The molecular regulation of ethylene in fruit ripening. *Small Methods* **2020**, *4*, 1900485. [CrossRef]
4. Itai, A.; Kawata, T.; Tanabe, K.; Tamura, F.; Uchiyama, M.; Tomomitsu, M.; Shiraiwa, N. Identification of 1-aminocyclopropane-1-carboxylic acid synthase genes controlling the ethylene level of ripening fruit in Japanese pear (*Pyrus pyrifolia* Nakai). *Mol. Gen. Genet.* **1999**, *261*, 42–49. [CrossRef] [PubMed]
5. Bapat, V.A.; Trivedi, P.K.; Ghosh, A.; Sane, V.A.; Ganapathi, T.R.; Nath, P. Ripening of fleshy fruit: Molecular insight and the role of ethylene. *Biotechnol. Adv.* **2010**, *28*, 94–107. [CrossRef]
6. Pattyn, J.; Vaughan-Hirsch, J.; Van de Poel, B. The regulation of ethylene biosynthesis: A complex multilevel control circuitry. *New Phytol.* **2021**, *229*, 770–782. [CrossRef]
7. Kende, H. Ethylene biosynthesis. *Annu. Rev. Plant Physiol. Plant Mol. Biol.* **1993**, *44*, 283–307. [CrossRef]
8. Liu, M.; Pirrello, J.; Chervin, C.; Roustan, J.; Bouzayen, M. Ethylene control of fruit ripening: Revisiting the complex network of transcriptional regulation. *Plant Physiol.* **2015**, *169*, 2380–2390. [CrossRef]
9. Sunako, T.; Sakuraba, W.; Senda, M.; Akada, S.; Ishikawa, R.; Niizeki, M.; Harada, T. An allele of the ripening-specific 1-aminocyclopropane-1-carboxylic acid synthase gene (*ACS1*) in apple fruit with a long storage life. *Plant Physiol.* **1999**, *119*, 1297–1304. [CrossRef]
10. Wang, A.; Yamakake, J.; Kudo, H.; Wakasa, Y.; Hatsuyama, Y.; Igarashi, M.; Kasai, A.; Li, T.; Harada, T. Null mutation of the *MdACS3* gene, coding for a ripening-specific 1-aminocyclopropane-1-carboxylate synthase, leads to long shelf life in apple fruit. *Plant Physiol.* **2009**, *151*, 391–399. [CrossRef]
11. Li, T.; Tan, D.; Liu, Z.; Jiang, Z.; Wei, Y.; Zhang, L.; Li, X.; Yuan, H.; Wang, A. Apple MdACS6 regulates ethylene biosynthesis during fruit development involving ethylene-responsive factor. *Plant Cell Physiol.* **2015**, *56*, 1909–1917. [CrossRef]
12. Zeng, W.; Pan, L.; Liu, H.; Niu, L.; Lu, Z.; Cui, G.; Wang, Z. Characterization of 1-aminocyclopropane-1-carboxylic acid synthase (*ACS*) genes during nectarine fruit development and ripening. *Tree Genet.* **2015**, *11*, 18. [CrossRef]
13. Li, X.; Liu, L.; Ming, M.; Hu, H.; Zhang, M.; Fan, J.; Song, B.; Zhang, S.; Wu, J. Comparative transcriptomic analysis provides insight into the domestication and improvement of pear (*P. pyrifolia*) fruit. *Plant Physiol.* **2019**, *180*, 435–452. [CrossRef]
14. Mudassar, A.; Li, J.Z.; Yang, Q.S.; Jamil, W.; Teng, Y.W.; Bai, S.L. Phylogenetic, molecular, and functional characterization of PpyCBF proteins in Asian pears (*Pyrus pyrifolia*). *Int. J. Mol. Sci.* **2019**, *20*, 2074.
15. Guo, L.Q.; Zhang, J.G.; Liu, X.X.; Rao, G.D. Polyploidy-related differential gene expression between diploid and synthesized allotriploid and allotetraploid hybrids of Populus. *Mol. Breed.* **2019**, *39*, 69.
16. Wang, L.; Feng, Z.; Wang, X.; Wang, X.; Zhang, X. DEGseq: An R package for identifying differentially expressed genes from RNA-seq data. *Bioinformatics* **2010**, *26*, 136–138. [CrossRef]
17. Qi, L.Y.; Chen, L.; Wang, C.S.; Zhang, S.L.; Yang, Y.J.; Liu, J.L.; Li, D.L.; Song, J.K.; Wang, R. Characterization of the auxin efflux transporter PIN proteins in pear. *Plants* **2020**, *9*, 349. [CrossRef]
18. Jung, S.; Lee, T.; Cheng, C.; Buble, K.; Zheng, P.; Yu, J.; Humann, J.; Ficklin, S.; Gasic, K.; Scott, K.; et al. 15 years of GDR: New data and functionality in the genome database for Rosaceae. *Nucleic Acids Res.* **2019**, *47*, D1137–D1145. [CrossRef] [PubMed]
19. Jung, S.; Bassett, C.; Bielenberg, D.G.; Cheng, C.-H.; Dardick, C.; Main, D.; Meisel, L.; Slovin, J.; Troggio, M.; Schaffer, R.J. A standard nomenclature for gene designation in the Rosaceae. *Tree Genet.* **2015**, *11*, 108. [CrossRef]
20. Chen, C.; Chen, H.; Zhang, Y.; Thomas, H.; Frank, M.; He, Y.; Xia, R. TBtools: An integrative toolkit developed for interactive analyses of big biological data. *Mol. Plant* **2020**, *13*, 1194–1202. [CrossRef] [PubMed]
21. Lescot, M.; Déhais, P.; Thijs, G.; Marchal, K.; Moreau, Y.; Peer, Y.; Rouzé, P.; Rombauts, S. PlantCARE, a database of plant *cis*-acting regulatory elements and a portal to tools for in silico analysis of promoter sequences. *Nucleic Acids Res.* **2002**, *30*, 325–327. [CrossRef]
22. Xu, L.; Dong, Z.; Fang, L.; Luo, Y.; Wei, Z.; Guo, H.; Zhang, G.; Gu, Y.; Coleman-Derr, D.; Xia, Q.; et al. OrthoVenn2: A web server for whole-genome comparison and annotation of orthologous clusters across multiple species. *Nucleic Acids Res.* **2019**, *47*, 52–58. [CrossRef] [PubMed]
23. Yamagami, T.; Tsuchisaka, A.; Yamada, K.; Haddon, W.; Harden, L.; Theologis, A. Biochemical diversity among the 1-amino-cyclopropane-1-carboxylate synthase isozymes encoded by the *Arabidopsis* gene family. *J. Biol. Chem.* **2003**, *278*, 49102–49112. [CrossRef] [PubMed]
24. Kumar, S.; Stecher, G.; Tamura, K. MEGA7: Molecular evolutionary genetics analysis version 7.0 for bigger datasets. *Mol. Biol. Evol.* **2016**, *33*, 1870–1874. [CrossRef] [PubMed]
25. Livak, K.J.; Schmittgen, T.D. Analysis of relative gene expression data using real-time quantitative PCR and the $2^{-\Delta\Delta Ct}$ method. *Methods* **2001**, *25*, 402–408. [CrossRef] [PubMed]
26. Wang, Y.; Dai, M.; Cai, D.; Shi, Z. Screening for quantitative real-time PCR reference genes with high stable expression using the mRNA-sequencing data for pear. *Tree Genet.* **2019**, *15*, 54. [CrossRef]

27. Li, H.; Huang, C.-H.; Ma, H. Whole-Genome Duplications in Pear and Apple. In *The Pear Genome*; Korban, S.S., Ed.; Springer International Publishing: Cham, Switzerland, 2019; pp. 279–299.
28. Yoshida, H.; Nagata, M.; Saito, K.; Wang, K.; Ecker, J. Arabidopsis ETO1 specifically interacts with and negatively regulates type 2 1-aminocyclopropane-1-carboxylate synthases. *BMC Plant Biol.* **2005**, *5*, 14. [CrossRef]
29. Zhu, J.-H.; Xu, J.; Chang, W.-J.; Zhang, Z.-L. Isolation and molecular characterization of 1-aminocyclopropane-1-carboxylic acid synthase genes in *Heveabrasiliensis*. *Int. J. Mol. Sci.* **2015**, *16*, 4136–4149. [CrossRef]
30. Shirasawa, K.; Itai, A.; Isobe, S. Chromosome-scale genome assembly of Japanese pear (*Pyrus pyrifolia*) variety 'Nijisseiki'. *DNA Res.* **2021**, *28*, dsab001. [CrossRef]
31. Yamane, M.; Abe, D.; Yasui, S.; Yokotani, N.; Kimata, W.; Ushijima, K.; Nakano, R.; Kubo, Y.; Inaba, A. Differential expression of ethylene biosynthetic genes in climacteric and non-climacteric Chinese pear fruit. *Postharvest Biol. Technol.* **2007**, *44*, 220–227. [CrossRef]
32. Lee, J.-H.; Kim, Y.-C.; Choi, D.; Han, J.-H.; Jung, Y.; Lee, S. RNA expression, protein activity, and interactions in the *ACC synthase* gene family in cucumber (*Cucumis sativus* L.). *Hortic. Environ. Biotechnol.* **2018**, *59*, 81–91. [CrossRef]
33. Li, T.; Tan, D.; Yang, X.; Wang, A. Exploring the apple genome reveals six *ACC synthase* genes expressed during fruit ripening. *Sci. Hortic.* **2013**, *157*, 119–123. [CrossRef]
34. Yuan, H.; Yue, P.; Bu, H.; Han, D.; Wang, A. Genome-wide analysis of *ACO* and *ACS* genes in pear (*Pyrus ussuriensis*). *In Vitro Cell. Dev. Biol. Plant* **2020**, *56*, 193–199. [CrossRef]
35. Paul, V.; Pandey, R.; Srivastava, G.C. The fading distinctions between classical patterns of ripening in climacteric and non-climacteric fruit and the ubiquity of ethylene-an overview. *J. Food Sci. Technol.* **2012**, *49*, 1–21. [CrossRef] [PubMed]
36. Itai, A.; Kotaki, T.; Tanabe, K.; Tamura, F.; Kawaguchi, D.; Fukuda, M. Rapid identification of 1-aminocyclopropane-1-carboxylate (ACC) synthase genotypes in cultivars of Japanese pear (*Pyrus pyrifolia* Nakai) using CAPS markers. *Theor. Appl. Genet.* **2003**, *106*, 1266–1272. [CrossRef]
37. Tan, D.; Li, T.; Wang, A. Apple 1-aminocyclopropane-1-carboxylic acid synthase genes, *MdACS1* and *MdACS3a*, are expressed in different systems of ethylene biosynthesis. *Plant Mol. Biol. Report* **2012**, *31*, 204–209. [CrossRef]
38. Huang, G.; Li, T.; Li, X.; Tan, D.; Jiang, Z.; Wei, Y.; Li, J.; Wang, A. Comparative transcriptome analysis of climacteric fruit of Chinese pear (*Pyrus ussuriensis*) reveals new insights into fruit ripening. *PLoS ONE* **2014**, *9*, e107562. [CrossRef]
39. Zhang, T.; Qiao, Q.; Zhong, Y. Detecting adaptive evolution and functional divergence in aminocyclopropane-1-carboxylate synthase (*ACS*) gene family. *Comput. Biol. Chem.* **2012**, *38*, 10–16. [CrossRef]
40. Barry, C.S.; Llop-Tous, M.I.; Grierson, D. The regulation of 1-aminocyclopropane-1-carboxylic acid synthase gene expression during the transition from system-1 to system-2 ethylene synthesis in tomato. *Plant Physiol.* **2000**, *123*, 979–986. [CrossRef]
41. Booker, M.A.; DeLong, A. Producing the ethylene signal: Regulation and diversification of ethylene biosynthetic enzymes. *Plant Physiol.* **2015**, *169*, 42–50. [CrossRef]
42. Gao, Y.; Ma, J.; Zheng, J.C.; Chen, J.; Chen, M.; Zhou, Y.B.; Fu, J.D.; Xu, Z.S.; Ma, Y.Z. The elongation factor GmEF4 is involved in the response to drought and salt tolerance in soybean. *Int. J. Mol. Sci.* **2019**, *20*, 3001. [CrossRef]
43. Wang, W.; Wu, H.; Liu, J.H. Genome-wide identification and expression profiling of copper-containing amine oxidase genes in sweet orange (*Citrus sinensis*). *Tree Genet. Genomes* **2017**, *13*, 31. [CrossRef]
44. Jakubowicz, M.; Sadowski, J. 1-aminocyclopropane-1-carboxylate synthase genes and expression. *Acta Physiol. Plant.* **2002**, *24*, 459–478. [CrossRef]
45. Van de Poel, B.; Van Der Straeten, D. 1-aminocyclopropane-1-carboxylic acid (ACC) in plants: More than just the precursor of ethylene! *Front. Plant. Sci.* **2014**, *5*, 640. [CrossRef] [PubMed]
46. Matarasso, N.; Schuster, S.; Avni, A. A novel plant cysteine protease has a dual function as a regulator of 1-aminocyclopropane-1-carboxylic acid synthase gene expression. *Plant. Cell* **2005**, *17*, 1205–1216. [CrossRef] [PubMed]

horticulturae

Review

Insights into Factors Controlling Adventitious Root Formation in Apples

Muhammad Mobeen Tahir [1,†], Jiangping Mao [1,†], Shaohuan Li [1], Ke Li [1], Yu Liu [1], Yun Shao [1], Dong Zhang [1,*] and Xiaoyun Zhang [1,2,*]

1 College of Horticulture, Yangling Sub-Center of National Center for Apple Improvement, Northwest A&F University, Yangling 712100, China; mubeentahir924@gmail.com (M.M.T.); mjp588@163.com (J.M.); lishaohuan@nwafu.edu.cn (S.L.); keli050035@nwafu.edu.cn (K.L.); liuyu18060098@nwafu.edu.cn (Y.L.); shaoyun@nwafu.edu.cn (Y.S.)

2 Agricultural College, The Key Laboratory of Special Fruits and Vegetables Cultivation Physiology and Germplasm Resources Utilization in Xinjiang Production and Construction Group, Shihezi University, Shihezi 832003, China

* Correspondence: afant@nwsuaf.edu.cn (D.Z.); zhangxiaoyun@stu.shzu.edu.cn (X.Z.)

† These authors contributed equally to this work.

Citation: Tahir, M.M.; Mao, J.; Li, S.; Li, K.; Liu, Y.; Shao, Y.; Zhang, D.; Zhang, X. Insights into Factors Controlling Adventitious Root Formation in Apples. *Horticulturae* **2022**, *8*, 276. https://doi.org/10.3390/horticulturae8040276

Academic Editor: Stefano Giovanni La Malfa

Received: 21 February 2022
Accepted: 21 March 2022
Published: 22 March 2022

Abstract: Adventitious root (AR) formation is required for the vegetative propagation of economically important horticultural crops, such as apples. Asexual propagation is commonly utilized for breeding programs because of its short life cycle, true-to-typeness, and high efficiency. The lack of AR formation from stem segments is a barrier to segment survival. Therefore, understanding the AR regulatory mechanisms is vital for the prolonged and effective use of biological resources. Several studies have been undertaken to comprehend the molecular and physiological control of AR, which has greatly extended our knowledge regarding AR formation in apples and other crops. Auxin, a master controller of AR formation, is widely used for inducing AR formation in stem cutting. At the same time, cytokinins (CKs) are important for cell division and molecular reprograming, and other hormones, sugars, and nutrients interact with auxin to control excision-induced AR formation. In this review, we discuss the present understandings of ARs' formation from physiological and molecular aspects and highlight the immediate advancements made in identifying underlying mechanisms involved in the regulation of ARs. Despite the progress made in the previous decades, many concerns about excision-induced AR formation remain unanswered. These focus on the specific functions and interactions of numerous hormonal, molecular, and metabolic components and the overall framework of the entire shoot cutting in a demanding environment.

Keywords: apples; adventitious root (AR); formation; asexual reproduction; phytohormones; phenolic compounds; sugars; polyamines; nutrients

1. Introduction

Asexual reproduction is a highly efficient and economical technique for generating homogenous forestry and horticultural plants when apomictic seeds are lacking [1]. Different methods are commonly used for vegetative propagation, including layering, budding, cutting, and grafting. Propagation by cuttings has been identified as the most productive and profitable technique for producing large quantities of homogeneous plants [2]. However, adventitious roots (ARs) formation from stem cuttings is a bottleneck for mass propagation, and the growth of new roots from stem segments restores the ability of water and nutrient acquisition. AR formation in stem cuttings is affected by a variety of factors, including hormones, phenolic compounds, sugars, polyamines, nutrients, etc. [3].

Plant hormones are essential to regulate ARs. They are important for cell fate determination and cell specialization because they can respond to changing environmental conditions and deliver a signaling network inside the plant body. Conversely, it is inadequate to understand the control and involvement of plant hormones and their sophisticated signaling networks during adventitious rooting. Similar to plant hormones, phenolic compounds, sugars, polyamines, and mineral nutrients also have essential and particular roles in regulating ARs. However, the detailed effect of these factors on forming ARs in cuttings is also limited. This review provides physiological and molecular insights into the regulation of ARs in apples. We also propose avenues for further investigations into the formation of ARs by hormonal and nutritional clues.

2. Developmental Phase of Excision Induced AR Formation in Stem Cuttings

Stem cuttings are produced by removing young shoots from the source plant and usually comprise a leafy stem having a terminal shoot apex with one completely formed leaf. Excision-induced AR may be determined by two factors, including wounding and separation from resources and signals from the mother plant. Wounding at the stem base initiates a new developmental program via specific responsive cells, resulting in the regrowth of a new root system. Different cell types, known as AR-forming cells, may produce ARs based on the plant and type of explant.

Anatomical studies on apples indicate that AR formation begins with the induction stage (3 days). At this stage, cell divisions in cambial cells have occurred and are considered necessary for cell reprogramming. Following the identification of root founder cells, the AR initiation stage begins with the fundamental variations in cell architecture, succeeded by cell division and differentiation of newly formed cell clusters, creating a dome-shaped AR primordium at 7 days. The last stage (emergence starts after 7 days) comes with primordia differentiation into a new root, which has distinct vascular bundles joined to the stem's vascular cylinder and extends with AR emergence [4,5]. Depending on these findings, the AR formation process may be classified into three phases: 0–3 day induction, 3–7 day initiation, and the emerging stage, which starts after 7 days.

3. Multiple Hormonal Pathways Mediate Adventitious Rooting

After separation from the donor plants, stem cuttings critically modify hormone homeostasis in the detached shoots [6]. Different growth regulators were examined to increase the rooting ability of stem cuttings. In the last few decades, several studies have been conducted on the formation of ARs in apples (Table 1). Different endogenous hormones play a key role in forming ARs in apples. Auxin, as a master controller, promotes AR induction and initiation stages while inhibiting the AR emergence stage. Auxin crosstalk with melatonin (MT) also promotes the AR induction stage. Cytokinin (CK) works antagonistically to auxin. Gibberellic acid 3 (GA_3) also plays a negative role in the initial stages. Moreover, ethylene (ET) and jasmonic acid (JA) can inhibit the induction stage, and their roles in other stages are unclear. Abscisic acid (ABA) is a negative regulator of AR formation at all stages. The role of various hormones in the multiple phases of AR formation is shown in Figure 1. The details of hormones are explained below.

Table 1. A summary of the described effects of exogenous hormones, sugars, polyamines, and nutrients on adventitious rooting in apple stem cuttings.

Application	Test Material	Culture	Reported Effect	Reference
IBA	M26 and M9	*In vitro* tissueculture	IBA promotes AR formation by increasing free IAA accumulation at stem basal parts	[1]
IBA	M26	*In vitro* tissueculture	IBA treatment increased endogenous IAA content at the primordia formation stage	[2]
IBA	T337	*In vitro* tissueculture	IBA-treated cuttings produced more ARs by hormone signaling and protein homeostasis	[3]
IBA and ethylene	M9-T337	*In vitro* tissueculture	IBA boosted ethylene and auxin production during AR formation, reducing AR elongation as indicated by ethylene	[4]

Table 1. *Cont.*

Application	Test Material	Culture	Reported Effect	Reference
IBA and NAA	M116	Mist chamber conditions	Increasing auxin levels correlated positively with rooting success	[5]
IAA, IBA, NAA	Jork9	*In vitro* tissueculture	IAA is the more preferred auxin for increasing rooting *in vitro*	[6]
IBA	T337	*In vitro* tissueculture	Endogenous and exogenous auxins both influence AR development through homologous signaling pathways	[7]
IAA	M9	Hydroponic culture in a growth chamber	IAA promotes AR founder cell division and elongation via upregulation of *PINs*	[8]
IBA	MP, M26, SH6, T337	*In vitro* tissueculture	The mode of IBA during AR formation is species-specific	[9]
IBA	M9-Jork	*In vitro* tissueculture	IBA induced more roots than IAA	[10]
IBA	M9-T337	*In vitro* tissueculture	Several miRNAs and their targets collaborated with hormone signaling pathways to contribute ARs	[11]
IBA	*Malus xiaojinesis*	Sand culture in a greenhouse	miR156 high expression is required for auxin-induced AR formation	[12]
6BA	M26	*In vitro* tissueculture	CK prevents AR primordia formation by increasing the expression of CK signaling pathway genes	[13]
IAA and 6BA	*Malus sieversii*	Hydroponic culture in a growth chamber	Auxin and CK are both important regulators of AR formation	[14]
6BA	Jork9	*In vitro* tissueculture	CK is essential for AR formation	[15]
Ethylene	Gala and Triple Red Delicious	*In vitro* tissueculture	AR in apples was not related to ethylene	[16]
ABA	*Malus pumila*	*In vitro* tissueculture	ABA negatively regulates AR formation	[17]
IAA and ABA	Jonathan	*In vitro* tissueculture	IBA increases adventitious rooting	[18]
Melatonin	*Malus prunifolia*	*In vitro* tissueculture	MT promotes ARs' initiation stage by IAA homeostasis	[19]
GA_3	M9 cv Jork	*In vitro* tissueculture	GA_3 treatments limit AR formation from the initial to final stages of AR formation	[20]
Phenolic compounds	Jork9	*In vitro* tissueculture	Phenolic compounds may act as antioxidants, preventing auxin oxidation and, thus, contributing to AR formation	[21]
Phenolic compounds	M9 and M26	*In vitro* tissueculture	Auxin and PG stimulated ARs more than auxin-alone controls	[22]
IBA and sugar	Jork9	*In vitro* tissueculture	Starch grains provide energy for AR growth	[23]
Carbohydrates	Jork9	*In vitro* tissueculture	Sugar promotes root regeneration, and sucrose content affects AR density	[24]
Sucrose	Jork9	*In vitro* tissueculture	Sucrose and auxin interacted to mediate AR formation	[25]
Sucrose and IBA	M7	*In vitro* tissueculture	Soluble saccharides are crucial for primordia formation but not essential for later stages	[26]
Spermidine	*Malus prunifolia*	*In vitro* tissueculture	Spd promotes AR formation by interacting with IAA and regulating different gene sets	[27]
Nitrate	B9	*In vitro* tissueculture	Nitrate promotes ARs at lower-medium levels and inhibits them at higher levels	[28]
Nitrate	B9	*In vitro* tissueculture	Nitrate promotes AR's length by upregulating the expression of different gene sets	[29]

Table 1. *Cont.*

Application	Test Material	Culture	Reported Effect	Reference
Nitrate	B9	*In vitro* tissueculture	High nitrate inhibited ARs by ABA signaling miRNA	[30]
Nitrate	B9	*In vitro* tissueculture	The high availability of nitrate delays AR initiation and emergence stages	[31]
Potassium	B9	*In vitro* tissueculture	KCl-treated cuttings produced more ARs than control cuttings	[32]

Abbreviations: AR: adventitious root; IBA: indole-3-butyric acid; IAA: indole-3-acetic acid; NAA, naphthaleneacetic acid; 6BA: 6-Benzylaminopurine; CK: cytokinin; ABA: abscisic acid; MT: melatonin; GA_3: gibberellic acid 3; PG: phloroglucinol; SPD: spermidine; and KCl: potassium chloride.

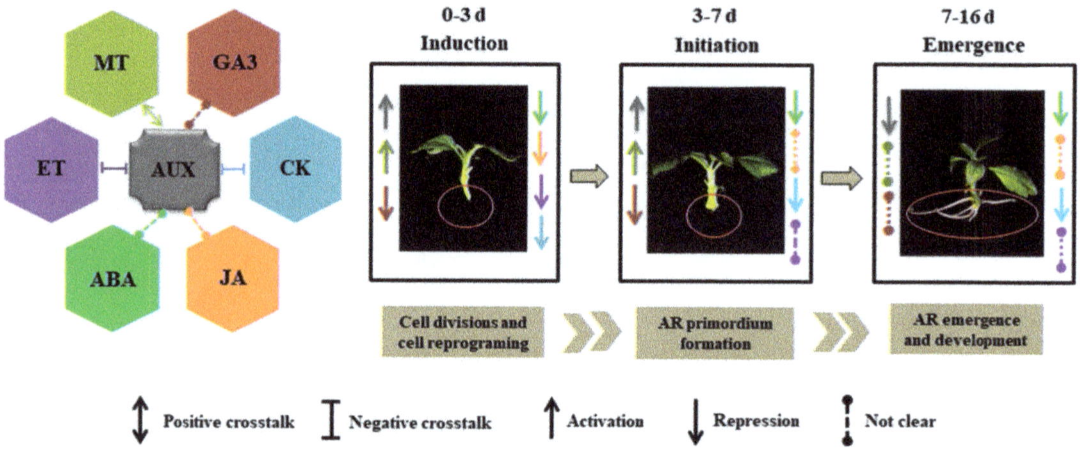

Figure 1. A suggested schematic diagram of adventitious root (AR) formation stages (0–3 day, AR induction phase; 3–7 day, AR initiation phase; and 7–16 day, AR emergence phase) influenced by endogenous hormones in apples: Auxin (AUX), abscisic acid (ABA), cytokinin (CK), melatonin (MT), jasmonic acid (JA), ethylene (ET) and gibberellic acid 3 (GA_3). Different colors show specific hormone activities. Different arrow heads show different functions.

3.1. Auxin: A Master Regulator for ARs

Auxin is involved in various physiological events, including vascular differentiation, cell expansion, lateral roots (LRs), and ARs. More auxin is needed to induce ARs, but this is not necessary for the AR emergence stage [33]. Auxin treatment was shown to boost assimilate translocation from the leaf and sugar content at the root growth site [34]. Indole-3-acetic acid (IAA) is perhaps the most prevalent natural auxin, but Indole-3-butyric acid (IBA) is the most frequently utilized exogenous auxin for increasing ARs in most species, particularly in difficult-to-root genotypes. Recently, a study was conducted on M 116 apple clonal rootstock to identify the effects of different concentrations of IBA and naphthaleneacetic acid (NAA) (IBA: T1, 1500; T2, 2000; T3, 2500; T4, 3000; T5, 3500; T6, 4000 ppm; NAA: T7, 500; T8, 1000; T9, 1500 ppm) compared with T10, control plants during adventitious rooting under mist chamber conditions. The results suggested that increasing auxin levels correlated positively with rooting success. IBA treatments significantly improved rooting. T5 had the highest rooting percentage (57.12%), AR numbers per cutting (7.33), AR length (34.85 cm), and AR diameter (5.27 mm) when compared to other treatments and control [5]. Furthermore, the Jork9 was also treated with auxins (IAA, IBA, and NAA). The cultures were kept in darkness during the initial time of rooting treatment. During this phase, the AR initials were made, then cultures were shifted to light. NAA and IAA or IBA treatments obtained the lowest ARs (8) and the highest ARs (15), respectively. The maximal AR numbers were recorded throughout a broad range of IAA levels (10–100 μM), although at only one level of IBA or NAA (10 μM and 3 μM, respectively) [6]. Studies indicated that the duration of exposure to auxins mainly controlled the AR formation. According to a previous study, the cuttings are not highly responsive to auxin within the first 24 h after being collected. It is believed that during this lag time, cells dedifferentiate and become capable of responding to the rhizogenic signal, auxin. The cells that give rise to the AR primordia are often seen between the vascular bundles and store starch over the first 24 h. The rhizogenic activity of auxin in the induction stage then commits previously activated cells to create AR primordia for up to 72 h. During this time, auxin pulses stimulate the highest AR numbers. Auxin is no longer necessary after 96 h, and auxin levels advantageous for establishing root meristemoids are restrictive during this period [35].

IBA promotes AR formation in M.9 and M.26 apple rootstocks, but higher AR formation was seen in M.26, which is due to the high amount of free IAA in the stem basal part compared to M.9. Furthermore, the conjugated IAA level was higher in M.9 as compared to M.26, suggesting that the difference between AR responses in both rootstocks may be associated with free IAA content in stem basal parts [1]. During the whole process of adventitious rooting, endogenous auxin, or applied externally, plays a critical function at each step. In the AR initial phases, a high level of endogenous auxin is generally associated with a high rooting rate [36]. When the auxin is applied exogenously to increase root formation, it influences the endogenous auxin concentration, which generally reaches a maximum after wounding [37]. Still, in some cases, the auxin peak was not detected [38]. The endogenous level of auxin induces the AR primordia formation, and the quantity of primordia formation is raised together with the elevation of the IAA level [39]. In apples, IBA was found to be highly important for inducing adventitious rooting in M.26 apple rootstock. The other hormones seem to be indirectly involved in the control of ARs by their interaction with the auxin, where the endogenous level of IAA was observed to be higher at the early stage and induces primordia. Then, the level of IAA decreases towards a later stage [2]. These variations in the IAA levels suggest that auxin plays a critical function at the early stages and may not be crucial after the AR primordia have been established. IBA-treated cuttings of *Malus prunifolia var. ringo* produced more ARs than cuttings treated with the auxin transport inhibitor NPA, which completely inhibited AR formation. This shows that IBA plays a role in AR formation [27]. In M9-T337 apple stem cutting, IBA inhibited AR elongation at later stages of development by decreasing cell length and by decreasing the expression of genes related to cell elongation [4]. Moreover, transcriptome sequencing studies on IBA-treated T337 cuttings showed that auxin, both endogenous and exogenous, regulates AR formation through homologous signaling pathways to some degree. AR formation is largely controlled via the auxin signaling pathway. Furthermore, various hormone, wounding, as well as sugar signaling pathways work together with the auxin signaling pathway to regulate adventitious rooting in T337 [7]. In M.9 apple rootstock, IAA stimulates the process of AR formation by the upregulation of PIN-FORMED (PIN). Auxin promotes the multiplication and extension of AR founder cells via starch grain hydrolysis, resulting in endomembrane system multiplication, lenticel dehiscence, and AR emergence. This effect was inhibited by the NPA application [8]. Nevertheless, in stem cuttings of Malus species, the IBA promotes more root formation than the IAA. At the same time, it was changed only at deficient IAA levels, implying that the IBA was operational or controlled IAA activity [10].

Recently, Meng and colleagues [9] applied 1 mg/L IBA to different apple rootstocks with different rooting abilities, including MP, SH6, T337, and M.26, to detect the effect of IBA on AR formation compared with control plants. The results showed that MP was easy to root and developed ARs in both groups. However, M.26 produced some ARs in IBA-treated groups. However, T337 and SH6 have poorly developed ARs in both groups. Moreover, microRNA160 (miR160) mediates AR formation; therefore, five members of miR160 were found in the apple with two target genes: auxin response factor 16 (*ARF16*) and *ARF17*. MiR160 was highly expressed in SH6 and T337 and lower in MP and M.26 in both groups. Following that, miR160a was cloned from M.26 and overexpressed in tobacco plants, causing severe defects in AR formation, which were released by IBA application to transgenic plants [9]. A comprehensive proteomic analysis was conducted on the role of IBA (1 mg/L) in T337 during AR formation. The study showed that AR rate and length were increased after IBA treatment compared with control cuttings. IBA treatment increases the content of IAA, ABA, and (brassinolide) BR and decreases ZR, GA, and JA content. The expression profiles of differentially expressed proteins were strictly related to phytohormone signaling, protein homeostasis, carbohydrate uptake and energy synthesis, reactive oxygen and nitric oxide signaling, as well as biological processes associated with cell wall remodeling. These are all the most important AR formation processes by IBA application [3].

Concerning auxin signaling pathways, the degradation of AUX/IAA stimulated *ARF* transcripts, which in turn stimulated the auxin-responsive gene expressions [40]. Numerous genetic investigations have shown the contribution of *ARFs* in plant developments. In Arabidopsis, *ARF7* and *ARF19* show specific and dynamic gene expression profiles throughout embryogenesis, seedling development, and rooting [41,42]. Single mutants *arf7* and *arf19* limit the formation of ARs and LRs, but double mutants are initiated to produce significantly fewer roots [42,43]. They are considered key players in the formation of ARs in apples, where they are more expressed at important time points [44]. Several miRNAs are key players in root growth, including miR160, which is considered necessary for the development of root tips and gravity sensing via the participation of their targets, such as *ARF10* and *ARF16* [45], which regulate auxin-responsive gene expression through binding with auxin response elements in the promoters. In addition, callus initiation was repressed by miR160 from pericycle-like cells but activated by *AFR10* [46]. Furthermore, smaller and gravitropic roots and tumor-like apexes were seen by the over-expression of miR160c [45]. This phenotype was related to unrestrained mitosis and a lack of columella cell differentiation in the root apical meristem (RAM) [45]. A very similar phenomenon was also perceived in double mutants of *arf10* and *arf16*, where miR160 and their target genes might be necessary to limit cell divisions in root tips and stimulate cell extension and differentiation. Moreover, the contribution of miR160 and *ARFs* in response to auxin signals during AR formation might play a critical role in regulating apple ARs [11]. In contrast with miR160, miR167, also an auxin-related miRNA whose target genes are *ARF6* and *ARF8*, is an adverse controller for adventitious rooting. Furthermore, *ARF8* and *ARF17* play opponents in auxin homeostasis in Arabidopsis [47,48].

On the other hand, miR167 in rice seemed to be a key controller of AR formation [49]. According to a former study on Arabidopsis, in which miR390 was regulated by auxin via the roles of *ARF2*, *ARF3*, and *ARF4*, conversely, miR390 expression was restricted by *ARF4* at the base of root primordia [50]. Moreover, AR development was positively regulated by miR390a in apple, which shows its regulatory ability over *ARFs*. MiR390 and miR160 are important in forming ARs. The *ARFs* (target genes) also act as a crucial controller in auxin signal transduction during the process of adventitious rooting in apple rootstocks [11]. Auxin-induced adventitious rooting largely depends on miR156 high expressions in *Malus xiaojinensis* via *MxSPL26*, independent of *PINs* and *ARFs* [12]. Conversely, it does not affect *Eucalyptus grandis* [51]. So, rooting regulated by miR156 might be species-specific. However, miR156 can be an essential factor for AR induction but not a determining element in woody plant species. Moreover, in apple cuttings, wounding is knotty with the induction of adventitious rooting [11].

The auxin gradient in root tips is maintained by PIN auxin polar transport, which stimulates root growth [52]. In apples, the *PIN* family consists of eight members. All were differentially expressed at different stages of AR formation. For example, *PIN8* and *PIN10* were upregulated and largely associated with the induction stage. Upregulation of all *PINs* was detected towards the initiation stage, whereas *PIN4*, *PIN5*, and *PIN8* were all upregulated and largely associated with the emergence stage, suggesting their regulatory roles in adventitious rooting [8]. Furthermore, CK repressed *PIN1* expression during adventitious rooting to regulate polar auxin transport [53]. The AUXIN RESISTANCE1 (*AUX1*) gene is recognized as an auxin influx carrier [54]. Moreover, the aux1 mutant developed few roots and showed reduced root gravitropism; it reduced IAA accumulation in the roots and acted as an auxin influx transporter to stimulate root growth via allocating IAA among root and shoot organs [55].

In Arabidopsis, LATERAL ORGAN BOUNDARIES-DOMAIN (LBD) *LBD16* and *LBD29* are positively controlled by *ARF7* and *ARF19* and induce rooting [43]. In a recent study, *ARF7* and *ARF19* stimulate *LBD16* and *LBD19*, which, in turn, activate adventitious rooting in apples [44]. The Wuschel-related homeobox gene *WOX11* was identified as a key regulator of Arabidopsis root development and apple ARs [19,56]. Short root (*SHR*) also plays a positive function in forming the LR stem cell niche as well as AR's meristem [57]. *WOX5* stimulates *WOX11* to endorse the beginning of root primordia and organogenesis [58], and *WOX5* is considered the initial indicator for adventitious rooting [59]. Furthermore, *WOX5*, *WOX11*, and *SHR* are all auxin-inducible and contribute positively to integrating numerous signals related to root induction [60,61]. Furthermore, in Arabidopsis, Liu and colleagues found that the auxins induced *WOX11* during the initial stage of cell fate transition, and *WOX11* increased the transcript of *LBD16* and *LBD29* during adventitious rooting [61]. Subsequently, harmonizing transcript abundance of these genes could promote rooting by stimulating cell cycle-related (*CYCD1;1* and *CYCP4;1*) gene expression in apples [13,44].

3.2. *Cytokinin: A Required Inhibitor*

CK is an essential hormone that plays many vital roles in regulating the cell cycle and several developmental processes. Root growth requires CK signaling and perception [62]. Several studies have linked CK with the inhibition of AR formation [63–66]. Exogenous treatment of CK inhibited the formation of ARs, and auxin is generally used to initiate ARs in different plant species [67–69]. In AR induction, the hypothesis of an inhibitory effect for high CK contents and a high CK to auxin ratio has garnered significant scientific evidence, based on the finding that low CK to auxin ratios favor root regeneration in explants. The crosstalk of CK-auxin also played several functions in controlling the size of root meristems, with auxin and CK presenting opposing positions in controlling root growth [70]. The low CK concentrations promoted AR initiation. M.26 apple cuttings were treated exogenously with 2 mg/L 6-benzyl adenine (6-BA) to 3 day and 7 day old cuttings by transferring them from IBA containing medium to 6-BA medium. Stem anatomy showed that 6-BA treatment limits primordia formation in 3 d-treated cuttings but not in 7 day treated cuttings. The endogenous IAA concentration and the IAA/CK and IAA/ABA ratios were seen to be lower in 3 d-treated cuttings than in 7 d-treated cuttings. The high abundance of CK in 3-d treated cuttings increased the expression levels of CK signaling pathway genes, *RR1*, *RR2*, *RR3*, and *AHK4*, which prevented the genes related to auxin synthesis and transport, *PIN1*, *PIN2*, *PIN3*, *AUX1*, *YUCCA1*, and *YUCCA10*, resulting in a decline in inner auxin content. Thus, reduced auxin content inhibits the expression levels of auxin signaling pathway genes *IAA23*, *ARF6*, *ARF7*, *ARF8*, and *ARF19*. Furthermore, the transcripts of the cell cycle and AR development-related genes were also reduced, which all have a negative effect on forming AR primordia (initiation stage). Contrarily, reduced CK content in 7 day treated cuttings reduces the negative effect on auxin content, increasing the gene expression recognized to stimulate the formation of AR primordia. These results showed that 3–7 day is the period of primordia formation. The decline in AR formation in the 3 day treated group is due to suspending the AR primordial formation stage in apples [13].

Di Zhao and colleagues [14] found the interaction between CK and auxin during AR formation in wild apple (*Malus sieversii* Roem). Overexpression of *MsGH3.5* (encoding IAA amido synthetase) developed fewer ARs than the empty vector by significantly decreasing the amount of free IAA and elevating specific IAA amino acid conjugates. This feature is similar to GH3's involvement in auxin balance, which involves conjugating the concentration of free active IAA to amino acids. CK content was increased and altered the gene expression involved in CK biosynthesis, absorption, and signaling in overexpressed plants. Moreover, external CK treatment promoted *GH3.5* expression via the action of RR1a (the CK type-B response regulator that regulates the CK main response). RR1a induced the *GH3.5* transcript via binding to its promoter, connecting auxin and CK signaling. Overexpressed *MsRR1* plants also showed some ARs, consistent with *MsRR1a*'s control of *MsGH3.5* expression. Taken together, *MsGH3.5* influences adventitious rooting by changing auxin and CK content and their sophisticated signaling pathways [14]. However, CK may have a stimulatory effect in the first few hours after cutting excision by activating the cell cycle earlier [71]. A previous study showed that CK is essential in the initial phase of adventitious rooting in the Jork 9 apple rootstock. Lovastatin and simvastatin are CK-synthesis inhibitors and inhibit AR formation. This negative effect was partially released by adding zeatin, confirming its role in forming ARs. Still, it stimulates cell divisions at low concentrations just after wounding [15].

3.3. *Ethylene: A Positive or Negative Regulator for ARs*

ET plays a crucial function in regulating ARs in many species. Many experiments were conducted to identify their roles in root formation. The findings of these experiments were extremely flexible for specific species. ET behaved as activators or inhibitors and did not influence the formation of ARs. Because of earlier findings, auxin affects ET synthesis [72]. Subsequently, many efforts have been made to identify how auxin interacts with ET during AR formation. High ET level tissues amplified the responsiveness of root developing tissues in response to endogenous IAA. The role of ET was studied in M9-T337 apple stem cutting during AR formation. The $AgNO_3$ (ethylene inhibitor) reduced the appearance of ET and promoted the AR's emergence and development. However, the ET precursor, 1 aminocyclopropane-1-carboxylic acid (ACC), was added to the MS medium, where it may convert into ET, inhibiting AR emergence and decreasing AR length in M9-T337 [4]. Harbage and Stimart [16] found that ET was not involved in AR formation in apple microcuttings of Gala and Triple Red Delicious. This study found that IBA-induced ET formation was reduced by aminoethoxyvinylglycine (AVG), although the AR number continued to be IBA-dependent. ACC restored the inhibitory effect of IBA+AVG on rooting, while ACC separately had little impact on the AR number. Unlike 2,3,5-triiodobenzoic acid (TIBA) and N-1-naphthylphthalamic acid (NPA), which impede polar auxin transport, ET production is increased without increasing the AR number [16]. ET inhibits or promotes the process of root development depending on the stage of the process. It has a stimulatory role at the initial stage but reflects an inhibitory role at later stages of ARs' development [35]. Root development can also be inhibited by ET, mainly by limiting cell expansion, but this does not affect root meristem activation [73].

3.4. *Abscisic Acid: A Negative Regulator for ARs*

The ABA mainly responds to various environmental factors, including water and salt stress, controlling root architecture and regulating root elongation by controlling cell division and elongation [74]. Root cell division is limited to meristems, a region with a quiescent center (QC) bordered by stem cells, which further split to form cells that differentiate into several root tissues, along with the formation of stem cells, which keep on isolating and allow for indeterminate root growth [75,76]. In comparison, high ABA content is well known to inhibit root formation and development.

Consequently, in Arabidopsis, the ABA regulated quiescence and inhibited differentiation in the root meristem, hence continuing the stem cell population [77]. ABA controlled root growth by regulating the dividing cell population in the root tips. Furthermore, ABA also held root elongation by controlling cell length, an intense effect on the development of the root. However, cell division continuously is crucial for root growth. ABA has been found to limit LR and AR proliferation in tobacco, apple, and Arabidopsis [17,78,79]. High ABA content in Jonathan apple microcuttings is related to root inhibition. It is thought that high ABA is the key factor involved in the inhibition of ARs in apples [18]. Recently, Qingzhen

1 apple plants were exogenously treated with 5 μM ABA, where root development was significantly inhibited compared to control plants by significantly increasing the endogenous content of ABA and by decreasing IAA, ZR, and JA concentrations. ABA treatment upregulates the expressions of ABA-related genes (*ABF3*, *ABI1*, *AREB2*, and *CYP707A2*) and downregulates the expressions of auxin-related genes (*ARF19*, *YUCCA8*, *PIN1*, *PIN2*, *PIN3*, and *YUCCA3*), root development, and cell cycle (*WOX5*, *WOX11*, *CYCD1;1*, and *CYCD3;1*) [80]. AR proliferation was strongly repressed by the treatment of high nitrate and polyethylene glycol (PEG) by crosstalk with endogenous ABA, whose content was considerably higher than control cutting at all sampling points, suggesting that high endogenous ABA content is harmful for AR formation in apples [31,81].

3.5. Jasmonic Acid: A Positive or Negative Regulator for ARs

JA, usually related to mechanical wounding, provides defense against plant pathogens. Studies in the last decade have focused primarily on JA's role in plant development and exposed JA as a critical hormone. It participates in different developmental mechanisms: hypocotyl elongation, primary root (PR) elongation, LR and AR formation, and flower development [82]. Recently, JA seemed higher in B9 apple rootstock cuttings when treated with high nitrate and inhibited AR formation by JA signaling pathways [31]. The endogenous concentration of Methyl jasmonate (MeJA) was higher in GL-3 apple plants when exposed to drought and inhibits AR formation [81]. In many plant species, wounding can increase the jasmonates [83]. JA is promptly induced after cutting or in the case of injury. It accumulates at the place of injury, bringing responses that defend plants from pathogens' attacks. At the base of petunia stem cuttings, JA rapidly accumulates during the induction stage of adventitious rooting. It appears to serve as a positive regulator for the formation of ARs [84]. Based on these findings, Lischewski and colleagues [85] tested JA's role in forming ARs in petunia stem cuttings. They found some ARs are formed due to the downregulation of allene oxide synthase, which represses a critical aspect in the synthesis of cis-12-oxo-phytodienoic acid [85]. However, the same writers demonstrated that when petunia cuttings were treated continuously with exogenous JA, they formed significantly fewer ARs than the control cuttings [85]. The findings mentioned above are in dispute with those suggesting that JA could be a positive controller for ARs initiation and that JA's continuous treatment inhibited the ARs initiation stage. Recent experiments have revealed that JA is an essential controller for the formation of ARs, possibly acting as an inhibitor at the AR initiation stage, while their functions may be more complicated and plant species sensitive. JA severely inhibited cell differentiation and expansion [86,87]; however, this information is very limited in identifying the action mode of JA at the particular stage of ARs formation and development.

3.6. Melatonin: A Positive Regulator

MT is thought to be necessary for the creation and growth of ARs in apples. [19]. In plants, MT served as a vital regulatory signal [88] and was essential for root formation, stress response, explants, and shoots [89–91]. A recent study has indicated that the exogenous treatment of MT stimulated AR formation in cuttings of *Malus prunifolia*, where MT mainly affects the AR induction stage by IAA homeostasis. *WOX11* was induced by MT, and apple plants overexpressing *MdWOX11* developed more ARs than the GL-3 WT plant, suggesting that *MdWOX11* promotes ARs by MT signaling [19]. A few studies have advocated that exogenous MT at low concentrations could increase the endogenous content of IAA, and it is supposed that this stimulating MT effect on growth and development might be triggered by this rise in IAA levels [92]. On the other hand, another study has suggested that the effect of MT on root formation and differentiation is IAA independent [93]. The IAA content was increased at the AR induction stage after the application of MT; however, it was reduced at the AR initiation stage and emergence stage in apple [19].

3.7. Gibberellic Acid and Brassinosteroids: A Positive or Negative Regulator for ARs

The specific roles of GA and BR are still largely unknown in the regulation of ARs. The effect of GA_3 was studied in M9 cv. Jork stem discs. First, the discs were cultured in darkness for 24 h on a root-inducing medium containing 24.6 μM IBA. Afterward, the discs were shifted to light exposure and cultured on a hormone-free medium and a medium containing 10 μM GA_3 for different timepoints of adventitious rooting. The results suggest that GA_3 treatments limit AR formation from the initial to final stages of AR formation [20]. Moreover, the concentration of GA_3 was significantly increased at the initiation of ARs, indicating that GA_3 plays an important role in forming apple ARs [27]. However, some shreds of evidence show that BR participated in AR formation. The availability of BRs triggers dual effects on the formation of ARs: the enhancing effect at low levels and the inhibitory effect at high levels [94]. Their high concentration inhibited AR formation in apple rootstock [27]. However, the above information is not enough to identify the specific role of GA and BR in forming ARs.

4. Role of Phenolic Compound in the Regulation of ARs

The involvement of phenolic compounds in AR formation has been well established for a long time [21,22,35]. These were shown to either synergize or inhibit the activity of auxin [95]. Phenolic compounds keep plants from oxidative stress [96]. Besides shielding auxins from oxidation, phenolic compounds have been associated with them in various ways. Flavonoids may also limit auxin transport [97]. Flavonoids either interact with PIN2 or have an impact on the extent of PIN proteins [98]. The role of phenolic compounds was tested on Jork 9 stem slices during AR formation. The results suggested that all orthodiphenols, paradiphenols, and triphenols investigated with IAA enhanced adventitious rooting from stem slices. The most effective treatment was ferulic acid (FA) (a methylated orthodiphenol), which raised the number of ARs from 0.9 to 5.8. There was little or no improvement with NAA following the inclusion of phenolics. FA and phloroglucinol (a triphenol) were studied in-depth. Based on their effects on the IAA dose–response curve and the duration of their activity, both acted as antioxidants, preventing IAA decarboxylation and oxidative stress in the tissue [21]. Auxin degradation includes oxidative decarboxylation via peroxidases, but since phenolics influence peroxidase activity, auxin catabolism at the cuttings' base may be prevented [21]. After wounding, there is an elevation in JA, auxin, and phenolic compounds at the cutting base during the AR induction stage, with a decrease in peroxidase activity. However, peroxidase activity increases to a peak during the initiation stage, and the auxin concentration decreases [35,84].

5. Role of Sugars in the Regulation of ARs

Sugars are both energy sources and signaling molecules that control plant growth. The published articles related to sugar and apple ARs are gathered in Table 1. The detailed studies of sugar metabolism focusing on the anatomical changes in apple rootstock Jork 9 were conducted by Jasik and De Klerk [23]. As highlighted in the initial developmental stage, a large number of starch grains were found in the cells that started AR primordium formation at the stem base, and thus the percentage of plastids grew dramatically, with the starch grains occupying a major share of all apparent plastids at the same stage. The utilization of sugars produced by the hydrolysis of all of these starch grains provides energy for adventitious rooting, which is associated with an increase in the number of cambial mitochondria, dictyosomes, and nuclei to the visible detriment of vacuolar expansion. Moreover, Pawlicki and colleagues [24] used different combinations and concentrations of carbohydrates and auxin to investigate AR differentiation. They found that sucrose (29–50 mM) stimulated adventitious rooting but also supported callus formation. In contrast, sorbitol, which is required in large quantities for optimal adventitious rooting, was not highly supportive of callus formation. Furthermore, the combinations of sucrose and mannitol (59/29 mM) or glucose and sorbitol (117/59 mM) ensued in 100% AR formation, with more than 6 ARs per disc. Then, in the presence of the sucrose-mannitol combination, the discs were treated with 49.2 μM of IBA for 540 min, obtaining an adventitious rooting of 100%.

The number of ARs was affected by sucrose concentration, although the effect was minor across a wide range of sucrose concentrations (1–9%). There was also a synergy between sucrose and auxin, which facilitated adventitious rooting. When slices were cultivated in a sucrose-free medium for 0–2 days, root formation was decreased. However, 2 days of cultivation minus sucrose had no impact or even a minor enhancer effect on later days, suggesting that sucrose is needed as a source of energy and a building block during adventitious rooting in apples [25]. The cuttings of M7 apple rootstock were treated with IBA and found to contain glucose, sorbitol, fructose, and inositol. In the initial 10 days of AR formation, the content of soluble saccharides improved significantly, and this was associated with the rapid cell divisions at the stem base in this stage. Specifically, the amount of fructose content in the stem basal part was related to rooting potential. At the AR primordia differentiation and AR emergence stages, the concentration of soluble saccharides in the cuttings had dropped to its initial level, indicating that soluble saccharides are crucial for the initial stages but not essential for later stages of AR formation in apples [26].

A recent study was conducted on T337, where the IBA treatment induces the expression of sucrose synthase4 (SUS4), sucrose phosphate synthase4 (SPS4), and polyol/monosaccharide transporter (PMT5) at the induction stage [7]. The application of KCl to B9 apple rootstock also upregulates the expression of various starch and sucrose metabolism-related genes [32]. These results show that IBA and KCl had some crosstalk with sugars during the AR induction stage, and sugar was transported into the stem basal parts by the activities of several genes, providing adequate energy and signal activity for ARs to begin.

6. Role of Polyamines in the Regulation of ARs

Polyamines (PAs), including putrescine, spermidine, and spermidine, are a class of organic compounds with two or more main amino acid groups found abundantly in plant cells. Additionally, PAs are important in controlling DNA duplication and cell proliferation, senescence, stress responses, and morphogenesis. They also play a key function in forming and developing root architecture [99]. Exogenous application of spermidine on *Malus prunifolia var. ringo* stem cuttings increased AR formation by interacting with IAA. Spermidine application upregulated the expression levels of spermidine-related genes (*SAMDC1*, *PAO*, and *SPDS6*) and auxin biosynthesis-related genes (*IAA7*, *IAA14*, and *IAA23*) during AR formation. In contrast, *WOX11* upregulated the expression levels of *LBD16* and *LBD29*, which prompted the transcripts of genes related to the cell cycle (*CYCD1;1* and *CYCP4;1*) [27]. Nonetheless, they have been shown to be inhibitors in a number of species, including poplar [100] and walnut [101]. PAs, in conjunction with auxin, modulate cell division and root primordia initiation during the induction stage [102]. Nonetheless, the relationship between PAs and auxin for the rooting process is still lacking.

7. Role of Nutrients in the Life of ARs

Mineral nutrients are necessary for plant metabolism. AR formation and nutrients are closely related. The published articles related to nutrients and apple ARs are gathered in Table 1. The details of nutrients are explained below.

7.1. Role of Nitrogen in the Formation of ARs

Nitrogen is regarded as an important macronutrient that is needed for plant development and higher yields [103]. Nitrogen assimilation in soil occurs in two ways, such as organic and inorganic [104]. Nitrate is a vital nitrogen source that also functions as a signaling molecule for controlling flowering time, AR, and LR formation, as well as prompting auxin-related gene expressions [28,105,106]. Nitrate levels in the soil are relatively low due to its high solubility, leaching capabilities, and fast absorption by bacteria and fungus [107]. In higher plants, such as apples, there are two types of nitrate transport mechanisms: low-affinity transport systems (LATS) and high-affinity transport systems

(HATS), which are responsible for uptake, distribution, and storage of nitrate [108]. The nitrate supply is immediately and strongly detected by the plant cells. Following this, the nitrate signaling system changes the relative expression levels of several gene sets to control cell and organ metabolism. Furthermore, the availability of nitrate has a significant effect on AR formation. In general, the external concentration of nitrate generated binary effects on the formation of AR depending on their concentrations, with an activated impact at low levels; however, a limiting effect at high levels was seen in Arabidopsis [109]. A similar phenomenon was also seen in a recent study, where different nitrate concentrations (9.4 mM/L, 18.8 mM/L, 28.1 mM/L, 46.9 mM/L, and 84.5 mM/L) were exogenously treated to B9 apple rootstock stem cuttings during adventitious rooting, and 28.1 mM/L was found to be the most favorable nitrate level for adventitious rooting, and 46.9 mM/L and 84.5 mM/L were found to be inhibitors [28]. High nitrate inhibits AR in apples by elevating the endogenous levels of ABA, ZR, JA, BR, and GA_3, which may create a hormonal imbalance in the plant. In addition, the high ratios of IAA/ABA and IAA/ZR promote ARs under nitrate treatments. Furthermore, transcriptome analysis showed that hormone signaling pathway-related genes were upregulated, and root development and cell cycle-related pathways were repressed by the application of high nitrate [31]. Moreover, auxin and ABA signaling miRNAs (miR390a, miR160a, miR167, miR169a, and miR394) were activated, and miRNAs related to cell fate transformation, expansion, and enlargement (miR166, miR171, miR319, miR156, and miR396) were repressed by high nitrate [30]. The same cited authors explained the mechanism of a different set of genes how 28.1 mM/L treatment promotes AR formation in B9 compared with 46.9 mM/L (inhibiting treatment). The results showed that treatment with 28.1 mM/L noticeably upregulated the relative expression levels of nitrate related genes (*NRT1.1*, *NRT2.1*, *NIA1*, and *ANR1*) and auxin biosynthesis (*IAA14* and *IAA23*), which enhances the AR development-related gene expression (*WOX11*, *ARRO1*, and *SHR*) and collectively induces the expression of cell cycle related genes (*CYCD1;1*, *CYCD3;1*, and *CYCP4;1*) in comparison with 46.9 mM/L nitrate treatment [29]. *NRT2.1* a high affinity nitrate transporter showed the highest response to nitrate availability, indicating that *NRT2.1* may play a key role in forming AR in apples, and the overexpression of *MdNRT2.1* gene in tobacco produced superior roots compared to WT plants.

Ammonium, such as nitrate, played an important role in root development in apple and other crops [110–112]. The effects of nitrate and ammonium were studied in apples, where significant differences were detected in root morphology within a week of application. The roots were thin and long in response to nitrate treatment, although they were thick and short in response to ammonium application, with prominent enlarged areas behind the tip. Furthermore, nitrate-treated roots were nearly devoid of root hairs, but those treated with ammonium were entirely covered in thick, long root hairs, and the root hair cylinder diameter in the ammonium treated was around three times that of the nitrate treated [111]. Hilo and colleagues [112] found that ammonium-treated cuttings (without nitrate) had a higher total nitrogen content, which was indicated by enhanced glutamine and asparagine levels. These authors point to faster ammonium assimilation in the stem base, which could have resulted in a lower expression level of N-regulated genes such as the ammonium transporter *AMT1* [112]. Moreover, the effects of ammonium nitrate (NH_4NO_3) and potassium nitrate (KNO_3) were also studied in three apple scion cultivars [110]. The percentage ARs of Gala and Royal Gala rose dramatically when the level of NH_4NO_3 in the medium was reduced from full strength to 1/4 strength, but not in Jonagold. A further decrease in NH_4NO_3 concentration from 1/4 strength to zero considerably decreased the rate of ARs in Gala but not in Royal Gala. However, Jonagold rooted optimally in the absence of NH_4NO_3. Moreover, without NH_4NO_3, adventitious rooting for all three cultivars was as high as 100% when KNO_3 was given at full strength. These results suggest that the effect of NH_4NO_3 was cultivar-specific, but KNO_3 treatment at full strength promoted ARs in all cultivars.

7.2. Role of Potassium in the Formation of ARs

Potassium (K) is an indispensable macronutrient, and it is the most abundant cation absorbed by higher plants. It comprises more than 10% of the plant's total dry weight [113]. Its decline to below 10 g/kg of dry weight leads to serious growth issues in various plant species. Indeed, despite not being an important element of any structural and functional molecules, it is involved in various key physiological processes, such as metabolism and plant growth and development [114], flowering [115], improved fruit quality [116], and AR formation in apples and other plants [32,117]. In addition, under K deficiency in soil, plant root growth is weakened. As a result, the yields and outcomes are often limited [118]. The uptake and translocation of K is mediated by several K channels and transporters [113,119].

The underlying physiological and molecular mechanisms regulating AR by KCl application were studied in B9 apple rootstock. KCl-treated cuttings produced a significantly higher number of ARs and increased AR length than control cuttings. At most time points during AR formation, KCl promoted several hormone levels, including IAA, ZR, JA, and GA, and decreased the ratios of ABA/JA, ABA/ZR, and ABA/IAA. Moreover, transcriptome analysis revealed that KCl promoted ARs through the auxin signaling pathway and sugar metabolism and by increasing the genes related to AR development and cell cycle [32]. Z.R. Zhao and colleagues found the promotive effect of K on the formation of ARs in cucumber cotyledons and mung bean hypocotyls, as well as in kidney beans [117]. The results of these studies indicated that the K application promotes adventitious rooting in apples and other crops.

The effect of nitrate and potassium treatments on the various endogenous hormones during adventitious rooting in apples is shown in Figure 2.

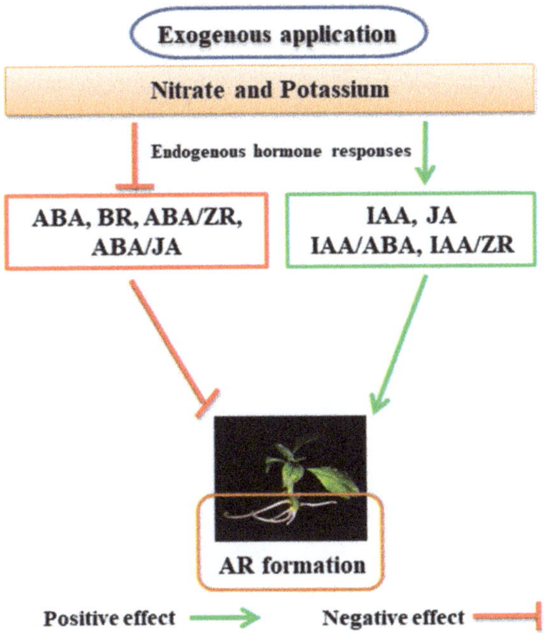

Figure 2. A suggested schematic diagram of how exogenous application of nitrate and potassium regulates different endogenous hormones during the formation of adventitious roots (ARs) in apples. Indole-3-acetic acid (IAA), zeatin ribosome (ZR), jasmonic acid (JA), abscisic acid (ABA), and brassinosteroid (BR). Ratios of hormone content, such as ABA/ZR, ABA/JA, IAA/ZR, and IAA/ABA. The green color shows a positive effect, and the red indicates a negative effect.

8. Summary and Future Perspectives

AR formation is a necessary step in a plant's asexual reproduction. Several apple genotypes have elite characteristics but are limited in their AR formation. Therefore, increasing rooting ability is the primary pursuit. AR formation is a complicated process, influenced by numerous factors. Hormones seem to be the most significant and key among them since they interact with one another, phenolic compounds, sugars, PAs, and mineral nutrients. Despite the widespread belief that auxin is the most critical hormone controller of AR formation in apples, there is complicated crosstalk with and between other plant hormones, as well as with other internal and external factors. Studies on the roles of GA, BR, and JA and their interactions with auxin are still lacking in apples. All these factors remain, heightening the mechanisms regulating ARs' formation throughout the rooting process. Therefore, more extensive investigations are required to address these interactions and evaluate various genotypes.

Recently, multiple advanced molecular techniques have been executed during the formation stages of ARs, and various genes directly or indirectly involved in AR formation have already been reported. AR induction stage is the most crucial, and many genes alter their expression levels after wounding, indicating their roles in the very initial stage. Unfortunately, due to the significant differences in rooting abilities across apple genotypes, evaluating data from different investigations is challenging and may not reveal the intricate rooting mechanisms. It would be fascinating to examine the genetic variations between high and poor AR formation abilities in order to determine the genes involved in these inconsistencies. Transcriptomic comparisons of extreme phenotypes might be a beginning for identifying putative candidate genes in apples that either stimulate or inhibit ARs. To overcome the functional studies of a single gene or whole process, the expression of a specific gene must be modulated using overexpression, RNAi, and mutation.

Author Contributions: M.M.T. and J.M. writing the original draft. S.L., K.L., Y.L. and Y.S. provided guidance. D.Z. and X.Z. supervision and editing. All authors have read and agreed to the published version of the manuscript.

Funding: This work was financially supported by Shaanxi Apple Industry Science and Technology Project (2020zdzx03-01-04), Studying Abroad Personnel Science and Technology Activity Fund Project of Shaanxi Province (2020-07), the National Key Research and Development Program of China (2019YFD1000803), Tang Scholar by Cyrus Tang Foundation (C200022002), The China Apple Research System (CARS-27), Tang Scholar by Cyrus Tang Foundation and Northwest A&F University.

Institutional Review Board Statement: Not applicable.

Informed Consent Statement: Not applicable.

Data Availability Statement: Not applicable.

Conflicts of Interest: The authors declare no conflict of interest.

References

1. Alvarez, R.; Nissen, S.J.; Sutter, E.G. Relationship between indole-3-acetic acid levels in apple (Malus pumila Mill) rootstocks cultured in vitro and adventitious root formation in the presence of indole-3-butyric acid. *Plant Physiol.* **1989**, *89*, 439–443. [CrossRef] [PubMed]
2. Meng, Y.; Xing, L.; Li, K.; Wei, Y.; Wang, H.; Mao, J.; Dong, F.; Ma, D.; Zhang, Z.; Han, M. Genome-wide identification, characterization and expression analysis of novel long non-coding RNAs that mediate IBA-induced adventitious root formation in apple rootstocks. *Plant Growth Regul.* **2019**, *87*, 287–302. [CrossRef]
3. Lei, C.; Fan, S.; Li, K.; Meng, Y.; Mao, J.; Han, M.; Zhao, C.; Bao, L.; Zhang, D. iTRAQ-based proteomic analysis reveals potential regulation networks of IBA-induced adventitious root formation in apple. *Int. J. Mol. Sci.* **2018**, *19*, 667. [CrossRef] [PubMed]
4. Bai, T.; Dong, Z.; Zheng, X.; Song, S.; Jiao, J.; Wang, M.; Song, C. Auxin and its interaction with ethylene control adventitious root formation and development in apple rootstock. *Front. Plant Sci.* **2020**, *11*, 574881. [CrossRef] [PubMed]
5. Patial, S.; Chandel, J.; Sharma, N.; Verma, P. Influence of Auxin on Rooting in Hardwood Cuttings of Apple (×Borkh.) Clonal Rootstock'M 116'under Malus domestica Mist Chamber Conditions. *Indian J. Ecol.* **2021**, *48*, 429–433.
6. De Klerk, G.-J.; Brugge, J.T.; Marinova, S. Effectiveness of indoleacetic acid, indolebutyric acid and naphthaleneacetic acid during adventitious root formation in vitro in Malus 'Jork 9'. *Plant Cell Tissue Organ. Cult.* **1997**, *49*, 39–44. [CrossRef]

7. Li, K.; Liang, Y.; Xing, L.; Mao, J.; Liu, Z.; Dong, F.; Meng, Y.; Han, M.; Zhao, C.; Bao, L. Transcriptome analysis reveals multiple hormones, wounding and sugar signaling pathways mediate adventitious root formation in apple rootstock. *Int. J. Mol. Sci.* **2018**, *19*, 2201. [CrossRef]
8. Guan, L.; Li, Y.; Huang, K.; Cheng, Z.-M.M. Auxin regulation and MdPIN expression during adventitious root initiation in apple cuttings. *Hortic. Res.* **2020**, *7*, 143. [CrossRef]
9. Meng, Y.; Mao, J.; Tahir, M.M.; Wang, H.; Wei, Y.; Zhao, C.; Li, K.; Ma, D.; Zhao, C.; Zhang, D. Mdm-miR160 participates in auxin-induced adventitious root formation of apple rootstock. *Sci. Hortic.* **2020**, *270*, 109442. [CrossRef]
10. Van der Krieken, W.; Breteler, H.; Visser, M. Uptake and metabolism of indolebutyric acid during root formation on Malus microcuttings. *Acta Bot. Neerl.* **1992**, *41*, 435–442. [CrossRef]
11. Li, K.; Liu, Z.; Xing, L.; Wei, Y.; Mao, J.; Meng, Y.; Bao, L.; Han, M.; Zhao, C.; Zhang, D. miRNAs associated with auxin signaling, stress response, and cellular activities mediate adventitious root formation in apple rootstocks. *Plant Physiol. Biochem.* **2019**, *139*, 66–81. [CrossRef] [PubMed]
12. Xu, X.; Li, X.; Hu, X.; Wu, T.; Wang, Y.; Xu, X.; Zhang, X.; Han, Z. High miR156 expression is required for auxin-induced adventitious root formation via MxSPL26 independent of PINs and ARFs in Malus xiaojinensis. *Front. Plant Sci.* **2017**, *8*, 1059. [CrossRef] [PubMed]
13. Mao, J.; Zhang, D.; Meng, Y.; Li, K.; Wang, H.; Han, M. Inhibition of adventitious root development in apple rootstocks by cytokinin is based on its suppression of adventitious root primordia formation. *Physiol. Plant* **2019**, *166*, 663–676. [CrossRef] [PubMed]
14. Zhao, D.; Wang, Y.; Feng, C.; Wei, Y.; Peng, X.; Guo, X.; Guo, X.; Zhai, Z.; Li, J.; Shen, X. Overexpression of MsGH3. 5 inhibits shoot and root development through the auxin and cytokinin pathways in apple plants. *Plant J.* **2020**, *103*, 166–183. [CrossRef] [PubMed]
15. De Klerk, G.-J.; Hanecakova, J.; Jasik, J. The role of cytokinins in rooting of stem slices cut from apple microcuttings. *Plant Biosyst. Int. J. Deal. All Asp. Plant Biol.* **2001**, *135*, 79–84. [CrossRef]
16. Harbage, J.F.; Stimart, D.P. Ethylene does not promote adventitious root initiation on apple microcuttings. *J. Am. Soc. Hortic. Sci.* **1996**, *121*, 880–885. [CrossRef]
17. Sriskandarajah, S. *Induction of Adventitious Roots in Some Scion Cultivars of Apple (Malus pumila Mill)*; University of Sydney: Sydney, Australia, 1984.
18. Noiton, D.; Vine, J.H.; Mullins, M.G. Endogenous indole-3-acetic acid and abscisic acid in apple microcuttings in relation to adventitious root formation. *Plant Growth Regul.* **1992**, *11*, 63–67. [CrossRef]
19. Mao, J.; Niu, C.; Li, K.; Chen, S.; Tahir, M.M.; Han, M.; Zhang, D. Melatonin promotes adventitious root formation in apple by promoting the function of MdWOX11. *BMC Plant Biol.* **2020**, *20*, 536. [CrossRef]
20. Pawlicki, N.; Welander, M. The effects of benzyladenine and gibberellic acid on adventitious root formation in apple stem discs. *Agronomie* **1992**, *12*, 783–788. [CrossRef]
21. De Klerk, G.-J.; Guan, H.; Huisman, P.; Marinova, S. Effects of phenolic compounds on adventitious root formation and oxidative decarboxylation of applied indoleacetic acid in Malus 'Jork 9'. *Plant Growth Regul.* **2011**, *63*, 175–185. [CrossRef]
22. James, D.J.; Thurbon, I.J. Phenolic compounds and other factors controlling rhizogenesis in vitro in the apple rootstocks M. 9 and M. 26. *Z. Für Pflanzenphysiol.* **1981**, *105*, 11–20. [CrossRef]
23. Jasik, J.; De Klerk, G.-J. Anatomical and ultrastructural examination of adventitious root formation in stem slices of apple. *Biol. Plant* **1997**, *39*, 79–90. [CrossRef]
24. Pawlicki, N.; Welander, M. Influence of carbohydrate source, auxin concentration and time of exposure on adventitious rooting of the apple rootstock Jork 9. *Plant Sci.* **1995**, *106*, 167–176. [CrossRef]
25. Calamar, A.; De Klerk, G.-J. Effect of sucrose on adventitious root regeneration in apple. *Plant Cell Tissue Organ. Cult.* **2002**, *70*, 207–212. [CrossRef]
26. Kromer, K.; Gamian, A. Analysis of soluble carbohydrates, proteins and lipids in shoots of M 7 apple rootstock cultured in vitro during regeneration of adventitious roots. *J. Plant Physiol.* **2000**, *156*, 775–782. [CrossRef]
27. Wang, H.; Tahir, M.M.; Nawaz, M.A.; Mao, J.; Li, K.; Wei, Y.; Ma, D.; Lu, X.; Zhao, C.; Zhang, D. Spermidine application affects the adventitious root formation and root morphology of apple rootstock by altering the hormonal profile and regulating the gene expression pattern. *Sci. Hortic.* **2020**, *266*, 109310. [CrossRef]
28. Tahir, M.M.; Wang, H.; Ahmad, B.; Liu, Y.; Fan, S.; Li, K.; Lei, C.; Shah, K.; Li, S.; Zhang, D. Identification and characterization of NRT gene family reveals their critical response to nitrate regulation during adventitious root formation and development in apple rootstock. *Sci. Hortic.* **2021**, *275*, 109642. [CrossRef]
29. Tahir, M.M.; Lu, Z.; Wang, C.; Shah, K.; Li, S.; Zhang, X.; Mao, J.; Liu, Y.; Shalmani, A.; Li, K. Nitrate Application Induces Adventitious Root Growth by Regulating Gene Expression Patterns in Apple Rootstocks. *J. Plant Growth Regul.* **2021**, 1–12. [CrossRef]
30. Tahir, M.M.; Li, S.; Mao, J.; Liu, Y.; Li, K.; Zhang, X.; Lu, X.; Ma, X.; Zhao, C.; Zhang, D. High nitrate inhibited adventitious roots formation in apple rootstock by altering hormonal contents and miRNAs expression profiles. *Sci. Hortic.* **2021**, *286*, 110230. [CrossRef]

31. Zhang, X.; Tahir, M.M.; Li, S.; Mao, J.; Nawaz, M.A.; Liu, Y.; Li, K.; Xing, L.; Niu, J.; Zhang, D. Transcriptome analysis reveals the inhibitory nature of high nitrate during adventitious roots formation in the apple rootstock. *Physiol. Plant* **2021**, *173*, 867–882. [CrossRef]
32. Tahir, M.M.; Chen, S.; Ma, X.; Li, S.; Zhang, X.; Shao, Y.; Shalmani, A.; Zhao, C.; Bao, L.; Zhang, D. Transcriptome analysis reveals the promotive effect of potassium by hormones and sugar signaling pathways during adventitious roots formation in the apple rootstock. *Plant Physiol. Biochem.* **2021**, *165*, 123–136. [CrossRef] [PubMed]
33. Druege, U.; Franken, P.; Hajirezaei, M.R. Plant hormone homeostasis, signaling, and function during adventitious root formation in cuttings. *Front. Plant Sci.* **2016**, *7*, 381. [CrossRef] [PubMed]
34. Agulló-Antón, M.Á.; Sánchez-Bravo, J.; Acosta, M.; Druege, U. Auxins or sugars: What makes the difference in the adventitious rooting of stored carnation cuttings? *J. Plant Growth Regul.* **2011**, *30*, 100–113. [CrossRef]
35. de Klerk, G.-J.; van der Krieken, W.; de Jong, J.C. Review the formation of adventitious roots: New concepts, new possibilities. *Vitr. Cell. Dev. Biol.-Plant* **1999**, *35*, 189–199. [CrossRef]
36. Caboni, E.; Tonelli, M.; Lauri, P.; Iacovacci, P.; Kevers, C.; Damiano, C.; Gaspar, T. Biochemical aspects of almond microcuttings related to in vitro rooting ability. *Biol. Plant* **1997**, *39*, 91–97. [CrossRef]
37. Gatineau, F.; Fouché, J.G.; Kevers, C.; Hausman, J.-F.; Gaspar, T. Quantitative variations of indolyl compounds including IAA, IAA-aspartate and serotonin in walnut microcuttings during root induction. *Biol. Plant* **1997**, *39*, 131–137. [CrossRef]
38. Label, P.; Sotta, B.; Miginiac, E. Endogenous levels of abscisic acid and indole-3-acetic acid during in vitro rooting of wild cherry explants produced by micropropagation. *Plant Growth Regul.* **1989**, *8*, 325–333. [CrossRef]
39. Li, X.; He, Y.; Tang, Z. Effects of IAA and stimulated microgravity on formation of adventitious roots of Chinese cabbage. *Shi Yan Sheng Wu Xue Bao* **2000**, *33*, 179.
40. Gray, W.M.; Kepinski, S.; Rouse, D.; Leyser, O.; Estelle, M. Auxin regulates SCF TIR1-dependent degradation of AUX/IAA proteins. *Nature* **2001**, *414*, 271–276. [CrossRef]
41. Okushima, Y.; Overvoorde, P.J.; Arima, K.; Alonso, J.M.; Chan, A.; Chang, C.; Ecker, J.R.; Hughes, B.; Lui, A.; Nguyen, D. Functional genomic analysis of the AUXIN RESPONSE FACTOR gene family members in Arabidopsis thaliana: Unique and overlapping functions of ARF7 and ARF19. *Plant Cell* **2005**, *17*, 444–463. [CrossRef]
42. Wilmoth, J.C.; Wang, S.; Tiwari, S.B.; Joshi, A.D.; Hagen, G.; Guilfoyle, T.J.; Alonso, J.M.; Ecker, J.R.; Reed, J.W. NPH4/ARF7 and ARF19 promote leaf expansion and auxin-induced lateral root formation. *Plant J.* **2005**, *43*, 118–130. [CrossRef] [PubMed]
43. Okushima, Y.; Fukaki, H.; Onoda, M.; Theologis, A.; Tasaka, M. ARF7 and ARF19 regulate lateral root formation via direct activation of LBD/ASL genes in Arabidopsis. *Plant Cell* **2007**, *19*, 118–130. [CrossRef]
44. Mao, J.; Niu, C.; Li, K.; Mobeen Tahir, M.; Khan, A.; Wang, H.; Li, S.; Liang, Y.; Li, G.; Yang, Z. Exogenous 6-benzyladenine application affects root morphology by altering hormone status and gene expression of developing lateral roots in Malus hupehensis. *Plant Biol.* **2020**, *22*, 1150–1159. [CrossRef] [PubMed]
45. Wang, J.-W.; Wang, L.-J.; Mao, Y.-B.; Cai, W.-J.; Xue, H.-W.; Chen, X.-Y. Control of root cap formation by microRNA-targeted auxin response factors in Arabidopsis. *Plant Cell* **2005**, *17*, 2204–2216. [CrossRef] [PubMed]
46. Liu, Z.; Li, J.; Wang, L.; Li, Q.; Lu, Q.; Yu, Y.; Li, S.; Bai, M.Y.; Hu, Y.; Xiang, F. Repression of callus initiation by the mi RNA-directed interaction of auxin–cytokinin in Arabidopsis thaliana. *Plant J.* **2016**, *87*, 391–402. [CrossRef] [PubMed]
47. Sorin, C.; Bussell, J.D.; Camus, I.; Ljung, K.; Kowalczyk, M.; Geiss, G.; McKhann, H.; Garcion, C.; Vaucheret, H.; Sandberg, G. Auxin and light control of adventitious rooting in Arabidopsis require ARGONAUTE1. *Plant Cell* **2005**, *17*, 1343–1359. [CrossRef] [PubMed]
48. Tian, C.E.; Muto, H.; Higuchi, K.; Matamura, T.; Tatematsu, K.; Koshiba, T.; Yamamoto, K.T. Disruption and overexpression of auxin response factor 8 gene of Arabidopsis affect hypocotyl elongation and root growth habit, indicating its possible involvement in auxin homeostasis in light condition. *Plant J.* **2004**, *40*, 333–343. [CrossRef]
49. Meng, Y.; Huang, F.; Shi, Q.; Cao, J.; Chen, D.; Zhang, J.; Ni, J.; Wu, P.; Chen, M. Genome-wide survey of rice microRNAs and microRNA–target pairs in the root of a novel auxin-resistant mutant. *Planta* **2009**, *230*, 883–898. [CrossRef]
50. Marin, E.; Jouannet, V.; Herz, A.; Lokerse, A.S.; Weijers, D.; Vaucheret, H.; Nussaume, L.; Crespi, M.D.; Maizel, A. miR390, Arabidopsis TAS3 tasiRNAs, and their AUXIN RESPONSE FACTOR targets define an autoregulatory network quantitatively regulating lateral root growth. *Plant Cell* **2010**, *22*, 1104–1117. [CrossRef]
51. Levy, A.; Szwerdszarf, D.; Abu-Abied, M.; Mordehaev, I.; Yaniv, Y.; Riov, J.; Arazi, T.; Sadot, E. Profiling microRNAs in Eucalyptus grandis reveals no mutual relationship between alterations in miR156 and miR172 expression and adventitious root induction during development. *BMC Genom.* **2014**, *15*, 524. [CrossRef]
52. Grieneisen, V.A.; Xu, J.; Marée, A.F.; Hogeweg, P.; Scheres, B. Auxin transport is sufficient to generate a maximum and gradient guiding root growth. *Nature* **2007**, *449*, 1008–1013. [CrossRef] [PubMed]
53. Ioio, R.D.; Nakamura, K.; Moubayidin, L.; Perilli, S.; Taniguchi, M.; Morita, M.T.; Aoyama, T.; Costantino, P.; Sabatini, S. A genetic framework for the control of cell division and differentiation in the root meristem. *Science* **2008**, *322*, 1380–1384. [CrossRef] [PubMed]
54. Bennett, M.J.; Marchant, A.; Green, H.G.; May, S.T.; Ward, S.P.; Millner, P.A.; Walker, A.R.; Schulz, B.; Feldmann, K.A. Arabidopsis AUX1 gene: A permease-like regulator of root gravitropism. *Science* **1996**, *273*, 948–950. [CrossRef] [PubMed]

55. Marchant, A.; Bhalerao, R.; Casimiro, I.; Eklöf, J.; Casero, P.J.; Bennett, M.; Sandberg, G. AUX1 promotes lateral root formation by facilitating indole-3-acetic acid distribution between sink and source tissues in the Arabidopsis seedling. *Plant Cell* **2002**, *14*, 589–597. [CrossRef]
56. Liu, B.; Wang, L.; Zhang, J.; Li, J.; Zheng, H.; Chen, J.; Lu, M. WUSCHEL-related Homeobox genes in Populus tomentosa: Diversified expression patterns and a functional similarity in adventitious root formation. *BMC Genom.* **2014**, *15*, 296. [CrossRef]
57. Tian, H.; Jia, Y.; Niu, T.; Yu, Q.; Ding, Z. The key players of the primary root growth and development also function in lateral roots in Arabidopsis. *Plant Cell Rep.* **2014**, *33*, 745–753. [CrossRef]
58. Hu, X.; Xu, L. Transcription factors WOX11/12 directly activate WOX5/7 to promote root primordia initiation and organogenesis. *Plant Physiol.* **2016**, *172*, 2363–2373. [CrossRef]
59. Mashiguchi, K.; Tanaka, K.; Sakai, T.; Sugawara, S.; Kawaide, H.; Natsume, M.; Hanada, A.; Yaeno, T.; Shirasu, K.; Yao, H. The main auxin biosynthesis pathway in Arabidopsis. *Proc. Natl. Acad. Sci. USA* **2011**, *108*, 18512–18517. [CrossRef]
60. Sozzani, R.; Cui, H.; Moreno-Risueno, M.; Busch, W.; Van Norman, J.; Vernoux, T.; Brady, S.; Dewitte, W.; Murray, J.A.H.; Benfey, P. Spatiotemporal regulation of cell-cycle genes by SHORTROOT links patterning and growth. *Nature* **2010**, *466*, 128–132. [CrossRef]
61. Liu, J.; Sheng, L.; Xu, Y.; Li, J.; Yang, Z.; Huang, H.; Xu, L. WOX11 and 12 are involved in the first-step cell fate transition during de novo root organogenesis in Arabidopsis. *Plant Cell* **2014**, *26*, 1081–1093. [CrossRef]
62. Mähönen, A.P.; Higuchi, M.; Törmäkangas, K.; Miyawaki, K.; Pischke, M.S.; Sussman, M.R.; Helariutta, Y.; Kakimoto, T. Cytokinins regulate a bidirectional phosphorelay network in Arabidopsis. *Curr. Biol.* **2006**, *16*, 1116–1122. [CrossRef]
63. Atkinson, J.A.; Rasmussen, A.; Traini, R.; Voß, U.; Sturrock, C.; Mooney, S.J.; Wells, D.M.; Bennett, M.J. Branching out in roots: Uncovering form, function, and regulation. *Plant Physiol.* **2014**, *166*, 538–550. [CrossRef] [PubMed]
64. Ramírez-Carvajal, G.A.; Morse, A.M.; Dervinis, C.; Davis, J.M. The cytokinin type-B response regulator PtRR13 is a negative regulator of adventitious root development in Populus. *Plant Physiol.* **2009**, *150*, 759–771. [CrossRef] [PubMed]
65. Jinxiang, W.; Xiaolong, Y.; Ruichi, P. Relationship between adventitious root formation and plant hormones. *Plant Physiol. Commun.* **2005**, *41*, 133–142.
66. Amiri, E.M.; Elahinia, A. Optimization of medium composition for apple rootstocks. *Afr. J. Biotechnol.* **2011**, *10*, 3594–3601.
67. Hutchison, C.E.; Li, J.; Argueso, C.; Gonzalez, M.; Lee, E.; Lewis, M.W.; Maxwell, B.B.; Perdue, T.D.; Schaller, G.E.; Alonso, J.M. The Arabidopsis histidine phosphotransfer proteins are redundant positive regulators of cytokinin signaling. *Plant Cell* **2006**, *18*, 3073–3087. [CrossRef]
68. Chen, L.; Tong, J.; Xiao, L.; Ruan, Y.; Liu, J.; Zeng, M.; Huang, H.; Wang, J.-W.; Xu, L. YUCCA-mediated auxin biogenesis is required for cell fate transition occurring during de novo root organogenesis in Arabidopsis. *J. Exp. Bot.* **2016**, *67*, 4273–4284. [CrossRef]
69. Zhao, Y.; Cheng, S.; Song, Y.; Huang, Y.; Zhou, S.; Liu, X.; Zhou, D.-X. The interaction between rice ERF3 and WOX11 promotes crown root development by regulating gene expression involved in cytokinin signaling. *Plant Cell* **2015**, *27*, 2469–2483. [CrossRef]
70. Chandler, J.W.; Werr, W. Cytokinin–auxin crosstalk in cell type specification. *Trends Plant Sci.* **2015**, *20*, 291–300. [CrossRef]
71. Da Costa, C.T.; De Almeida, M.R.; Ruedell, C.M.; Schwambach, J.; Maraschin, F.D.S.; Fett-Neto, A.G. When stress and development go hand in hand: Main hormonal controls of adventitious rooting in cuttings. *Front. Plant Sci.* **2013**, *4*, 133. [CrossRef]
72. Zimmerman, P.W. Several chemical growth substances which cause initiation of roots and other responses in plants. *Contrib. Boyce Thompson Inst.* **1935**, *7*, 209–229.
73. Růžička, K.; Ljung, K.; Vanneste, S.; Podhorská, R.; Beeckman, T.; Friml, J.; Benková, E. Ethylene regulates root growth through effects on auxin biosynthesis and transport-dependent auxin distribution. *Plant Cell* **2007**, *19*, 2197–2212. [CrossRef] [PubMed]
74. Harris, J.M. Abscisic acid: Hidden architect of root system structure. *Plants* **2015**, *4*, 548–572. [CrossRef] [PubMed]
75. Petricka, J.J.; Benfey, P.N. Root layers: Complex regulation of developmental patterning. *Curr. Opin. Genet. Dev.* **2008**, *18*, 354–361. [CrossRef]
76. Heidstra, R.; Sabatini, S. Plant and animal stem cells: Similar yet different. *Nat. Rev. Mol. Cell Biol.* **2014**, *15*, 301–312. [CrossRef] [PubMed]
77. Zhang, H.; Han, W.; De Smet, I.; Talboys, P.; Loya, R.; Hassan, A.; Rong, H.; Jürgens, G.; Paul Knox, J.; Wang, M.H. ABA promotes quiescence of the quiescent centre and suppresses stem cell differentiation in the Arabidopsis primary root meristem. *Plant J.* **2010**, *64*, 764–774. [CrossRef]
78. Pelese, F.; Megnegneau, B.; Sotta, B.; Sossountzov, L.; Caboche, M.; Miginiac, E. Hormonal characterization of a nonrooting naphthalene-acetic acid tolerant tobacco mutant by an immunoenzymic method. *Plant Physiol.* **1989**, *89*, 86–92. [CrossRef]
79. De Smet, I.; Signora, L.; Beeckman, T.; Inzé, D.; Foyer, C.H.; Zhang, H. An abscisic acid-sensitive checkpoint in lateral root development of Arabidopsis. *Plant J.* **2003**, *33*, 543–555. [CrossRef]
80. Zhang, X.; Tahir, M.M.; Li, S.; Tang, T.; Mao, J.; Li, K.; Shao, Y.; Yang, W.; Niu, J.; Zhang, D. Effect of exogenous abscisic acid (ABA) on the morphology, phytohormones, and related gene expression of developing lateral roots in 'Qingzhen 1'apple plants. *Plant Cell Tissue Organ. Cult. (PCTOC)* **2022**, *148*, 23–34. [CrossRef]
81. Li, S.; Tahir, M.M.; Wu, T.; Xie, L.; Zhang, X.; Mao, J.; Ayyoub, A.; Xing, L.; Zhang, D.; Shao, Y. Transcriptome Analysis Reveals Multiple Genes and Complex Hormonal-Mediated Interactions with PEG during Adventitious Root Formation in Apple. *Int. J. Mol. Sci.* **2022**, *23*, 976. [CrossRef]
82. Wasternack, C.; Song, S. Jasmonates: Biosynthesis, metabolism, and signaling by proteins activating and repressing transcription. *J. Exp. Bot.* **2017**, *68*, 1303–1321. [CrossRef] [PubMed]

83. Schilmiller, A.L.; Howe, G.A. Systemic signaling in the wound response. *Curr. Opin. Plant Biol.* **2005**, *8*, 369–377. [CrossRef]
84. Ahkami, A.H.; Lischewski, S.; Haensch, K.T.; Porfirova, S.; Hofmann, J.; Rolletschek, H.; Melzer, M.; Franken, P.; Hause, B.; Druege, U. Molecular physiology of adventitious root formation in Petunia hybrida cuttings: Involvement of wound response and primary metabolism. *New Phytol.* **2009**, *181*, 613–625. [CrossRef]
85. Lischweski, S.; Muchow, A.; Guthörl, D.; Hause, B. Jasmonates act positively in adventitious root formation in petunia cuttings. *BMC Plant Biol.* **2015**, *15*, 229. [CrossRef]
86. Światek, A.; Lenjou, M.; Van Bockstaele, D.; Iné, D.; Van Onckelen, H. Differential effect of jasmonic acid and abscisic acid on cell cycle progression in tobacco BY-2 cells. *Plant Physiol.* **2002**, *128*, 201–211. [CrossRef] [PubMed]
87. Pauwels, L.; Morreel, K.; De Witte, E.; Lammertyn, F.; Van Montagu, M.; Boerjan, W.; Inzé, D.; Goossens, A. Mapping methyl jasmonate-mediated transcriptional reprogramming of metabolism and cell cycle progression in cultured Arabidopsis cells. *Proc. Natl. Acad. Sci. USA* **2008**, *105*, 1380–1385. [CrossRef]
88. Park, W.J. Melatonin as an endogenous plant regulatory signal: Debates and perspectives. *J. Plant Biol.* **2011**, *54*, 143–149. [CrossRef]
89. Arnao, M.B.; Hernández-Ruiz, J. Melatonin promotes adventitious-and lateral root regeneration in etiolated hypocotyls of *Lupinus albus* L. *J. Pineal Res.* **2007**, *42*, 147–152. [CrossRef]
90. Sarropoulou, V.; Dimassi-Theriou, K.; Therios, I.; Koukourikou-Petridou, M. Melatonin enhances root regeneration, photosynthetic pigments, biomass, total carbohydrates and proline content in the cherry rootstock PHL-C (Prunus avium × Prunus cerasus). *Plant Physiol. Biochem.* **2012**, *61*, 162–168. [CrossRef]
91. Li, C.; Liang, B.; Chang, C.; Wei, Z.; Zhou, S.; Ma, F. Exogenous melatonin improved potassium content in Malus under different stress conditions. *J. Pineal Res.* **2016**, *61*, 218–229. [CrossRef] [PubMed]
92. Chen, Q.; Qi, W.-b.; Reiter, R.J.; Wei, W.; Wang, B.-m. Exogenously applied melatonin stimulates root growth and raises endogenous indoleacetic acid in roots of etiolated seedlings of Brassica juncea. *J. Plant Physiol.* **2009**, *166*, 324–328. [CrossRef] [PubMed]
93. Pelagio-Flores, R.; Muñoz-Parra, E.; Ortiz-Castro, R.; López-Bucio, J. Melatonin regulates Arabidopsis root system architecture likely acting independently of auxin signaling. *J. Pineal Res.* **2012**, *53*, 279–288. [CrossRef]
94. Müssig, C.; Shin, G.-H.; Altmann, T. Brassinosteroids promote root growth in Arabidopsis. *Plant Physiol.* **2003**, *133*, 1261–1271. [CrossRef] [PubMed]
95. Gorter, C.J. Further Experiments on Auxin-Synergists. *Physiol. Plant* **1962**, *15*, 88–95. [CrossRef]
96. Jaleel, C.A.; Riadh, K.; Gopi, R.; Manivannan, P.; Ines, J.; Al-Juburi, H.J.; Chang-Xing, Z.; Hong-Bo, S.; Panneerselvam, R. Antioxidant defense responses: Physiological plasticity in higher plants under abiotic constraints. *Acta Physiol. Plant* **2009**, *31*, 427–436. [CrossRef]
97. Murphy, A.; Peer, W.A.; Taiz, L. Regulation of auxin transport by aminopeptidases and endogenous flavonoids. *Planta* **2000**, *211*, 315–324. [CrossRef]
98. Buer, C.S.; Imin, N.; Djordjevic, M.A. Flavonoids: New roles for old molecules. *J. Integr. Plant Biol.* **2010**, *52*, 98–111. [CrossRef]
99. Couée, I.; Hummel, I.; Sulmon, C.; Gouesbet, G.; El Amrani, A. Involvement of polyamines in root development. *Plant Cell Tissue Organ. Cult.* **2004**, *76*, 1–10. [CrossRef]
100. Hausman, J.F.; Gevers, C.; Gaspar, T. Involvement of putrescine in the inductive rooting phase of poplar shoots raised in vitro. *Physiol. Plant* **1994**, *92*, 201–206. [CrossRef]
101. Kevers, C.; Hausman, J.-F.; Faivre-Rampant, O.; Evers, D.; Gaspar, T. Hormonal control of adventitious rooting: Progress and questions. *J. Appl. Bot.* **1997**, *71*, 71–79.
102. Geiss, G.; Gutierrez, L.; Bellini, C. Adventitious root formation: New insights and perspectives. *Annu. Plant Rev. Online* **2009**, *37*, 127–156. [CrossRef]
103. Wang, Y.-Y.; Hsu, P.-K.; Tsay, Y.-F. Uptake, allocation and signaling of nitrate. *Trends Plant Sci.* **2012**, *17*, 458–467. [CrossRef] [PubMed]
104. Masclaux-Daubresse, C.; Daniel-Vedele, F.; Dechorgnat, J.; Chardon, F.; Gaufichon, L.; Suzuki, A. Nitrogen uptake, assimilation and remobilization in plants: Challenges for sustainable and productive agriculture. *Ann. Bot.* **2010**, *105*, 1141–1157. [CrossRef]
105. O'Brien, J.A.; Vega, A.; Bouguyon, E.; Krouk, G.; Gojon, A.; Coruzzi, G.; Gutiérrez, R.A. Nitrate transport, sensing, and responses in plants. *Mol. Plant* **2016**, *9*, 837–856. [CrossRef] [PubMed]
106. Tahir, M.M.; Zhang, X.; Shah, K.; Hayat, F.; Li, S.; Mao, J.; Liu, Y.; Shao, Y.; Zhang, D. Nitrate application affects root morphology by altering hormonal status and gene expression patterns in B9 apple rootstock nursery plants. *Fruit Res.* **2021**, *1*, 1–11. [CrossRef]
107. López-Bucio, J.; Cruz-Ramırez, A.; Herrera-Estrella, L. The role of nutrient availability in regulating root architecture. *Curr. Opin. Plant Biol.* **2003**, *6*, 280–287. [CrossRef]
108. Crawford, N.M.; Glass, A.D. Molecular and physiological aspects of nitrate uptake in plants. *Trends Plant Sci.* **1998**, *3*, 389–395. [CrossRef]
109. Zhang, H.; Jennings, A.; Barlow, P.W.; Forde, B.G. Dual pathways for regulation of root branching by nitrate. *Proc. Natl. Acad. Sci. USA* **1999**, *96*, 6529–6534. [CrossRef]
110. Sriskandarajah, S.; Skirvin, R.; Abu-Qaoud, H. The effect of some macronutrients on adventitious root development on scion apple cultivars in vitro. *Plant Cell Tissue Organ. Cult.* **1990**, *21*, 185–189. [CrossRef]
111. Bhat, K. Nutrient inflows into apple roots. *Plant Soil* **1983**, *71*, 371–380. [CrossRef]

112. Hilo, A.; Shahinnia, F.; Druege, U.; Franken, P.; Melzer, M.; Rutten, T.; von Wirén, N.; Hajirezaei, M.-R. A specific role of iron in promoting meristematic cell division during adventitious root formation. *J. Exp. Bot.* **2017**, *68*, 4233–4247. [CrossRef] [PubMed]
113. Véry, A.-A.; Nieves-Cordones, M.; Daly, M.; Khan, I.; Fizames, C.; Sentenac, H. Molecular biology of K+ transport across the plant cell membrane: What do we learn from comparison between plant species? *J. Plant Physiol.* **2014**, *171*, 748–769. [CrossRef]
114. Römheld, V.; Kirkby, E.A. Research on potassium in agriculture: Needs and prospects. *Plant Soil* **2010**, *335*, 155–180. [CrossRef]
115. Yamasaki, A.; Yano, T. In Effect of Supplemental Application of Fertilizers on Flower Bud Initiation and Development of Strawberry; Possible Role of Nitrogen. In Proceedings of the VI International Strawberry Symposium 842, Huelva, Spain, 3 March 2008; pp. 765–768.
116. Hartz, T.; Johnstone, P.; Francis, D.; Miyao, E. Processing tomato yield and fruit quality improved with potassium fertigation. *HortScience* **2005**, *40*, 1862–1867. [CrossRef]
117. Zhao, Z.; Li, G.; Huang, G. Promotive effect of potassium on adventitious root formation in some plants. *Plant Sci.* **1991**, *79*, 47–50. [CrossRef]
118. Kannan, C.G.; Perumalsamy, P.; Thangavelu, M. Influences of potassium chloride fertilization on mycorrhizal formation in a tropical alfisol. *Commun. Soil Sci. Plant Anal.* **2017**, *48*, 524–538. [CrossRef]
119. Wang, Y.; Wu, W.-H. Potassium transport and signaling in higher plants. *Annu. Rev. Plant Biol.* **2013**, *64*, 451–476. [CrossRef]

MDPI
St. Alban-Anlage 66
4052 Basel
Switzerland
www.mdpi.com

Horticulturae Editorial Office
E-mail: horticulturae@mdpi.com
www.mdpi.com/journal/horticulturae

www.ingramcontent.com/pod-product-compliance
Lightning Source LLC
Chambersburg PA
CBHW061514210926
51850CB00008B/1919
9783036587202